Precast Concrete Structures

Kim S. Elliott

OXFORD   AMSTERDAM   BOSTON   LONDON   NEW YORK   PARIS
SAN DIEGO   SAN FRANCISCO   SINGAPORE   SYDNEY   TOKYO

Butterworth-Heinemann
An imprint of Elsevier Science
Linacre House, Jordan Hill, Oxford OX2 8DP
225 Wildwood Avenue, Woburn, MA 01801-2041

First published 2002

British Library Cataloguing in Publication Data
Elliott, Kim S.
  Precast concrete structures
  1. Precast concrete construction
  I. Title
  693.5'22

Library of Congress Cataloguing in Publication Data
A catalogue record for this book is available from the Library of Congress

ISBN 0 7506 5084 2

For information on all Butterworth-Heinemann publications
visit our website at www.bh.com

Typeset by Integra Software Services Pvt. Ltd, Pondicherry 605 005, India
www.integra-india.com

Printed in Great Britain by Antony Rowe Ltd

contents

preface

In 1990, the chairman of the British Precast Concrete Federation (BPCF), Mr Geoff Brigginshaw, asked me what level of teaching was carried out in British universities in precast concrete construction for multi-storey buildings. The answer, of course, was very little, and remains that way today in spite of considerable efforts by the BPCF and sections of the profession to broadcast the merits, and pitfalls of precast concrete structures. Having given lectures at about 25 UK universities in this subject, I estimate that less than 5 per cent of our civil/structural engineering graduates know about precast concrete, and less than this have a decent grounding in the design of precast concrete structures. Why is this?

The precast concrete industry commands about 25 per cent of the multi-storey commercial and domestic building market if frames, floors and cladding (facades) are all included. In higher education (one step away from the market), precast education commands between zero and (about) 5 per cent of the structural engineering curriculum. This in turn represents only about 1/8 of a civil engineering course. The 5 per cent figure claimed above could indeed be an over estimate.

The reasons are two-fold:

1 British lecturers are holistic towards structural engineering.

2 British lecturers have no information in this subject.

This book aims to solve these suggestions simultaneously. Suggestion no. 2 is more readily solved. This book is, unfortunately, one of very few text books in this subject area aimed at students at a level which they can assimilate in their overall structural engineering learning process. It does this by considering design both at the macro and micro levels – global issues such as structural stability, building movement and robustness are dissected and analysed down to the level of detailed joints, localized stress concentrations and bolts and welds sizes.

Suggestion no. 1 is more complex. Having been acquainted with members of the FIB* (formerly FIP†) Commission on Prefabrication, it has come to my notice

---

*FIB (Federation International du Beton), born from a merger of FIP with CEB. An international, but predominantly European organization for the welfare and distribution of information on structural concrete.

†FIP (Federation International de la Prefabrication) is an international, but predominantly European, organization for the welfare and distribution of information on prefabricated concrete.

the differing attitudes towards the education of students in certain forms of building construction – precast concrete being one of them (timber another). In continental Europe, leading precast industrialists and/or consultants hold academic posts dedicated to precast concrete construction. Chairs are even sponsored in this subject. In South America, lecturers, students and practitioners hold seminars where precast concrete is a major theme. It is not uncommon for as many as 10 Masters students to study this subject in a civil engineering department. In the United States of America collaborative research between consultants, precast manufacturers and universities is common, as the number of papers published in the *PCI JOURNAL* testifies.

The attitude in Britain is more holistic and less direct. Firstly, basic tuition is given in solid mechanics, structural analysis and material properties. Students are required to be capable of dealing with structural behaviour – independent of the material(s) involved. Secondly, given the fundamental principles of design (and a reminder that code equations are often simulations and their data conservative) students can assimilate any design situation, with appropriate guidance. This may be true for structural steelwork and cast in situ concrete structures where the designer may (if he wishes) divorce themselves from the fabricator and contractor. It is not true for precast concrete (and timber) structures where the fabricator and site erector form part of the 'design team'.

Precast concrete design is an iterative procedure, linking many aspects of architecture, design, detailing, manufacture and site erection together in a 5-point lattice.

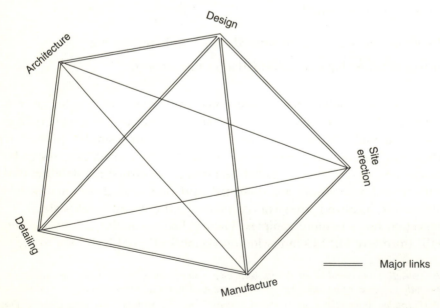

Figure i

Many students will be familiar with these names, but few will see or hear them in a single lecture. Some of the links are quite strong. Note the central role of 'designing' (this does not mean $wL^2/8$, etc.) in establishing relationships with architectural requirements, detailing components and connections, etc., manufacturing and erecting the said components at their connections. Could similar diagrams be drawn for structural steelwork or cast in situ concrete structures?

Further, there are a number of secondary issues involving precast concrete construction. Prefabrication of integrated services, automation of information, temporary stability and safety during erection, all result from the primary links.

Some of these are remote from 'designing'. The illustration reminds us of their presence in the *total* structure. The design procedure will eventually encompass all of these aspects.

This book is aimed at providing sufficient information to enable graduates to carry out structural design operations, whilst recognizing the role of the designer in precast concrete construction. Its content is in many parts similar to but more fundamental than the author's book 'Multi-Storey Precast Concrete Framed Structures' (Blackwell Science 1996). The Blackwell book assumed a prior knowledge of

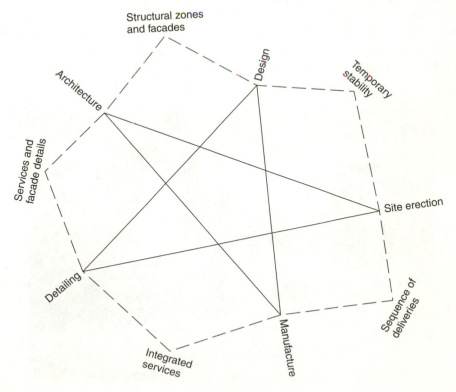

Figure ii

the building industry and some experience in designing concrete structures. This present book takes a backward step to many of the design situations, and does not always uphold the hypotheses given. Reference to the Blackwell book may therefore be necessary to support some of the design solutions.

The design examples are carried out to BS8110, and not EC2 as might be expected from a text book published today. The reason for this is that the clauses relevant to precast concrete in EC2 have yet to find a permanent location. Originally Part 1.3 was dedicated to precast concrete, but this was withdrawn and its content merged into the general code Part 1.1. For this reason specific design data relevant to precast concrete is not available.

The author is grateful to the contributions made by the following individuals and organizations: to members of the FIB Commission on Prefabrication, in particular Arnold Van Acker (Addtek Ltd., Belgium), Andre Cholewicki (BRI Warsaw), Bruno Della Bella (Precompressi Centro Nord, Italy), Ruper Kromer (Betonwerk + Fertigteil-Technik Germany), Gunner Rise (Stranbetong, Sweden), Nordy Robbens (Echo, Belgium) and Jan Vambersky, (TU Delft & Corsmit, Netherlands); to Trent Concrete Ltd (UK), Bison Ltd (UK), SCC Ltd (UK), Tarmac Precast Ltd (UK), Tarmac Topfloor Ltd (formerly Richard Lees) (UK), Techcrete Ltd (UK and Ireland), Composite Structures (UK), British Precast Concrete Federation, British Cement Association & Reinforced Concrete Council (UK), Betoni (Finland), Bevlon (Netherlands), C&CA of Australia, Cement Manufacturers Association of Southern Africa, CIDB (Singapore), Andrew Curd and Partners (USA), Echo Prestress (South Africa), IBRACON (Brazil), Grupo Castelo

Author visiting one of his structures in Malaysia 1998.

(Spain), National Precast Concrete Association of Australia, Nordimpianti-Otm (Italy), Hume Industries (Malaysia), Prestressed Concrete Institute (USA), Spaen-com Betonfertigteile GmbH (Germany), Varioplus (Germany), Spancrete (USA), AB Stranbetong (Sweden), Tammer Elementti Oy (Finland); to his research assist-ants Wahid Omar, Ali Mahdi, Reza Adlparvar, Dennis Lam, Halil Gorgun, Kevin Paine, Aziz Arshad, Adnan Altamimi, Basem Marmash and Marcelo Ferreira, and to his secretarial assistant Caroline Dolby.

1 What is precast concrete?

1.1 Why is precast different?

What makes precast concrete different to other forms of concrete construction? After all, the concrete does not know it is precast, whether statically reinforced or pretensioned (=prestressed). It is only when we consider the role that this concrete will play in developing structural characteristics that its precast background becomes significant. The most obvious definition for precast concrete is that it is concrete which has been prepared for casting, cast and cured in a location which is not its final destination. The distance travelled from the casting site may only be a few metres, where on-site precasting methods are used to avoid expensive haulage (or VAT in some countries), or maybe thousands of kilometres, in the case of high-value added products where manufacturing and haulage costs are low. The grit blasted architectural precast concrete in Figure 1.1 was manufactured 600 km from the site, whereas the precast concrete shown in Figure 1.2 travelled less than 60 m, having been cast adjacent to the final building.

What really distinguishes precast concrete from cast in situ is its stress and strain response to external (=load induced) and internal (=autogenous volumetric changes) effects. A precast concrete element is, by definition, of a finite size and must therefore be joined to other elements to form a

Figure 1.1: Architectural-structural precast concrete structure (courtesy Trent Concrete, UK).

Figure 1.2: Site cast precast concrete.

complete structure. A simple bearing ledge will suffice, as shown in Figure 1.3. But when shrinkage, thermal, or load induced strains cause volumetric changes (and shortening or lengthening) the two precast elements will try to move apart (Figure 1.4a). Interface friction at the mating surface prevents movement, but in doing so creates a force $F = \mu R$

Figure 1.3: Simple bearing nib.

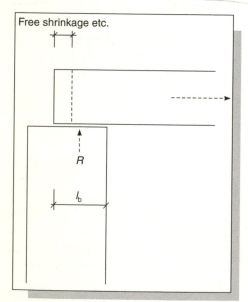

Figure 1.4a: Unrestrained movement between two precast concrete elements.

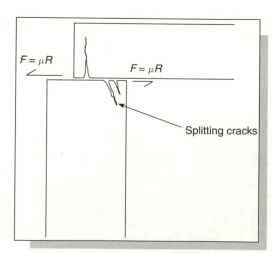

Figure 1.4b: Restrained movement but without tensile stress prevention.

which is capable of splitting both elements unless the section is suitably reinforced (Figure 1.4b). Flexural rotations of the suspended element (=the beam)
reduces the mating length l_b (=bearing length) creating a stress concentration until
local crushing at the top of the pillar (=the column) occurs, unless a bearing
pad is used to prevent the stress concentration forming (Figure 1.4c). If the
bearing is narrow, dispersal of stress from the interior to the exterior of the pillar

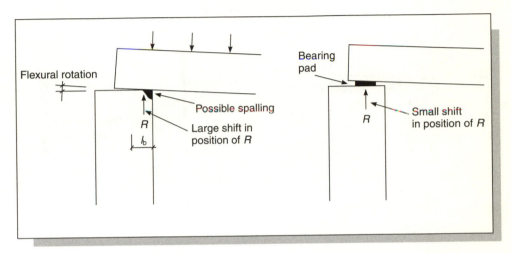

Figure 1.4c: Reduced bearing length and stress concentrations due to flexural rotation.

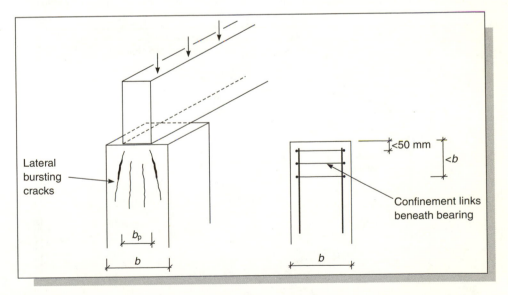

Figure 1.4d: Lateral splitting due to narrow bearings (left) and preventative rebars (right).

causes lateral tensile strain, leading to bursting of the concrete at some distance below the bearing unless the section is suitably reinforced (Figure 1.4d).

If the column is disturbed by an accidental, or a structural force H such that $H > \mu R$, the displacement is not elastically recoverable and may lead to instability

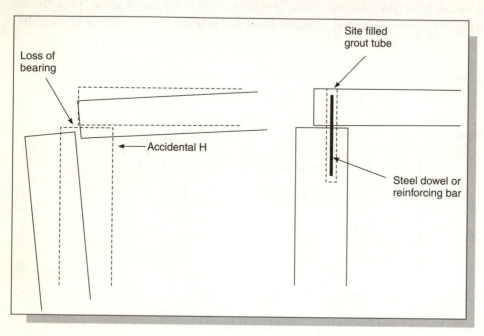

Figure 1.4e: Loss of bearing due to accidental load (left) and preventative dowel bar (right).

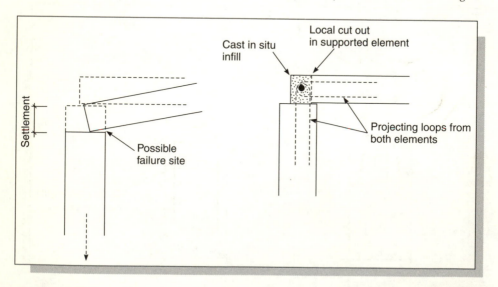

Figure 1.4f: Loss of bearing due to settlement (left) and preventative rebar mechanism (right).

or even loss of bearing altogether unless the bearing possesses shear capacity (Figure 1.4e). Should the column's foundation fail, loss of bearing will occur unless the bearing has tensile capacity (Figure 1.4f).

These are the factors which distinguish precast concrete from other forms of construction.

Figure 1.5: Precast concrete skeletal structure – The 'Green Apple' retail centre near Helsinki, Finland.

Figure 1.6: Precast skeletal structure with integrated architectural features (courtesy Trent Concrete, UK).

1.2 Precast concrete structures

A precast concrete structure is an assemblage of precast elements which, when suitably connected together, form a 3D framework capable of resisting gravitation and wind (or even earthquake*) loads. The framework is ideally suited to buildings such as offices, retail units, car parks, schools, stadia and other such buildings requiring minimal internal obstruction and multi-functional leaseable space. The quantity of concrete in a precast framework is less than 4 per cent of the gross volume of the building, and 2/3 of this is in the floors. In the case of the shopping centre and car park (completed in the year 2001) shown in Figure 1.5, the precast concrete elements are columns, beams, floor slabs, staircases and diagonal bracing. The framework in Figure 1.6 was built using similar elements, but thanks to some creative surface finishes and more expensive mouldings this building appears to have a completely different

* This book does not address seismic actions.

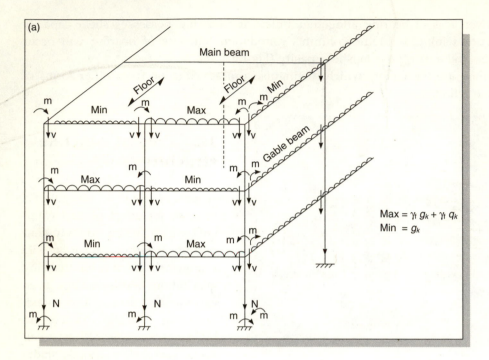

$Max = \gamma_f\, g_k + \gamma_f\, q_k$
$Min\ = g_k$

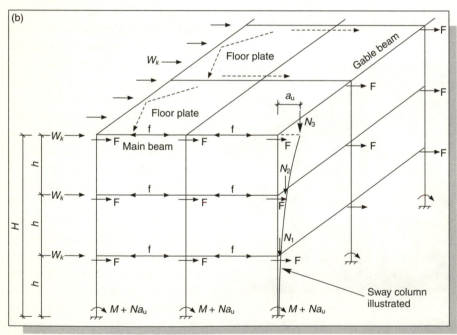

Figure 1.7: Load transfer in a skeletal unbraced or sway frame: (a) gravity loads; and (b) horizontal loads.

function, both architecturally and structurally. Such frameworks are called 'skeletal' structures.

In Figure 1.7 gravitation forces g_k (=pressure of dead loads [kN/m^2]) and q_k (=pressure of live or imposed loads [kN/m^2]) are carried by floors in bending and shear, and sometimes torsion. Floor loads are then transmitted to the beams (as shear v [kN/m]); from the beam (as shear V [kN]) to the column; and from the column to the foundation (as compression N[kN]). Bending moments m [kNm] may arise in the column due to beam–column interaction, but these will be ignored for now because, when compared with the moments M arising from horizontal forces W_k, they are rather small.

The forces w_k (=pressure due to horizontal wind loads [kN/m^2]) act over the full facade of the building. They are absorbed by the structure at each floor level. If the storey height is h [m] and the length of the facade is l [m] the total force absorbed at each floor level is $W_k = w_k hl$ [kN]. W_k is transmitted through the floors, now acting as plates (as horizontal shear f[kN]) to the beams; from the beams (as horizontal shear F[kN]) to the columns; and from the columns (as bending M [kNm] and shear Σf[kN]) to the foundations. As sway deflections a_u [m or mm] increase, the moment at the foundation $= M + m + Na_u$ [kNm]. Because a_u is proportional to h^3, a_u increases rapidly with height and quickly becomes dominant. (To check for frames built out of plumb, W_k[kN] has a minimum value of 0.015 times G_k in BS8110, where G_k is the dead load [kN] for each floor in turn.)

The sway profile of the column also depends on the degree of fixity (=resistance against bending) between the column and the foundation, and between the column and beams. At best, the foundation and beams are rigidly connected to the column and the column's sway profile is as shown in Figure 1.8a. At worst the foundation and beams are pinned, and the structure is a mechanism (Figure 1.8b). In practice, the foundation is very rigid (=not infinitely so) and the beams are nearly pinned (=not zero stiffness) so that the column's profile is as shown in Figure 1.8c. Acting as a free standing cantilever, the capacity of the column is exhausted by the combined actions of N, M, m and Na_u when the total height of the structure is $H = H_{crit}$. When $H < H_{crit}$ the skeletal structure is called a 'sway frame' or an 'unbraced structure'.

When $H > H_{crit}$, walls are introduced to the skeletal structure, which then becomes a 'no sway frame' or 'braced structure' (Figure 1.9). The walls (or other types of bracing), which are aligned lengthways with the direction of W_k, replace the function of the columns in resisting overturning moments. Because of their massive strength and stiffness, walls will not be required to replace every column, in fact 1 in 20 is more likely, as they are positioned at strategic positions, mostly at the ends of buildings and around staircases or lift shafts.

Because they are few in number, walls have little effect on the transfer of g_k and q_k, and so Figure 1.7a applies. However, their effect on horizontal forces is considerable as shown in Figure 1.9. W_k is transmitted through the floor, now

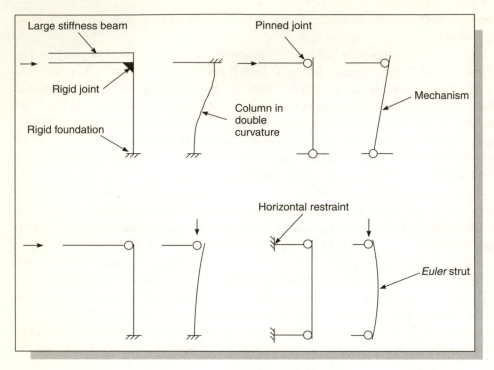

Figure 1.8: Column sway profiles for different end conditions.

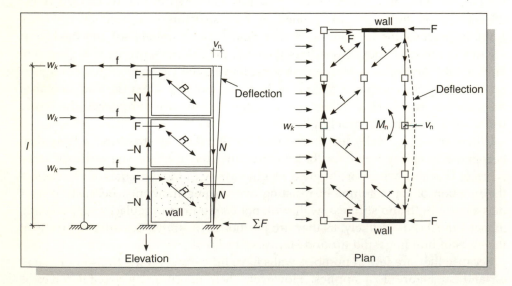

Figure 1.9: Load transfer in a braced frame.

acting as a horizontal plate (as horizontal shear f[kN] and horizontal bending m_h[kNm]) to the beams at the position of the walls; from the beam (as shear F[kN]) to the walls; through the walls (as diagonal force R[kN]) to the foundation (as shear ΣF[kN] and axial reactions N[kN]).

Though shear deflections v occur, the walls are made sufficiently stiff to ensure that the limit of $v/h < 1/500$ will eliminate a_u from further consideration. The sway profile of the columns is no longer dependent on the fixity of the foundation or beams if the connections are pinned. The sway profile is as shown in Figure 1.8d (=Euler strut). Second order deflections a_u only occur if the length of each column between storey heights to greater than critical values imposed by codes. (The minimum value of $W_k = 1.5\% \, G_k$ applies as before.)

Variations in the number and position of walls is endless, but the principles given above do not change. 'Partially braced' structures are an amalgam of unbraced, above or below a certain level, and braced structures, 'Undirectionally braced' structures are braced in one direction and unbraced in the other.

Having resolved the moments, shears, axial forces (and possibly torsion) in the precast elements, and taken note of the effects of shrinkage, bursting, accidental damage, loss at bearing etc. discussed in Section 1.1, element design follows the principles of traditional reinforced or prestressed concrete design. However, the manufacture, transportation, hoisting and temporary stability of the elements and of the structure during construction must be considered. Precast concrete frame designers cannot diverse themselves from the manufacture and site erection procedures. Feedback from the factory (as casting site) to the design office, and from the construction site to the factory and office is essential.

1.3 Why choose a precast structure?

The very existence of a precast concrete industry and the numerous successful building projects achieved using precast concrete, for the whole or just a part of the structure, is proof that the technique is practical and economical. In global terms, the market share of precast 'grey' frames (=structural concrete with no architectural qualities) is probably around 5 per cent of the multi-storey business. However, for precast structures with an integrated facade or other decorative features (Figure 1.6) the global market share is closer to 15 per cent, being as high as 70 per cent in the colder climates and/or where site labour is expensive. Figure 1.10 shows a building in a 'prefabricated concrete town', just outside Helsinki, Finland.

Many people believe that in certain countries, especially where the ratio of labour-to-materials or plant is low, say one man-hour pay is less than 1/50 tonne of cut-and-bent rebar, or one man-hour pay is less than 1/500 daily hire of a large mobile crane, precast cannot compete with cast in situ concrete. Where local labour policies demand high levels of unskilled site labour, heavy concrete prefabricates create the potential for new safety hazards due to transportation, handling and temporary stability. Similarly in countries with a strong steel industry and widespread education in steelwork design, the popular opinion is that precast cannot compete with structural steelwork frames. In many circumstances however precast concrete is the only economical and practical solution, and if the designer is unwilling to consider precast concrete as a total solution the result of this is the so-called 'mixed' solution. Figures 1.11 and 1.12

Figure 1.10: Precast concrete dwellings in a prefabricated concrete town.

show two such cases of using precast prestressed concrete for long spans, whilst vertical supports are in structural steelwork and cast in situ concrete, respectively.

Providing that the stability of the structure is not impaired, the substitution of a steel girder, truss or portal frame for a precast element is taken care of in the connection detail. In the structure shown in Figure 1.11 shallow steel beams were used as direct substitution for precast beams, which would otherwise have had a downstand below the floor. Savings in headroom are the obvious advantage of this method as the shallow steel beam cannot be as economical as a deeper steel beam or prestressed concrete beam.

Mixed construction is now being used in more than 50 per cent of new multi-storey buildings in the western world, where the increased use of precast concrete over the past 10–15 years is due to the move towards greater offsite prefabrication of structural elements. Some of the limitations found in precast concrete inevitably lead to it being used with other materials in a cost effective manner, e.g. to provide

Figure 1.11: Mixed precast and structural steelwork construction.

structural continuity using small quantites of cast in situ reinforced concrete, or to form long span steel or timber roofs. Structurally, combinations may work

Figure 1.12: Mixed cast in situ and precast concrete, with composite steel decking floors (courtesy Trent Concrete, UK).

together or independently but together they can provide many advantages over the use of a single material.

The 'key' to success using any form of mixed or precast concrete is to be able to offer the client, architect and consulting engineer, a solution that is:

1. Buildable

   (a) the construction sequence is sensible enabling other trades to dovetail into programmes;

   (b) the construction is safe and temporary stability is guaranteed;

   (c) economic cranes are used; and

   (d) labour is skilled.

2. Cost effective

   (a) building components cost-per-structural capacity is comparable to other materials, e.g. cost/bending moment or cost/axial load capacity;

   (b) the overall building costs, inclusive of transport, fixing finishing, main-tenance and repair, are competitive;

 (c) production is Quality Assured – lowest standard deviation on the population of all elements, materials, and methods;

 (d) the solution uses factory engineered concrete; and

 (e) destructive load testing or non-destructive testing assurances are given.

3. Fast to erect

 (a) although the precast manufacturing period may be several weeks, once started, construction proceeds rapidly;

 (b) following trades (bricklayers, electricians, joiners) move in quickly;

 (c) hand-over or possession can be phased;

 (d) clear heights and floor zones are satisfied;

 (e) service routes are not interrupted;

 (f) beam and column sizes are satisfied, so too the bracing positions; and

 (g) decorative concrete, both internally and externally may be exploited.

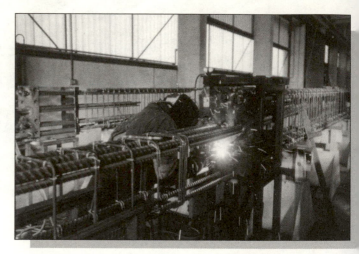

Figure 1.13: Fabrication of a beam end connection (courtesy Trent Concrete, UK).

Prefabrication factories are not cheap to run. In some instances, it is certainly economical to establish a prefabrication facility on site, or in an adjacent field, and work from there. Whichever method is chosen it is necessary to be able to do things in the factory that are impossible to achieve on site. For example, the welded steel connector being fabricated in Figure 1.13, has a shear capacity of around 400 kN. The two end connections alone probably cost as much as the rest of the element, but in being able to transmit this shear force in such a shallow depth

Figure 1.14: Polishing concrete to create pseudo marble finishes (courtesy Trent Concrete, UK).

Figure 1.16: Precast spirals adorn Braynston School, UK (courtesy Techcrete Ltd, Ireland and UK).

Figure 1.15: Polished columns resulting from the actions in Figure 1.14 (courtesy Trent Concrete, UK).

the advantages to the structure as a whole are obvious. Surface finishes are achieved in controlled environments using skilled labour and/or automated machines. The surface preparation in Figure 1.14 leads to a psuedo marble finish to the columns shown in Figure 1.15. Alternative grey concrete plus bolted-on marble panels would have been much more expensive, less durable and of greater size.

So what makes precast concrete different to other forms of construction? The answer is quite obvious when you look at the precast concrete achieved in the columns in Figure 1.16.

2 Materials used in precast structures

2.1 Concrete

Precast concrete is of the highest possible quality, both in terms of strength and durability. The essence of this statement is captured in Figure 2.1, where concrete is accurately delivered to every part of the mould, ensuring zero segregation, honey-

Figure 2.1: Precise delivery of concrete to the mould (courtesy Tammer Elementti Oy, Finland).

combing and minimal vibration. However, contrast this view with the precast concrete shown in Figure 2.2 to easily understand the problems of workmanship, rather than materials defects, that certain parts of the industry endure. Using materials that have passed strict quality control procedures, rapid hardening cement is mixed with excellent quality aggregates of known source and purity, often in computer controlled batching and mixing plant, to produce concrete of specified workability and strength. Even the introduction of small quantities of uncontaminated recycled concrete, usually from the factory's own waste production, superplasticizers and pozzolanic materials (such as pulverized fuel ash) has not reduced this standard. To quote standard deviations of less than 2N/mm^2 on concrete of 28-day compressive cube strength of between $f_{cu} = 50$ and 80N/mm^2 would not be inaccurate.

Figure 2.2: Compaction problems in a precast concrete column.

When precasting in a modern factory it is vital to achieve the dual requirements of good workability and early strength. Concrete is often transported automatically from an automatic batching plant to the mould leaving no opportunity for workability tests. It is unlikely that such tests are required, thanks to the experience of batching staff. Producers do not vary the mix specification unnecessarily – for example Grade C40 concrete ($f_{cu} = 40 \text{N/mm}^2$) might be used for a staircase even though the design calls for a lower strength of say $f_{cu} = 25 \text{N/mm}^2$.

For the production of standard elements such as columns and beams, concrete is cast into clean steel (sometimes timber, concrete) moulds, accurate to $\pm 3 \text{mm}$, or less, in cross-section. The use of clamped vibrators (Figure 2.3) tuned to the correct oscillations for the size and weight of the filled mould ensures correct compaction to a density of around 2400kg/m^3 (excluding reinforcement). The resulting surface-

Figure 2.3: Production of precast concrete structural elements in steel moulds (courtesy Trent Concrete, UK).

Figure 2.4: Production of precast concrete facade elements in timber moulds (courtesy Trent Concrete, UK).

finish, results in minimum porosity for maximum durability. Concrete strengths are made to match the optimum performance of each element, such that flexural members are produced in grade C40 concrete, whilst compression members are in C50 (or C60 if prestressed).

Non-standard elements, such as the cladding panel in preparation in Figure 2.4, are cast in timber or fibre glass moulds (epoxy-based materials may be remoulded after use). Although the surface-finish from a timber mould is not equal to that from a steel mould, it is nonetheless better than Class B to BS8110, clause 6.10.3. The grade of concrete is usually C40.

Machine extruded concrete is not, by definition cast and compacted. It is instead forced through rotating augers to pass out through steel formers, possibly with mandrels to create voids, as shown in Figures 2.5 and 2.6. Slip-formed concrete is similarly produced through a machine that compacts concrete by rows of small hammers – known as the 'shear-compaction' method. To enable both these processes to happen, the concrete is 'earth-dry', having a water: cement (or water–cementitious material) ratio of around 0.3. Aggregates are carefully selected, especially the sand grading and content, and the shape and size of these coarse aggregates, typically 14 mm down. Although not exclusive to these production methods, concrete produced in this way is usually prestressed for flexural members such as floor slabs (Figure 2.7). Because prestressed concrete benefits from additional strength in flexural compression, it is advantageous to use grade C60 concrete, although for certain geometry the requirement is only grade C50. Table 2.1 gives the relevant information. Mix design details are given by Elliott.[1]

Each producer will modify their mix according to variations in local supplies. The grading of fine aggregate is often the most critical mix

Figure 2.5: Rotating augers create forced extrusion of concrete.

Figure 2.6: *Circular mandrels create circular voids in extruded hollow core floor slabs.*

parameter, especially in the dry mixes used in slip-formed and extruded concrete. Although the 28-day characteristic strength specified in the design must be achieved, the controlling factor in mix design is the strength at demoulding, or in the case of prestressed concrete, its strength at detensioning (=transfer). Demoulding and

Figure 2.7: *Long line prestressed extrusion of hollow core floor slabs (courtesy Spaencom Betonfertigteile GmbH & Co. KG., Germany).*

detensioning of *cast* concrete takes place at around 18 hours. Detensioning of machine slip-formed or extruded concrete takes place between 10 and 18 hours. Various methods to accelerate the early strength of concrete include rapid hardening Portland cement, chemical accelerators (calcium chloride must not be used) and by external heating, such as steam curing and electrical heating. Even micro-

Table 2.1: Concrete strengths and elastic moduli used in precast elements

Component	Type	Grade	f_{cu} at 28 days (N/mm^2)	Demould cube strength (N/mm^2)	Design strength (N/mm^2)	Tensile strength (N/mm^2)	E_c at 28 days (kN/mm^2)	E_{ci} (kN/mm^2)
Beams, shear walls staircases, wet cast floors	RC	C40	40	20–25	18.0	N/A	28	N/A
Columns, load-bearing walls	RC	C50	50	25–30	22.5	N/A	30	N/A
Beams, columns, dry cast slabs	PSC	C50	50	30–35	22.5	3.2	30	27
Dry cast slabs	PSC	C60	60	35–40	27.0	3.5	32	28

Table 2.2: Strengths and short-term elastic modulus for typical concrete used in composite construction, from BS8110

Type of concrete	f_{cu} (N/mm^2)	f_t (N/mm^2)	E_c (kN/mm^2)
In situ	25	–	25.0
In situ	30	–	26.0
Precast reinforced	40	–	28.0
Prestressed	50	3.2	30.0
Prestressed	60	3.5	32.0

wave oven techniques have been tried. Care must be taken in thin walled section, such as hollow core floor units, to avoid thermal restraint due to large temperature differentials.

Where site concrete is added to precast elements to form composite sections (see Sections 4.4.4 and 5.5) the relative strengths and stiffness of the two materials must be considered, namely, the modular ratio method. Typical data are shown in Table 2.2. The most common combinations of in situ-to-precast strengths are C25-to-C40, and C30-to-C50/C60.

2.2 Steel reinforcement

Precast concrete elements can, if necessary, be heavily reinforced because they are cast horizontally. BS8110 permits up to 10 per cent of the cross-section to be reinforced, although this amount is rarely used in favour of higher concrete strengths.

High tensile hot rolled ribbed bar (HT rebar) is used in 95 per cent of cases, even in shear links where mild steel would be suitable. The small cost difference compared to the additional strength, i.e. 460 vs $250\,\text{N/mm}^2$, and the need for consistency of habit when assembling cages, makes it more economical. Tying wire is more secure around ribbed bar making the cage more robust. Mild steel is often used for projecting loops, etc., because it is easier to hand-bend on site. Bar diameters commonly used are 8 and 10 mm for

Figure 2.8: 7-wire helical strand in a hollow core floor unit.

column stirrups, 10 and 12 mm for beam stirrups and other distribution or anti-crack bars, and 16, 20, 25, 32 and 40 mm for main flexural bars.

Welded fabric, or 'mesh', is used in flat units such as slabs and walls. It is also used to reinforce structural toppings to floor slabs on site. The popular mesh size for flat panels, walls, etc., is A142 or A193 (6 or 7 mm bars at 200 mm centres in both directions). A rectangular mesh C283 (6 mm at 100 mm centres × 5 mm at 400 mm centres) is often used in one-way spanning units such as the flanges of double-tee slabs. For rebar and mesh the characteristic strength $f_y = 460\,\text{N/mm}^2$. The design strength $f_d = f_y / \gamma_m = 460/1.05 = 438\,\text{N/mm}^2$. Young's modulus is taken as $200\,\text{kN/mm}^2$.

Two main types of steel are used for pretensioning: (1) plain or indented (or crimped) wire; and (2) 7-wire helical strand. Figure 2.8 shows 12.5 mm diameter strand in a hollow core floor unit. Note the excellent compaction of concrete around the strand. This is vital for good transfer of stress to the concrete. The choice of tendon is often a matter of the arrangement of tendons and the correct distribution of pretensioning force in a section. Large tendons should not be placed in thin wall sections – to avoid localized splitting and bond failure, the edge cover to tendons should be at least twice the diameter. For this reason helical strand is preferred in larger units or where the level of prestress is high.

There is a long-term loss of force in all pretensioning tendons, called 'relaxation'. It is due to a stress relieving heat treatment process. Today tendons are classified as 'Class 2 : 5 per cent low relaxation' meaning that the final stress after 1000 hours relaxation is 95 per cent of the original. (Class 1 relaxation is greater but is rarely used today.) Indeed the relaxation of some 'standard' and 'super strand' is 2.5 per cent, although manufacturers may quote characteristic relaxation as low as 1.6 per cent at 1000 hours. When calculating relaxation losses BS8110

Table 2.3: Types and specification of pretensioning tendons

Type	Diameter (mm)	Cross-section area (mm²)	Characteristic load (kN)	Nominal characteristic strength f_{pu} (N/mm²)	Elastic modulus (kN/mm²)
Wire	5.0	19.6	30.8	1570	205
	7.0	38.5	60.4	1570	
Standard strand	10.9	71.0	125	1760	195
	12.5	94.2	165	1750	
	15.2	138.7	232	1670	
Super strand	12.9	100.0	186	1860	195
	15.7	149.7	265	1770	
Drawn strand	12.7	112.4	209	1860	195
	15.2	164.8	300	1820	

applies a factor of 1.2 to pretensioned Class 2 tendons. Specified strengths and elastic moduli are given in Table 2.3.

2.3 Structural steel and bolts

Structural steelwork sections are used in many types of precast elements, especially at the connections. These include rolled rectangular and square hollow sections (RHS, SHS), solid billets, channels and angles, plates and welded-tees, etc. Details of how these are used in practice are given in Sections 8.4 and 9.2. Structural sections such as Universal beams and columns (UB, UC) may be cast into precast elements to enhance strength where the reinforced concrete capacity is exhausted. However, this may have severe cost implications that must be carefully examined.

Rolled steel sections and bent or flat steel plates are welded to form steel connectors in many highly stressed support situations where direct contact between concrete surfaces is to be avoided. The steel used is grade 43 (mostly) or grade 50. Welded electrodes are mostly grade E43. When used to join grade 43 steel, the yield strength of the weld, $p_w = 215\,\text{N/mm}^2$. When used in combination with grade 50, steel grade E51 electrodes gives $p_w = 255\,\text{N/mm}^2$. The usual rules for lap lengths $(4t_w)$, throat thickness $(0.7t_w)$ and returns and run-outs $(2t_w)$ apply, where t_w is the leg length. Intermittent fillet welds are rarely used as weld lengths tend to be short.

Hot dipped galvanized steel is used for exposed connections, usually of secondary structural significance, such as dovetail channels for brick ties. The basic plate is grade 43 steel, and grade 50 is used in the more highly stressed plates.

Black bolts grade 4:6 ($p_y = 195\,\text{N/mm}^2$ in tension and $160\,\text{N/mm}^2$ in shear) and 8:8 ($p_y = 450:375$) are used in many connections. High strength friction grip bolts are used in special circumstances where the integrity and safety (both temporary and permanent) of connections made with ordinary bolts in clearance holes cannot be guaranteed.

2.4   Non-cementitious materials

Epoxy-based mortars are used to make, either partially or completely, connections where a rapid gain in strength is required, e.g. up to 40 N/mm^2 in 2–3 hours. Care is taken to ensure that these materials have not exceeded their shelf life and are being used correctly for the right application. The thermal expansion of epoxy materials is seven times that of concrete, and this should be accounted for in design. Epoxy compounds comprise two parts to be site mixed: (1) epoxy resins; and (2) hardeners. They are occasionally used as pressure injections for crack filling or to restore tensile strength. Manufacturer's procedures should be strictly adhered to.

Neoprene, rubbers and mastics are used for soft bearings, backing strips, etc. The PCI Manual on Architectural Precast Cladding[2] gives extensive guidance on the use of these materials. Although they are not used extensively in precast structures, a typical range of applications is given in Table 2.4.

Table 2.4: Non-cementitious materials in precast construction

Material	Application	Data (at 20 °C)
Elastomeric bearings, e.g. neoprene, rubber	Bearing pads	Comp. strength = 7–10 N/mm^2 Shear strength = 1 N/mm^2 Compressive strain = 15%
Bitumen impregnated sealing strip	Backing strip to concrete joints	Compressibility = 85%
Polysulphide sealants	Expansion joints	Elongation strain <50%
Epoxy resin mortars	Compression, shear or tension joints	Comp strength = 55–110 N/mm^2 Tensile strength = 9–20 N/mm^2 Elongation strain <15%
Polyester resin mortars	Compression, shear or tension joints	Comp strength = 55–110 N/mm^2 Tensile strength = 6–15 N/mm^2 Elongation strain <2%
Polystyrene	Filler, back-up blocks	
PTFE (Teflon)	Frictionless bearings	Compression = 8 N/mm^2 virgin and 15 N/mm^2 reinforced. Coefficient of friction 0.04

References

1   Elliott, K. S., *Multi-storey Precast Concrete Framed Structures*, Blackwell Science, Oxford, UK, 1996, 624p.
2   Prestressed Concrete Institute, *Manual for the Structural Design of Architectural Precast Concrete*, 2nd edn, PCI, Chicago, USA, 1989, 340p.

3 Precast frame analysis

3.1 Types of precast concrete structures

Preliminary structural design, in what many people refer to as the feasibility stage, is more often a recognition of the type of structural frame, which is best suited to the form and function of a building, than the structural design itself. The creation of large 'open plan' accommodation giving the widest possible scope for room utilization clearly calls for a column and slab structure, e.g. in Figure 3.1, where internal partitions could be erected to suit any client's needs. The type of structure used in this case is often referred to as 'skeletal' – resembling a skeleton of rather small but very strong components of columns, beams, floors, staircases, and sometimes structural (as opposed to partition) walls. Of course a skeletal structure could be designed in cast in situ concrete and structural steelwork, but here we will consider only the precast concrete version.

The basis for the design of precast skeletal structures has been introduced in Figure 1.7. The major elements (=the precast components) in the structure are shown in Figure 3.2. Note that the major connections between beams and floors are designed and constructed as 'pinned joints' and therefore the horizontal elements (slabs, staircases, beams) are all simply supported. They need not

Figure 3.1: Precast skeletal structure showing large unobstructed spaces for the benefit of both the construction workers and client.

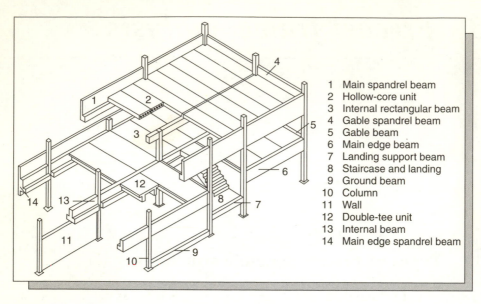

1	Main spandrel beam
2	Hollow-core unit
3	Internal rectangular beam
4	Gable spandrel beam
5	Gable beam
6	Main edge beam
7	Landing support beam
8	Staircase and landing
9	Ground beam
10	Column
11	Wall
12	Double-tee unit
13	Internal beam
14	Main edge spandrel beam

Figure 3.2: Definitions in a precast skeletal structure.

always be pinned (in seismic zones the connections are made rigid and very ductile) but in terms of simplicity of design and construction it is still the pre-ferred choice. Vertical elements (walls, columns) may be designed as continuous, but because the beam and slab connections are pinned there is no global frame action and no requirement for a frame stiffness analysis, apart from the distribution of some column moments arising from eccentric beam reactions. The stiff bracing elements such as walls are designed either as a storey-height element, bracing each storey in turn, or as a continuous element bracing all floors as tall cantilevers.

In office and retail development distances between columns and beams are usually in the range 6 m to 12 m, depending on the floor loading and intended use. In multi-storey car parks where the vehicle loading is always about the same it is around 16 m. The exterior of the frame – the building's weatherproof envelope, could also be a skeletal structure, in which case the spaces between the columns would be clad in brickwork, sheeting etc. Alternatively, the envelope might be constructed in solid precast bearing walls, which dispenses with the need

Figure 3.3: Wall frame with 3.6 m wide hollow core floors (courtesy A. Curd and Partners, USA).

Figure 3.4: Exterior facade to wall frame.

for beams, and is referred to as a 'wall frame'.

An example of a building where a precast wall frame would be the obvious choice is shown in Figure 3.3 – the walls are all load bearing and they support one-way spanning floor slabs. There is less architectural freedom compared to the skeletal frame, e.g. walls should (preferably) be arranged on a rectangular grid and a fixed modular distance, usually 300 mm, between walls is quite important economically. A wall frame may be more economical and may often be faster to build especially if the external walls are furnished with thermal insulation and a decorative finish at the factory. Figure 3.4 is a good example of this. Distances between walls may be around 6 m for hotels, schools, offices and domestic housing, and 10 m to 15 m in commercial developments. Given this description, wall frames appear to be very simple in concept, but in fact are quite complicated to analyse because the walls have very large in-plane rigidity whilst the connections between walls and floors are more flexible. Differential movement between wall panels and between walls and floors has resulted in major serviceability problems for more than a 25-year life, often leading to a breakdown in the weatherproof envelope and the eventual condemnation of buildings which are nevertheless structurally adequate.

A third category of precast building is the 'portal frame', used for industrial buildings and warehouses where clear spans of some 25 m to 40 m I or T section prestressed rafters are necessary (Figures 3.5 and 3.6). Although portal frames are nearly always used for single-storey buildings they may actually be used to form the roof structure to a skeletal frame, and as this book is concerned with multistorey structures it gives us a reason to mention them. The portal frame looks simple enough and in fact is quite rudimentary in design, providing that the flexural rotations at the end of the main rafters, which we can assume will always cause cracking damage to the bearing ledge, are catered for by inserting a flexible pad (e.g. neoprene) at the bearing. As mentioned earlier in this section, pinned connections between the rafter and column are the preferred choice – they are easy to design and construct. But the columns must be designed as moment resisting cantilevers – which might cause a problem in some structures as explained later in Section 6.2. A moment resisting connection is equally possible allowing moment distribution in the column. However, unless the columns are particularly tall, say more than about 8 m it is not worth the extra effort.

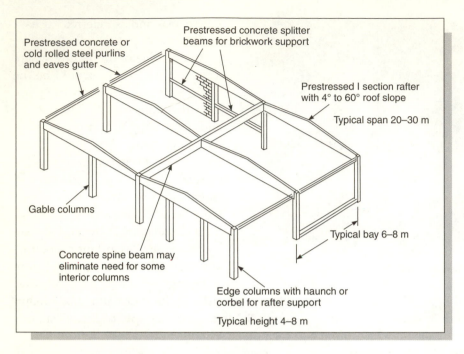

Prestressed concrete or cold rolled steel purlins and eaves gutter

Prestressed concrete splitter beams for brickwork support

Prestressed I section rafter with 4° to 60° roof slope

Typical span 20–30 m

Gable columns

Concrete spine beam may eliminate need for some interior columns

Typical bay 6–8 m

Edge columns with haunch or corbel for rafter support

Typical height 4–8 m

Figure 3.5: Definitions in a precast portal frame.

Figure 3.6: Examples of a precast portal frame (courtesy Crendon Ltd, UK).

Table 3.1: Application and types of precast concrete frames

Use of building	Number of storeys*	Interior spans (m)	Skeletal frame	Wall frame	Portal frame
Office	2–20	6–15	✓		
	2–50	6–15		✓	
Retail, shopping complex	2–10	6–10	✓	✓	
Cultural	2–10	6–10	✓		
Education	2–5	6–10	✓	✓	
Car parking	2–10	15–20	✓		
Stadia	2–4	6–8	✓		
Hotel	2–30	6–8		✓	
Hospital	2–10	6–10		✓	
Residential	1–20	4–6		✓	
Industrial	1	25–40			✓
Warehouse with office	2–3	6–8	✓		
		25–40			✓

Note:
*Typical values, depending on location, terrain, requirements etc.

Table 3.1 reviews the various types of precast structures with respect to their possible applications.

3.2 Simplified frame analysis

One of the most frequently asked questions is…*how is a precast concrete structure analysed compared to a monolithic cast in situ one?* The first response is to say that a precast concrete structure is not a cast in situ structure cut up into little pieces making it possible to transport and erect. It was mentioned in Chapter 1 that the passage of forces through the prefabricated and assembled components in a precast structure is quite different to a continuous (=monolithic) structure. This is certainly true near to connections. It is therefore possible to begin a global analysis by first considering the behaviour of a continuous frame and identifying the positions where suitable connections in a precast frame may be made. A 2D in-the-plane simplification is appropriate in the first instance. This is defined in Figure 3.7 where there are no structural *frame* components, only simply supported floor units, connecting the 2D in-plane frames together.

Figure 3.8 shows the approximate bending moments and deflected shape in a three-storey continuous beam and column frame subject to vertical (gravity) patch loads and horizontal (wind) pressure. Call this frame F1. The beam–column connections have equal strength and stiffness as the members. The stability of

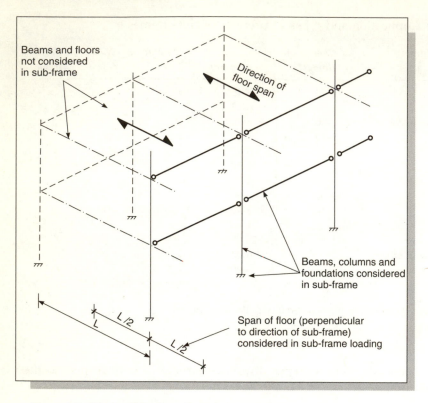

Figure 3.7: 2D simplification of 3D skeletal structure.

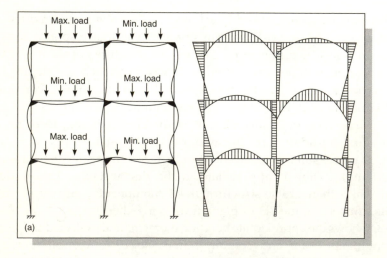

Figure 3.8a: Deformations and bending moment distributions in a continuous structure due to gravity loads.

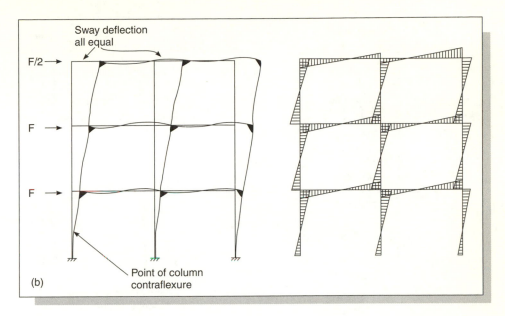

Figure 3.8b: Deformations and bending moment distributions in a continuous structure due to sway loads.

F1 is achieved through the combined action of the beams, columns and beam–column connections in bending, shear and axial. This is called an 'unbraced' frame. There are points of zero moment ('contraflexure') in F1 which depend on the relative intensity of the two load cases. If gravity load is dominant beam contraflexure is near to the beam–column connection, typically 0.1 times the span of the beam, if horizontal load is dominant (more rare) contraflexure is at mid-span. In the column, contraflexure is always at mid-storey height, and this is a good place to make a pinned (notionally = small moment capacity) connection between two precast columns.

Now, if the strength and stiffness of the connection at the end of the beam is reduced to zero, whilst the column and the foundation are untouched, the resulting moments and deflections in this frame, called F2, are as shown in Figure 3.9. The columns alone achieve the stability of F2 – the beams transfer no moments, only axial forces and shear. The foundations must be moment resisting (=rigid). This is the principle of a pin jointed unbraced skeletal frame. In taller structures, >3 storeys or about 10 m, the large sizes of the columns become impractical and uneconomic leading to bracing. The bracing may be used in the full height, called a 'fully braced' frame, or up to or from a certain level, called a 'partially braced' frame. The differences are explained in Figure 3.10. The bracing could be located in the upper storeys providing the columns in the unbraced part below the first floor are sufficiently stable to carry horizontal forces and any second order moments resulting from slenderness.

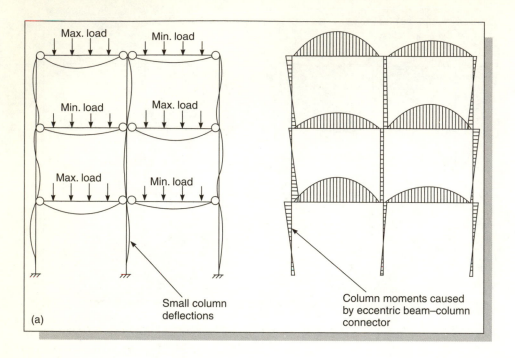

(a)

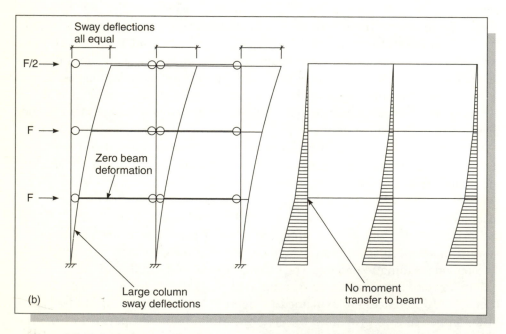

(b)

Figure 3.9: Deformations and bending moment distributions in a pinned jointed structure due to (a) gravity loads; and (b) sway loads.

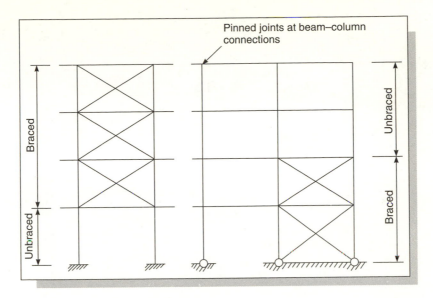

Figure 3.10: Partially braced structures.

Pinned connections may be formed at other locations. Referring back to frame F1, if the flexural stiffness of the members at the lower end of a column is greater than that at the upper end, the point of contraflexure will be near to the lower (=stiffer) end of the column. If the strength and stiffness of the lower end of the column is reduced to zero, whilst the beam and beam–column connections are untouched, the resulting moments and deflections in this frame, called F3, are as shown in Figure 3.11a. The stability of F3 is achieved by the portal frame action of inverted U frames – clearly not a practical solution for factory cast large spans so that this method is used for repetitious site casting. Therefore, a practical solution is to prefabricate a series of L frames as shown in Figure 3.11b. Foundations to F3 may be pinned, although most contractors prefer to use a fixed base for safety and immediate stability.

The so-called 'H-frame' is a variation on F3. Referring back to frame F1, if pinned connections are made at the points of column contraflexure structural behaviour is similar to a continuous frame as explained in Figure 3.12. Connections between frames are made at mid-storey height positions. Although in theory the connection is classed as pinned, in reality there will be some need for moment transfer, however small. Therefore, H-frame connections are designed with finite moment capacity, this also gives safety and stability to the H-frames which by their nature tend to be massive. The foundation to half-storey height ground floor columns must be rigid. The connection at the upper end of the column may be pinned if it is located at a point of contraflexure. If not the connection must possess flexural strength as shown in Figure 3.13 where the H-frame has been used in a number of multi-storey grandstands.

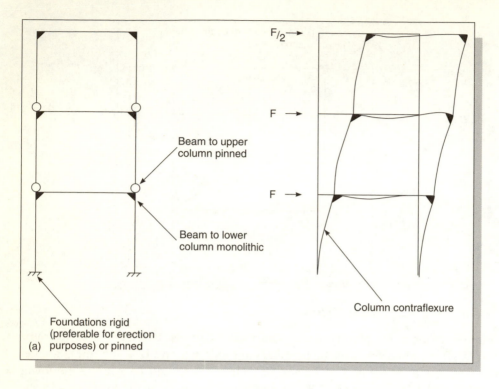

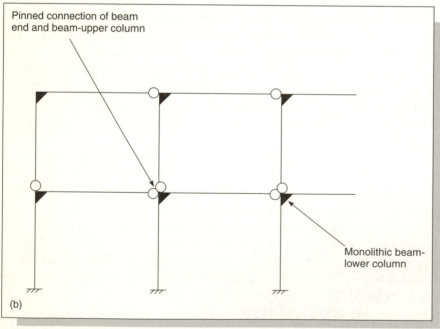

Figure 3.11: Structural systems for (a) portal U-frames; and (b) portal L-frames.

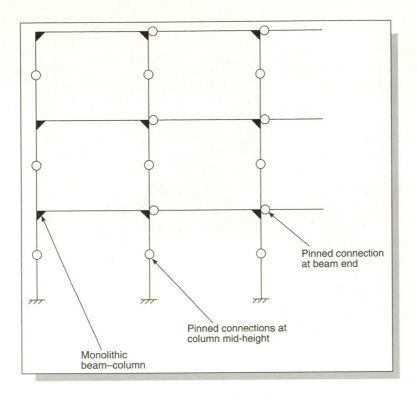

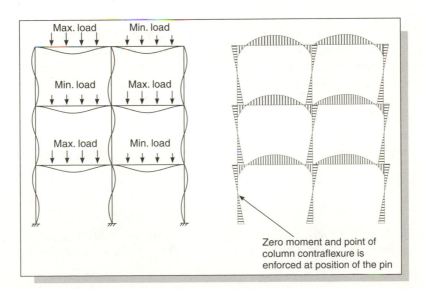

Figure 3.12: Structural system, deformations and bending moment distributions in an H-frame.

3.3 Substructuring methods

The object of analysis of a structure is to determine bending moments, shear and axial forces throughout the structure. Monolithic 2D plane frames are analysed using either rigorous elastic analysis, e.g. moment distribution or stiffness method, either manually or by computer program. Moment redistribution may be included in the analysis if appropriate. However, often it is only required to determine the moments and forces in one beam or one column, so codes of practice allow simplified substructuring techniques to be used to obtain these values. Fig-

Figure 3.13: Example of H-frame used in stadia (courtesy Tarmac Precast, UK).

ure 3.14 gives one such substructure, called a 'subframe' – refer to Ref. 1 for further details. If the frame is fairly regular, i.e. spans and loads are within 15 per cent of each other, substructuring gives 90–95 per cent agreement with full frame analysis.

Substructuring is also carried out in precast frame analysis, except that where *pinned* connections are used no moment distribution or redistribution is permitted. Figure 3.15 shows subframes for internal beam and upper and ground floor columns where all beam–column connections are pinned. (For rigid connections refer to Figure 3.14 etc.) Horizontal wind loads are not considered in subframes because the bending moments due to wind loads in an unbraced frame (there are no column moments due to wind in a braced frame) are additive to those derived from subframes. Elastic analysis is used to determine moments, forces and deflections, but a plastic (=ultimate) section analysis is used for the design of the components. Clearly some inaccuracies must be accepted.

The critical loading combinations with their associated partial safety factors for load γ_f are:

1   all spans loaded with the maximum ultimate load $w_{max} = \gamma_f g_k + \gamma_f q_k$; and

2   alternate (='pattern') spans loaded with the maximum ultimate load, $\gamma_f g_k + \gamma_f q_k$ on one span and the minimum $w_{min} = 1.0 g_k$ on the adjacent.

Where g_k and q_k are characteristic dead and imposed (=live) uniformly distributed loads.

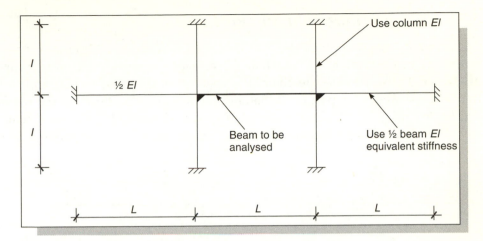

Figure 3.14: Substructuring method for internal beam in a continuous frame.

3.3.1 Beam subframe

The subframe consists of the beam to be designed of span L_1, and half of the adjacent beams of span L_2 and L_3 (Figure 3.15a). The eccentricity of the beam end reaction from the centroidal axis of the column is e. Alternate pattern loading is used. The height of the column above and below the beam is actually of no consequence to the beam. It is assumed that the cross-section and flexural stiffness of the column is constant.

The maximum moment in the beam is

$$M_1 = \frac{w_{\max}(L_2 - 2e)^2}{8} \qquad 3.1$$

The beam end reaction is

$$R_1 = \frac{w_{\max}L_2}{2} \qquad 3.2$$

(Note the shear force in the beam is $V = w_{\max}\dfrac{(L_2 - 2e)}{2}$.)
End reactions in the adjacent beams are

$$R_1 = \frac{w_{\min}L_1}{2} \quad \text{and} \quad R_3 = \frac{w_{\min}L_3}{2} \qquad 3.3$$

The resulting maximum bending moment in the column is given by:

$$M_{\text{col}} = \frac{(R_2 - R_1)eh_2}{(h_2 + h_3)} \qquad 3.4$$

assuming that $R_1 < R_3$ and $h_3 > h_2$. Figure 3.16a shows the final moments.

3.3.2   *Upper floor column subframe*

The subframe consists of the column to be designed of height (=distance between centres of beam bearing) h_2, and half the adjacent columns of heights h_1 and h_3 (Figure 3.15b). Because the column is continuous the cross-section and flexural stiffness EI of each part of the column is considered as shown in the figure. The beams are pattern loaded as above, of span $L_4/2$ and $L_5/2$, and the eccentricity of each beam end reaction from the centroidal axis of the column is e_4

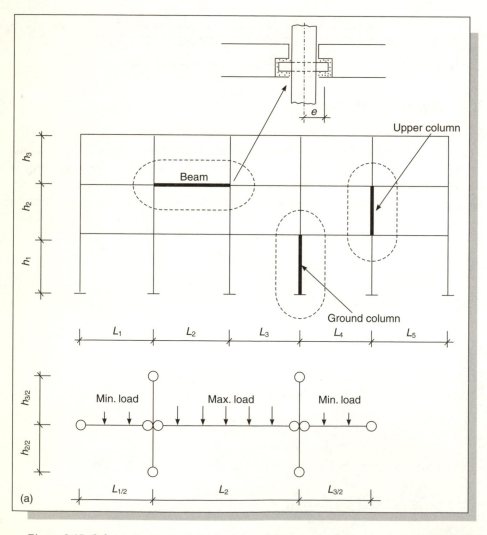

Figure 3.15: *Substructuring methods for internal beam and columns in a pinned-jointed frame.*

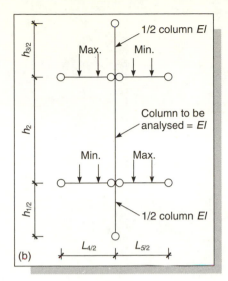

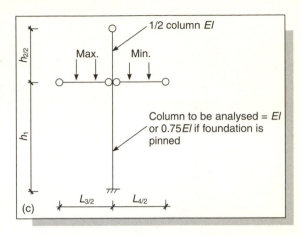

(b)

(c)

Figure 3.15 (continued): Substructuring methods for internal beam and columns in a pinned-jointed frame.

and e_5, respectively. The moment at the upper end of the designed column is given by:

$$M_{\text{col, upper}} = (R_4 e_4 - R_5 e_5) \frac{\dfrac{EI_2}{h_2}}{\dfrac{EI_2}{h_2} + \dfrac{EI_3}{h_3}}$$

3.5

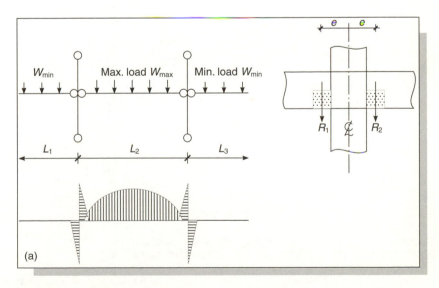

(a)

Figure 3.16a: Bending moments in a pinned-jointed frame for internal beams.

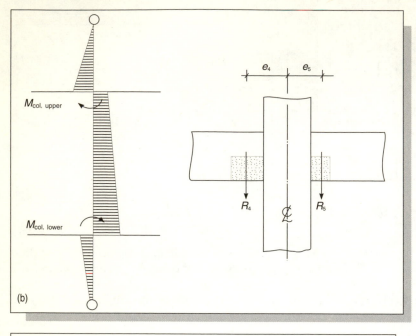

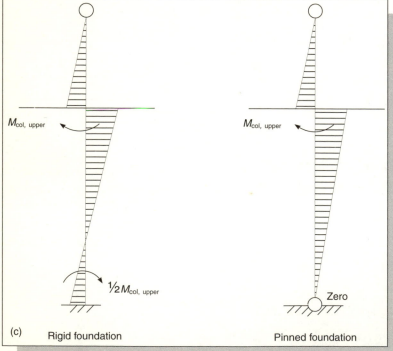

Figure 3.16 (continued): Bending moments in a pinned jointed frame for (b) upper floor columns and (c) ground floor columns.

and at the lower end is:

$$M_{\text{col, lower}} = (R_4 e_4 - R_5 e_5) \frac{\dfrac{EI_2}{h_2}}{\dfrac{EI_1}{h_1} + \dfrac{EI_2}{h_2}} \qquad 3.6$$

where R_4 and R_5 are given in Eqs 3.2 and 3.3. Figure 3.16b shows the final moments. Note that patch loading produces single curvature in the columns.

3.3.3 Ground floor column subframe

The subframe consists of the column to be designed of height (=distance between centre of first floor beam bearing and 50 mm below top of foundation (see Section 9.4)) h_1, and half the adjacent column of height h_2 (Figure 3.15c). All other details are as before.

If the foundation is rigid (=moment resisting) the moment at the upper end of the designed column is given by Eq. 3.5 with appropriate notation. The carry-over moment at the lower end is equal to 50 per cent of the upper end moment.

If the foundation is pinned, the upper end moment is given by:

$$M_{\text{col, upper}} = (R_4 e_4 - R_5 e_5) \frac{0.75\dfrac{EI_1}{h_1}}{0.75\dfrac{EI_1}{h_1} + \dfrac{EI_2}{h_2}} \qquad 3.7$$

and the lower end moment is zero. Figure 3.16c shows the final moments. Patch loads produce single curvature in the columns.

Example 3.1
Determine, using substructuring techniques, the bending moments in the beam X and columns Y and Z identified in Figure 3.17. The beam–column connections are pinned and the foundation is rigid. The distance from the edge of the column to the centre of the beam end reaction is 100 mm. Characteristic beam loading is $g_k = 40\,\text{kN/m}$ and $q_k = 30\,\text{kN/m}$. Adopt BS8110 partial safety factors of 1.4 and 1.6 for dead and live load, respectively.

Solution
$w_{\text{max}} = 1.4 \times 40 + 1.6 \times 30 = 104\,\text{kN/m}$; $w_{\text{min}} = 40\,\text{kN/m}$.
Beam subframe

$$e = 450/2 + 100 = 325\,\text{mm}$$

$$M_1 = \frac{104 \times (8.000 - 2 \times 0.325)^2}{8} = 702.3\,\text{kNm} \qquad (using\ Eq.\ 3.1)$$

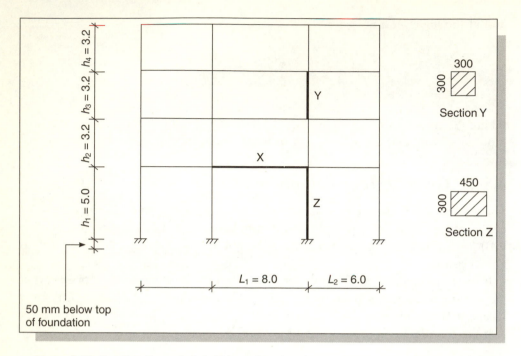

Figure 3.17: Detail to Example 3.1. (Dimensions in m).

Column Y subframe
Beam end reactions $R_1 = 104 \times 8.000/2 = 416\,\text{kN}$; $R_2 = 40 \times 6.000/2 = 240\,\text{kN}$
$e_1 = e_2 = 300/2 + 100 = 250\,\text{mm}$
but $EI_3/h_3 = EI_2/h_2$
At upper and lower ends

$$M_{\text{col}} = (416 - 240) \times 0.250 \times 0.5 = 22.0\,\text{kNm} \qquad (\text{using Eq. 3.5})$$

Column Z subframe
Beam end reactions as before. $e_1 = e_2 = 450/2 + 100 = 325\,\text{mm}$
Given that E is constant;

$$I_1/h_1 = (300 \times 450^3)/(12 \times 5050) = 451 \times 10^3\,\text{mm}^3;$$
$$I_2/h_2 = (300 \times 300^3)/(12 \times 3200) = 211 \times 10^3\,\text{mm}^3$$

At upper end

$$M_{\text{col, upper}} = (416 - 240) \times (0.325 \times 451)/(451 + 211) = 39.0\,\text{kNm} \qquad (\text{using Eq. 3.5})$$

At lower end $M_{\text{col, lower}} = 50\% \times 39.0 = 19.5\,\text{kNm}.$

3.4 Connection design

Connections form *the* vital part of precast concrete design and construction. They alone can dictate the type of precast frame, the limitations of that frame, and the erection progress. It is said that in a load bearing wall frame the rigidity of the connections can be as little as 1/100 of the rigidity of the wall panels – 200 N/mm^2 per mm length for concrete panels vs 2.7 to 15.0 N/mm^2 per mm length for joints.[2] Also the deformity of the bedding joint, i.e. the invisible interface where the panel is wet bedded onto mortar, between upper and lower wall panels can be 10 times greater than that of the panel.

The previous paragraph contained the words *connections* and *joints* to describe very similar things. Connections are sometimes called 'joints' – the terminology is loose and often interposed. The definition adopted in this book is:

    Connection: is the total construction between two (or more) connected components: it includes a part of the precast component itself and may comprise of several joints.

    Joint: is the part of a connection at individual boundaries between two elements (the elements can be precast components, in situ concrete, mortar bedding, mastic sealant, etc.).

For example, in the beam–column assembly shown in Figure 3.18, a bearing joint is made between the beam and column corbel; a shear joint is made between the dowel and the angle, and a bolted joint is made between the angle and column. When the assembly is completed by the use of in situ mortar/grout the entire construction is called a connection. This is because the overall behaviour of the assembly includes the behaviour of the precast components plus all of the interface joints between them. Engineers prove the capacity of the entire connection by assessing the behaviour of the individual joints.

Structurally joints are required to transfer all types of forces – the most common of these being compression and shear, but also tension, bending and (occasionally torsion). The combinations of forces at a connection can be resolved into components of compressive, tensile and shear stress and these can be assessed according to limit state design. Steel (or other materials) inserts may be included if the concrete stresses are greater than permissible values. The effects of localized stress concentrations near to inserts and geometric discontinuities can be assessed and proven at individual joints. However, connection design is much more important than that because of the sensitivity of connection behaviour to manufacturing tolerances, erection methods and workmanship.

It is necessary to determine the force paths through connections in order to be able to check the adequacy of the various joints within. Compared with cast in situ construction there are a number of forces which are unique to precast connections, namely frictional forces due to relative movement causes by shrinkage etc.,

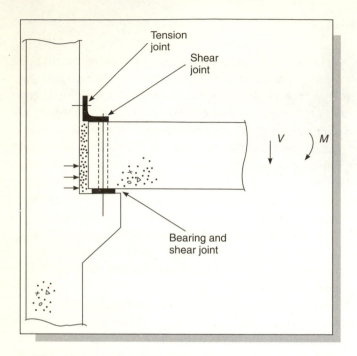

Figure 3.18: Moment and shear transfer at a bearing corbel.

pretensioning stresses in the concrete and steel, handling and self weight stresses. In the example shown in Figure 3.19, a reinforced concrete column and corbel support a pretensioned concrete beam. The figure shows there are 10 different force vectors in this connection as follows:

A: diagonal compression strut in corbel
B: horizontal component reaction to force at *A*
C: vertical component reaction to force at *A*
D: internal diagonal resultant to forces *B* and *C*
E: diagonal compression strut in beam
F & *G*: horizontal component reaction to force *E*
H: tension field reaction to forces *E* & *F*
J: horizontal friction force caused by relative movement of beam and corbel
K: horizontal membrane reaction to beam rotation due to eccentric prestressing.

The structural behaviour of the frame can be controlled by the appropriate design of connections. In achieving the various structural systems in Section 3.2 it may be necessary to design and construct either/both rigid and/or pinned connections.

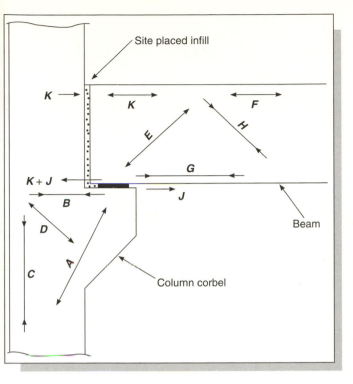

Figure 3.19: Force paths in beam to column corbel connection.

Rigid monolithic connections can only truly be made at the time of casting, although it is possible to site cast connections that have been shown to behave as monolithic, e.g. cast in situ filling of prefabricated U-beams shown in Figure 3.20. The advantages lost to in situ concreting work (cold climates in particular), the delayed maturity, the increase in structural cross-section, and the reliance on correct workmanship etc. detracts this solution in favour of bolted or welded mechanical devices. Rigid connections may be made at the foundation where there is less restriction on space as shown in Figure 3.21. In very simple terms, a bending moment is generated by the provision of a force couple in rigid embedment, i.e. no slippage

Figure 3.20: Precast U-beams awaiting cast in situ concrete filling.

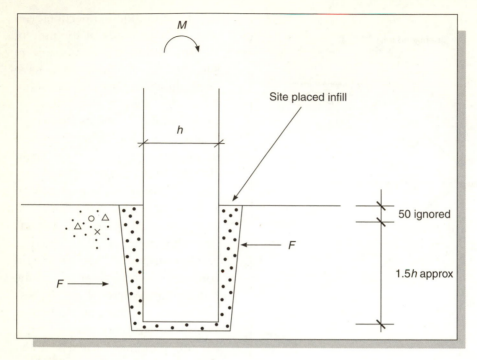

Figure 3.21: Precast column to pocket connection.

when the force is generated. Pinned connections are designed by an absence of this couple, although many connectors designed in this way inadvertently contain a force couple, giving rise to spurious moments which often cause cracking in a region of flexural tension.

To gain an overview of the various types, Figure 3.22 and Table 3.2 show the locations, classification and basic construction of connections in a precast structure.

In theory no connection is fully rigid or pinned – they all behave in a semi-rigid manner especially after the onset of flexural cracking. Using a 'beam-line' analysis (Figure 3.23), we can assess the structural classification of a connection. (Although the beam-line approach was developed for structural steelwork in c1936 recent research has shown that the method is appropriate to precast connections.[3])

The moment–rotation (M–θ) diagram in Figure 3.23 is constructed by considering the two extremes in the right-hand part of Figure 3.23. The hogging moment of resistance of the beam at the support is given by $M_R > wL^2/12$ and the rotation of a pin-ended beam subjected to a UDL of w is $\theta = wL^3/24EI$. The gradient of the beam line is 2EI/L. The M–θ plot for plots 1 and 2 give the monolithic and pinned connections, respectively. In reality, the behaviour of a connection in precast concrete will follow plots 3, 4 or 5 etc. If the M–θ plot for the connection fails to

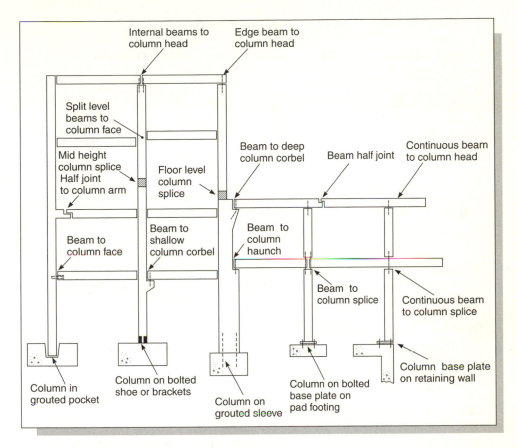

Figure 3.22: Types of connections in a precast structure.

pass through the beam-line, i.e. plot 5, the connection is deemed not to possess sufficient ductility and should be considered in design as 'pinned'. Furthermore its inherent stiffness (given by the gradient of the $M–\theta$ plot) is ignored. Conversely, if the $M–\theta$ behaviour follows plot 3 (the gradient must lie in the shaded zone and the failure take place outside the shaded zone) the effect of the connection will not differ from a monolithic by more than 5 per cent.

3.5   Stabilizing methods

Structural stability and safety are necessary considerations at all times during the erection of precast concrete frames. The structural components will not form a stabilizing system until the connections are completed – in some cases this can involve several hours of maturity of cast in situ concrete/grout joints, and several

Table 3.2: Types of connections in precast frames

Connection type	Location in Figure 3.22	Classification	Method of jointing
Beam–column head	1	Pinned	Dowel
Beam–column head	2	Rigid	Dowel plus continuity top steel
Rafter–column head	3	Pinned	Dowel
Rafter–column head	4	Rigid	Bolts (couple)
Column splice	5	Pinned	Bolts/Dowel
		Rigid	Bars in grouted sleeve (couple)
			Threaded couplers
			Steel shoes
Beam–column face	6	Pinned	Bolts
			Welded plates
			Notched plates
			Dowels
Beam–column corbel	7	Pinned	Dowel
Beam–column corbel	8	Rigid	Dowel plus continuity top steel
Beam–beam	9	Pinned	Bolts
			Dowels
Slab–beam	10	Pinned	Tie bars
Slab–wall	11	Pinned	Tie bars
Column–foundation	12	Pinned	Bolts
Column–cast in situ beam or retaining wall	13	Pinned or rigid*	Bolts
			Rebars in grouted sleeve

Note:
*Depending on the design of the cast in situ substructure.

days if structural cast in situ toppings are used to transfer horizontal forces. A stabilizing system must comprise two things as shown in Figure 3.24:

1 a horizontal system, often called a 'floor diaphragm' because it is extremely thin in relation to its plan area; and

2 a vertical system in which the reactions from the horizontal system are transferred to the ground (or other sub-structure).

The horizontal system is considered in detail in Chapter 7 where reference is also made to the many code regulations on this topic. When subjected to horizontal wind or lack-of-plumb forces the floor slab acts as a deep beam and is subjected to bending moments M_h and shear forces V_h (h being the subscript used for horizontal diaphragms). The basic design method is shown in Figure 3.25. The design is a three stage approach:

1 The floor diaphragm is analysed as a long, deep beam which is supported by a number of shear walls, shear cores, deep columns (=wind posts), or other kinds of bracing such as cross bracing (Figure 3.25a).

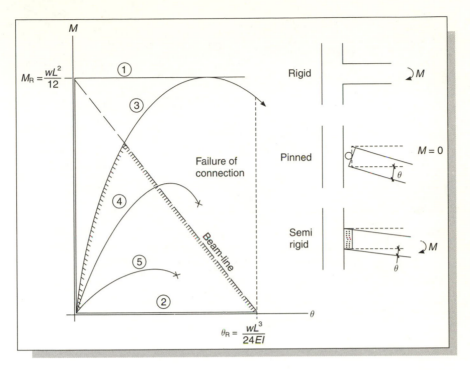

Figure 3.23: Definition of moment–rotation characteristics.

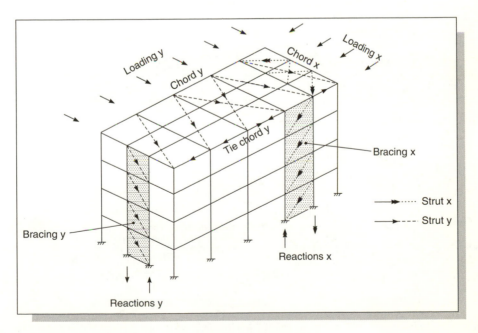

Figure 3.24: Stabilizing systems in braced frames.

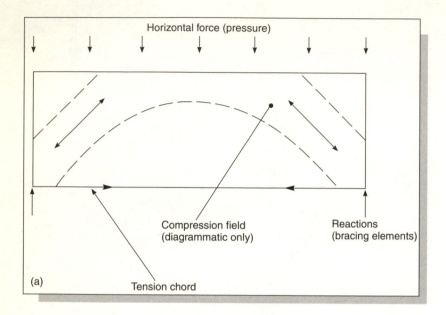

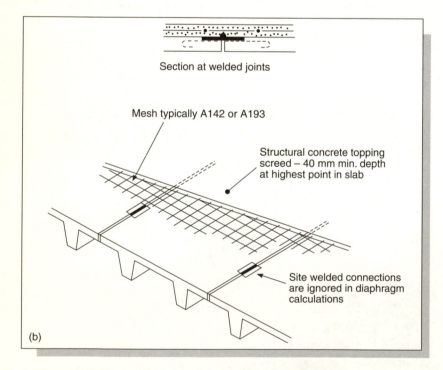

Figure 3.25: Diaphragm floor action: (a) deep beam analogy; (b) reinforced structural topping in double-tee floors.

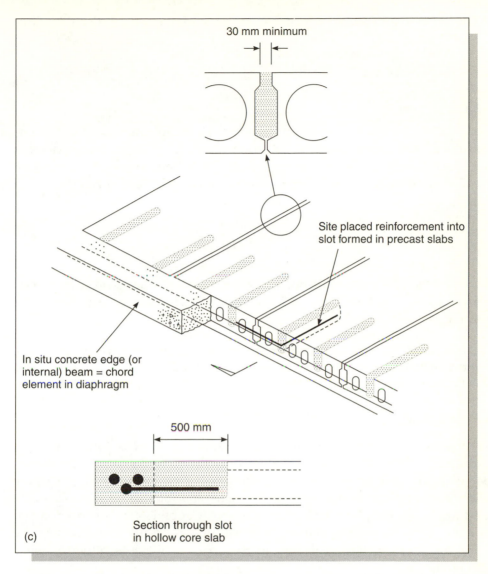

30 mm minimum

Site placed reinforcement into slot formed in precast slabs

In situ concrete edge (or internal) beam = chord element in diaphragm

500 mm

Section through slot in hollow core slab

(c)

Figure 3.25 (continued): Diaphragm floor action: (c) perimeter reinforcement in hollow core floors.

2 If there are only two supports (=bracing) the analysis is statically determinate and M_h and V_h may be calculated directly. If there are more than two supports, irrespective of where they are positioned, the analysis is statically indeterminate. The support reactions must first be found by a technique which considers the relative stiffness and position of each support, and the horizontal (e.g. wind load) pressure distribution. The derivation is given in Section 6.4 after which M_h and V_h may be calculated.

3 The area of reinforcement required to resist M_h and V_h is determined as follows:

$$A_{sh} = \frac{M_h \gamma_m}{0.8B \, f_y}$$

3.8

where 0.8B is the assumed lever arm between the compression zone and the tie steel (the assumption is known to be conservative), f_y/γ_m is the design stress in the tie steel. High tensile rebar with $f_y = 460–500 \, \text{N/mm}^2$ or helical strand with $f_y = 1750–1800 \, \text{N/mm}^2$ is used – the reasons are given in Section 7.4.

$$A_{svh} = \frac{V_h \gamma_m}{0.6 \mu f_y}$$

3.9

where μ is the coefficient of friction along the interface between adjacent precast units. According to BS8110, $\mu = 0.7$ units with no special, i.e. ex-factory, edge preparation (see Section 7.2).

4 The tie steel A_{sh} must be placed everywhere moments occur. The tie steel A_{svh} must be placed only where the shear force is greater than a certain value. This is found by checking that the interface shear stress $\tau = V_h/B(D - 30 \, \text{mm})$ does not exceed code values – the BS8110 value is $0.23 \, \text{N/mm}^2$. (The reason for the deduction of 30 mm is explained in Section 7.4.1.)

Diaphragms may be reinforced in several ways. In Figure 3.25b, a reinforced cast in situ topping transfers all horizontal forces to the vertical system – the precast floor plays no part but for restraining the topping against buckling. In Figure 3.25c, there is no cast in situ topping. Perimeter and internal tie steel resists the chord forces resulting from horizontal moments. Coupling bars are inserted into the ends of the floor units, and together with the perimeter steel provides the means for shear friction generated in the concrete filled longitudinal joints between the units.

Example 3.2
Determine the shear wall reactions and diaphragm reinforcement in the floor shown in Figure 3.26a. The precast units are 150 mm deep hollow cored and have an ex-factory edge finish. The characteristic wind pressure on the floor is 3 kN/m. Tie steel is high tensile grade 460. Suggest some reinforcement details.

Solution
Design ultimate wind load $= 1.4 \times 3.0 = 4.2 \, \text{kN/m}$. From Figure 3.26b, support reaction $R_1 = 47.2 \, \text{kN}$; $R_2 = 78.8 \, \text{kN}$.

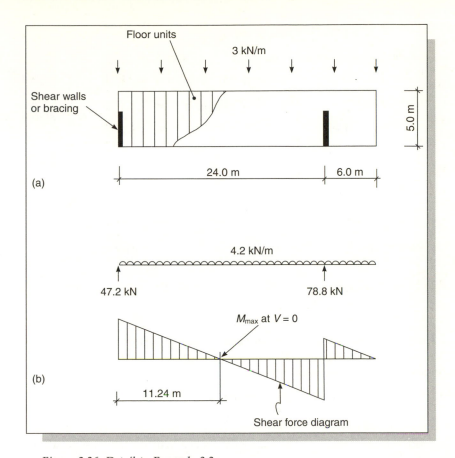

Figure 3.26: Detail to Example 3.2.

Shear span from LHS (=distance to zero shear and hence point of maximum moment) = 47.2/4.2 = 11.24 m.

$$M_{h,\,max} = 265 \text{ kNm}; \quad V_{h,\,max} = 53.6 \text{ kN}.$$
$$A_{sh} = (265 \times 10^6 \times 1.05)/(0.8 \times 5000 \times 460) = 151 \text{ mm}^2.$$

Use 2 no. T12 bars.

Interface shear stress $\tau = 53.6 \times 10^3/5000 \times (150 - 30) = 0.09 \text{ N/mm}^2 < 0.23 \text{ N/mm}^2$ allowed. No shear reinforcement needed.

Vertical stabilizing systems are dictated by the necessary actions of the structural system, i.e. skeletal, wall or portal frame. Column effective lengths depend on the type and direction of the bracing. However, there is a broad classification as the structure is either:

(i)   unbraced frame (Figure 3.27), where horizontal force resistance is provided either by moment resisting frame action, cantilever action of columns, or cantilever action of wind posts (=deep columns);

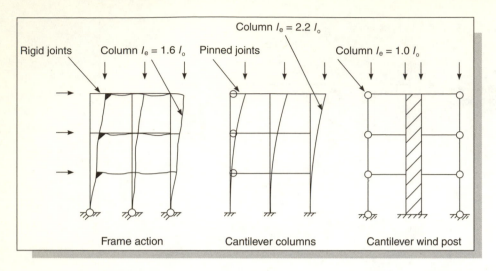

Figure 3.27: Alternative sway mechanisms and resulting column effective length factors.

(ii)   braced frame (Figure 3.28), where horizontal force resistance is provided either by cantilever action of walls or cores, in-plane panel action of shear walls or cores, infill walls, cross bracing etc.; and

(iii)   partially braced frame (Figure 3.29), which is some combination of (i) and (ii).

The type of stabilizing system may be different in other directions. The floor plan arrangement and the availability of shear walls/cores will dictate the solution. The simplest case is a long narrow rectangular plan where, as shown in Figure 3.30a, shear walls brace the frame in the y direction only, the x direction being unbraced. In other layouts, shown for example in Figure 3.30b, it is nearly always possible to find bracing positions. Precast skeletal frames of three or more storeys in height are mostly braced or partially braced. This is to avoid having to use deep columns to cater for sway deflections, which give rise to large second order bending moments. Section 6.2.4 refers in more detail.

It is not wise to use different stabilizing systems acting in the same direction in different parts of a structure. The relative stiffness of the braced part is likely to be much greater than in the unbraced part, giving rise to torsional effects due to the large eccentricity between the centre of external pressure and the centroid of the stabilizing system, as explained in Figure 3.30c. The different stabilizing systems should be structurally isolated – Figure 3.30d.

In calculating the position of the centroid of a stabilizing system the stiffness of each component of thickness t and length L is given by EI, where $E =$ long-term Young's modulus (usually taken as $15 \, kN/mm^2$) and $I = tL^3/12$. First moments of stiffness are used to calculate the centroid as explained in Example 3.3.

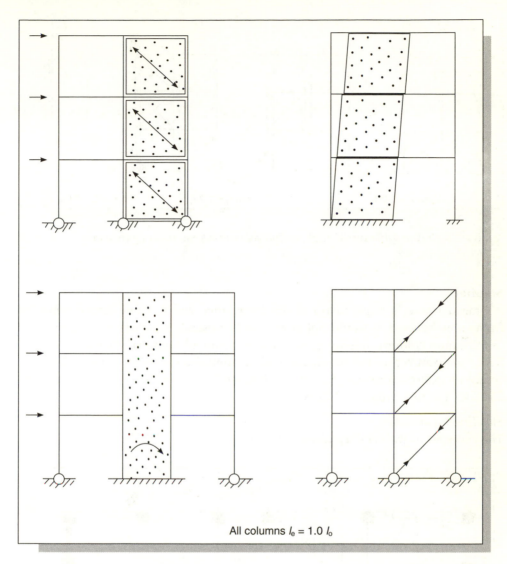

All columns $l_e = 1.0 \, l_o$

Figure 3.28: Alternative full height bracing mechanisms and resulting column effective length factors.

Example 3.3

Propose stabilizing systems for the five-storey skeletal frame shown in Figure 3.31a. The beam–column connections are all pinned, and the columns should be the minimum possible cross-section to cater for gravity loads. Wind loading may be assumed to be uniform over the entire facade. Use only shear walls for bracing.

Hint: The grid dimensions around the stairwell may be taken as 4 m × 3 m, and at the lift shaft 3 m × 3 m.

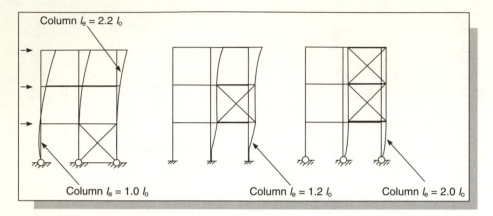

Figure 3.29: Alternative partial height bracing mechanisms and resulting column effective length factors.

Solution

A braced frame is required up to 4th floor, after which a one-storey unbraced frame may be used. It would not otherwise be possible to satisfy the requirement of minimum column sizes for gravity loads. To avoid torsional effects (see Figure 3.30c) the centroid of the stabilizing system should be as close as possible to the centre of external pressure, i.e. at $x \approx 24$ m and $y \approx 16$ m. It is necessary to first consider the two orthogonal directions.

Stability in y-direction
The centroid of the stability walls $x' \approx 24$ m.

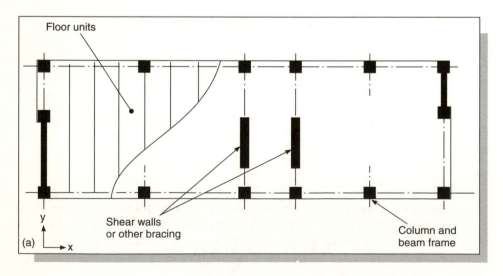

Figure 3.30: Positions of shear walls and cores in alternative floor plan layouts.

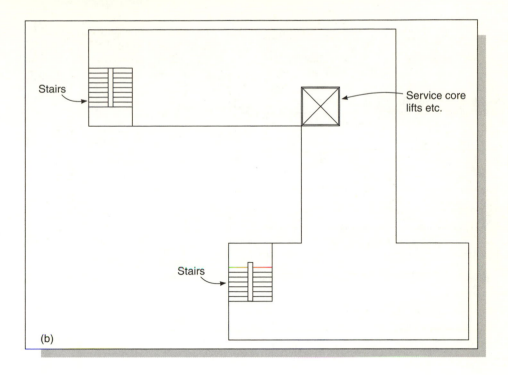

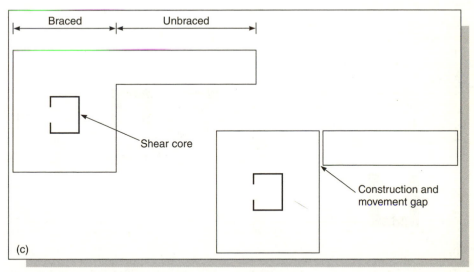

Figure 3.30 (continued): Positions of shear walls and cores in alternative floor plan layouts.

Select walls as shown in Figure 3.31b. On the assumption that the material and construction of all walls are the same, Young's modulus and thickness of wall is common to all walls and need **not** be used in the calculation.

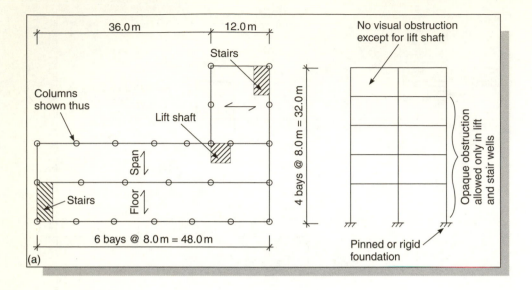

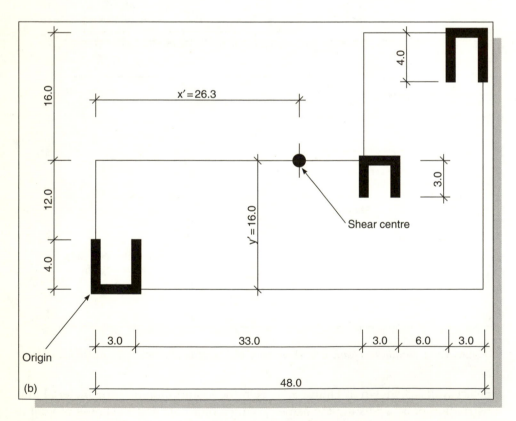

Figure 3.31: Detail to Example 3.3 (Figure 3.31b, dimensions in m).

Centroid of stiffness $x' = (4^3 \times 0) + (4^3 \times 3) + (3^3 \times 36) + (3^3 \times 39) + (4^3 \times 45) + (4^3 \times 48)/(4 \times 4^3) + (2 \times 3^3) = 26.3$ m, which is sufficiently close to the required point to eliminate significant torsional effects.

Stability in x-direction
The centroid of the stability walls $y' \approx 16$ m.

Centroid of stiffness $y' = (3^3 \times 0) + (3^3 \times 16) + (3^3 \times 32)/3 \times 3^3 = 16.0$ m, which is at the correct point.

References

1   Macginley, T. J. and Choo, B. S., *Reinforced Concrete – Design, Theory and Examples*, 2nd edn, Spon, 1990, 520p.
2   Straman, J. P., Precast Concrete Cores and Shear Walls, Prefabrication of Concrete Structures, International Seminar, Delft, October 25–26, 1990, pp. 41–54.
3   Elliott, K. S., Davies, G., Mahdi, A. A., Gorgun, H., Virdi, K. and Ragupathy, P., Precast Concrete Semi-rigid Beam-to-Column Connections in Skeletal Frames, Control of the Semi-rigid Behaviour of Civil Engineering Structural Connections, COST C1 International Conference, Liege, September 1998, pp. 45–54.

4 Precast concrete floors

4.1 Precast concrete flooring options

Precast concrete flooring offers an economic and versatile solution to ground and suspended floors in any type of building construction. Worldwide, approximately half of the floors used in commercial and domestic buildings are of precast concrete. It offers both design and cost advantages over traditional methods such as cast in situ concrete, steel-concrete composite and timber floors. There are a wide range of flooring types available to give the most economic solution for all loading and spans. The floors give maximum structural performance with minimum weight and may be used with or without structural toppings, non-structural finishes (such as tiles, granolithic screed), or with raised timber floors.

Precast concrete floors offer the twin advantages of:

1 off site production of high-strength, highly durable units; and

2 fast erection of long span floors on site.

Figure 4.1 shows some 12 m long × 1.2 m wide floors positioned at the rate of 1 unit every 10 to 15 minutes, equivalent to covering an area the size of a soccer field in 15 days. Each vehicle carries about 20 tonnes of flooring, approximately 6 units, and so erection rates are slowed down more by the problems of getting vehicles onto site than in erecting the units. These particular units are called 'hollow core floor units', or hollow-core *planks* in Australia and the United States of America. Figure 4.2 shows the moment when a hollow core unit is lifted from the steel casting bed, and illustrates the principle of a voided unit. Consequently, the self weight of a hollow core unit is about one-half of a solid section of the same depth. It is said to have a 'void ratio' of 50 per cent. Deeper hollow core units, such as the 730 mm units shown in Figure 4.3 from Italy have void ratios nearer to

Figure 4.1: Hollow core floor slabs (courtesy Bevlon, Association of Manufacturers of Prefabricated Concrete Floor Elements, Woerden, Netherlands).

Figure 4.2: Hollow core unit lifted from casting bed (courtesy Tarmac Precast, Tallington, UK).

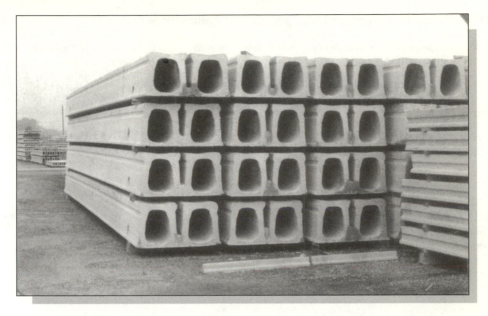

Figure 4.3: 700 mm deep hollow core floor units (courtesy Nordimpianti-Otm, Italy).

60 per cent. Units this deep have a limited market. The most common depths range from 150 to 300 mm. Most units are 1200 and 600 mm wide, although Figure 3.3 showed some 11 ft (3.66 m) wide units.

Table 4.1 lists a range of hollow core units manufactured under different trade names, e.g. *Spancrete*, or according to the type of machine used in their manufacture, e.g. *Roth* is the name of the machine for which *Bison Floors* happen to be the producer in this case. Variations in void ratio accounts for the different self weight for units of equal depth. Details of how to calculate the moment and shear resistances are given in Section 4.3.

The height of voids should not exceed $h - 50$ mm, where h is the overall depth of the unit. The diameter of circular voids is usually $h - 75$ mm. The minimum flange thickness depends on the overall depth of the unit h, given by $1.6\sqrt{h}$. However, because of cover requirements it is usually necessary for the bottom flange to be at least 30 mm thick. The minimum width of a web should not be less than 30 mm.

Hollow core units were developed in the 1950s when the dual techniques of long line prestressing and concrete production through machines were being developed by companies such as *Spiroll* in the United States of America and *Roth* in Europe. Precast concrete engineers continued to optimize the cross-section of the units leading to the so called 'double-tee' unit, achieving even greater spans and reduced mass compared with hollow core units. The 1.2 m deep double-tee

Table 4.1: Structural properties of prestressed hollow core floor units

Depth	Type or manufacturer	Country of origin of data	Self weight (kN/m²)	Service moment of resistance (kNm/m width)	Ultimate moment of resistance (kNm/m width)	Ultimate shear resistance (kNm/m width)
110 mm	Roth	UK	2.1	24	39	103
120 mm	Echo VS	Belgium	2.3	28	46	83
150 mm	Partek	Belgium	1.9	–	71	88
	Partek	Belgium	2.1	–	75	87
	Spiroll	UK	2.3	45	72	97
	Roth	UK	2.4	43	66	96
	Echo VS	Belgium	2.6	47	80	107
6 in	Spancrete	USA	–	–	71	–
165 mm	Varioplus	Germany	2.4	61	–	–
180 mm	Echo VS	Belgium	2.9	72	127	130
200 mm	Spiroll	UK	2.7	74	117	94
	Roth	UK	2.9	67	105	135
	Partek	Belgium		–	133	93
	Echo EP	Belgium	2.9	71	122	95
	Varioplus	Germany		78	–	–
	Stranbetong	Sweden	2.95	78	135	80
	Echo VS	Belgium	3.2	96	171	144
8 in	Spancrete	USA	–	–	153	–
250 mm	Roth	UK	3.5	92	148	162
	Echo VS	Belgium	3.9	132	231	179
10 in	Spancrete	USA	–	–	249	–
260 mm	Spiroll	UK	4.0	136	226	133
265 mm	Stranbetong	Sweden	3.65	133	230	84
	Partek	Belgium	3.7	–	275	161
	Varioplus	Germany	4.0	172	–	–
270 mm	Echo EP	Belgium	3.9	160	287	145
	Echo VS	Belgium	4.3	166	294	191
300 mm	Stranbetong	Sweden	3.9	186	318	96
	Echo VS	Belgium	4.5	192	335	213
12 in	Spancrete	USA	–	–	365	–
320 mm	Stranbetong	Sweden	3.95	204	344	96
	Roth	UK	4.3	–	–	–
	Partek	Belgium	4.0	–	412	–
	Echo EP	Belgium	4.3	202	349	172
	Varioplus	Germany	4.5	219	–	185
	Echo VS	Belgium	5.0	213	363	224
380 mm	Stranbetong	Sweden	4.6	307	525	158
400 mm	Roth	UK	4.7	–	–	–
	Spiroll	UK	4.8	–	–	–
	Partek	Belgium	5.0	–	626	247
	Varioplus	Germany	5.2	273	–	–
	Echo EP	Belgium	5.2	278	453	212
	Echo VS	Belgium	5.5	316	527	233
500 mm	Partek	Belgium	–	–	–	–

Note:

Cover or average cover to pretensioning tendons = 40 mm approx.

Figure 4.4: Double-tee floor slabs at a Missouri conference centre (courtesy PCI Journal, USA).

unit shown in Figure 4.4 spans 39.0 m. Although the finer points of detail of double-tees vary in many different countries, the unit comprises two deep webs, reinforced for strength, joined together by a relatively thin flange, for stability. Deflected or debonded tendons are used in some cases to overcome transfer stress problems in long span units. The cross-section profile is shown in Figure 4.5. Typical widths are 2.4 m to 3.0 m and depths range from 400 to 1200 mm. The void ratio is about 70 per cent, allowing the unit to span over longer spans and with less weight per area than the hollow core unit. The rate of erection is comparable to hollow core units, but most double-tee floors require a structural topping (see Section 4.4) to be site cast, together with a reinforcing mesh, thus reducing the overall benefit gained by the greater spans and reduced weight.

Table 4.2 lists the types of prestressed double-tee floors used, together with their moment and shear resistances – comparison with Table 4.1 is interesting. Unlike hollow core units double-tees do not have 'trade names' as their manufacture is not a proprietary method.

Both hollow core unit and double-tee floors are restricted, certainly in economical terms, to a rectangular plan shape. It is possible to make trapezoidal or splayed ended units to suit non-rectangular building grids, but the detailing of these units would be difficult and not economical. Some companies quote 20–50 per cent surcharges for manufacturing non-standard units. A precast flooring method which enables non-rectangular layouts is the 'composite beam and plank floor' shown in Figure 4.6. This is a tertiary system in which a composite floor is produced as shown in Figure 4.7; primary beams (r.c., precast, steel etc.) support long span

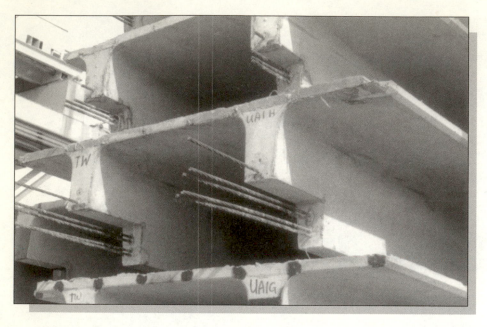

Figure 4.5: Double-tee end profile – the half joint is to raise the bearing level and reduce structural depth.

Table 4.2: Structural properties of double-tee floor units (without structural toppings)

Overall Depth (mm)	Flange depth (mm)	Web breadth* (mm)	Self weight (kN/m²)	Service moment of resistance (kNm/m width)	Ultimate moment of resistance (kNm/m width)	Ultimate shear resistance (kN/m width)
400	50	195	2.6	100	201	67
500	50	195	2.9	149	299	85
425	75	195	3.2	114	220	69
525	75	195	3.5	167	325	86
350	50	225	2.6	90	173	68
400	50	225	2.8	115	234	80
500	50	225	3.2	172	340	100
600	50	225	3.6	235	461	120
375	75	225	3.2	104	204	71
425	75	225	3.4	131	254	81
525	75	225	3.8	193	369	101
625	75	225	4.2	262	523	123

Note:

*Web breadth refers to breadth near to centroidal axis.

Source: Data based on $f_{cu} = 60\,\text{N/mm}^2$, $f_{ci} = 40\,\text{N/mm}^2$, 25% final losses of pretensioning force.

Figure 4.6: Composite beam and plank floor under construction.

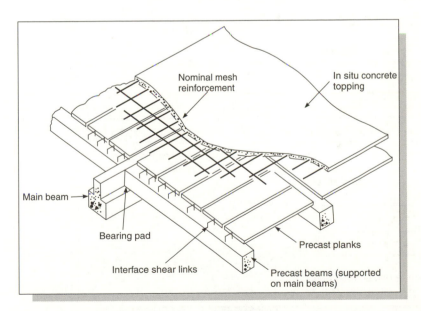

Figure 4.7: Composite beam and plank floor comprising precast beams, precast soffit units and cast in situ topping.

beams, reinforced or prestressed depending on structural requirements and manufacturing capability. These carry precast concrete planks that may be shaped to suit
non-rectangular, even curved, building layouts. The planks are relatively inexpensive to produce in a range of moulds of different sizes. It is usual for a structural
topping to be applied to the floor, and this is reinforced using a mesh. The final
constructed floor resembles a double-tee floor in structural form, and has a similar
void ratio of about 70 per cent, but the way in which each of these has been
achieved may be tailored to suit the building requirements.

The precast planks described above may be used in isolation of the precast
beams, spanning continuously between brick walls, steel or r.c. beams. The cross-
section of composite planks is as shown in Figure 4.8a. To speed erection rates the
planks may be up to 3 m wide (1.2 m and 2.4 m are common). The floor is ideal for
making both floors and beams continuous, for as shown in Figure 4.8b the tops of
the beams may be provided with interface shear loops to make a composite beam.
Lightweight infill blocks (e.g. dense polystyrene) are sometimes placed on to the
tops of the planks to reduce weight by about 25 per cent, but the weight saving

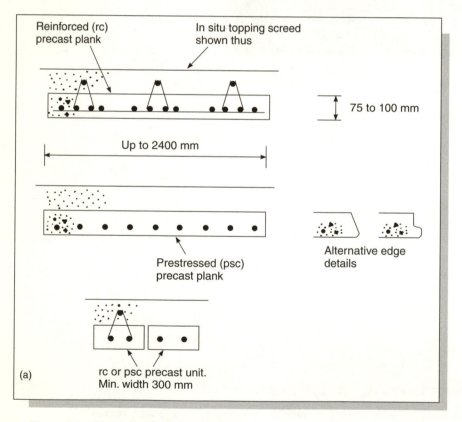

Figure 4.8a: Composite plank floor profiles.

Figure 4.8b: Practical layout of composite plank flooring (courtesy Pfeifer Seil und Hebetechnik, Memmingen, Germany).

blocks may cost more than the displaced concrete. It is relatively easy to form large size voids in this floor and to add site reinforcement to cater for stress raisers at corners etc.

A variation on this theme is the aptly named *bubble* floor (*BubbleDeck* is the trade name), shown in Figure 4.9a, where plastic spheres (about the size of footballs) are the weight saving medium. The spheres are fixed at the factory between two

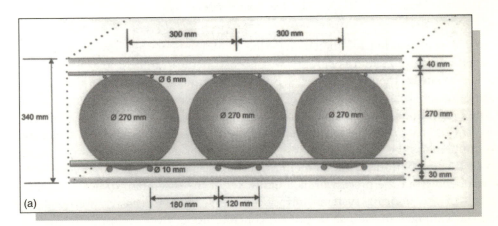

Figure 4.9a: Typical cross-section of bubble floor.

Figure 4.9b: BubbleDeck erected at Millennium Tower, Rotterdam (courtesy BubbleDeck GmbH, Darmstadt, Germany & BubbleDeck AG, Zug, Switzerland).

layers of spot welded reinforcement – the reinforcement cage can be manufactured robotically. A thin concrete soffit is cast at the factory and the units are trucked to site on their edges. Precast bubble floor units may be manufactured to a wide range of sizes, the maximum being about 6 × 3 m, which weighs only 2.2 ton at the crane hook. Figure 4.9b shows large floor panels erected at Millennium Tower, Rotterdam, in 2000. The depth of the floor is tailored to suit structural requirements as the floor may be designed as continuous by the addition of in situ top (and some bottom) reinforcement prior to in situ concreting. Large voids may be removed from the precast units, but always at the discretion of the designer.

Each of the flooring systems introduced above, has successively eroded the major advantages in the use of precast concrete floors over competitors such as timber or cast in situ floors. The advantages with precast are:

1   to manufacture units simply and economically;

2   to erect the floor as safely and as rapidly as possible;

3   to create a structurally complete precast floor; and

4   to use minimum amounts of in situ reinforcement and wet concrete.

However, these may be in competition with other criteria depending on site access, structural design requirements, interface with other trades, availability of expensive or cheap labour, services requirements, etc. Specifiers must therefore study all available options.

4.2   Flooring arrangements

A floor slab may comprise of a large number of individual *units*, each designed to cater for specified loads, moments etc., or it may comprise a complete slab *field* where the loads are shared between the precast units according to the structural response of each component. It is first necessary to define the following:

'Floor unit': a discrete element designed in isolation of other units (e.g. Figure 4.2).

'Floor slab': several floor units structurally tied together to form a floor area, with each unit designed in isolation (Figure 4.1).

'Floor field': a floor slab where each floor unit is designed as part of the whole floor. See Figures 4.10 and 4.11 later.

Most floor units, e.g. hollow core unit and double-tee, are one-way spanning, simply supported units. Composite plank and bubble floor may be designed to span in two directions, but the distances between the supports in the secondary direction may be prohibitively small to suit manufacturing or truck restrictions of about 3 m width. Structural toppings will enable slabs to span in two-directions, although this is ignored in favour of one-way spans. Hollow core units may be used without a topping because the individual floor units are keyed together over the full surface area of their edges – the longitudinal joints between the units shown in Figure 4.1 are site filled using flowable mortar to form a floor slab. Vertical and horizontal load transfer is effective over the entire floor area. This is not the case with all the other types of precast floor where a structural (i.e. containing adequate reinforcement) topping *must* be used either for horizontal load transfer, flexural and shear strength, or simply to complete the construction.

The most common situation is a uniformly distributed floor load acting on one-way spanning units with no secondary supports. Each unit will be equally loaded and there is no further analysis required of the slab field, only the design of each unit according to Section 4.3. Where line loads or point loads occur, unequal deflections of individual units will cause interface shear in the longitudinal joints between them, and load sharing will result as shown in Figure 4.10.

Hollow core units are not provided with transverse reinforcement in the precast units or in the joints between the units. The line load produces a shear reaction in the longitudinal edge of the adjacent units, and this induces torsion in the next slab. The capacity of the hollow core slab to carry the torsion is limited by the tensile capacity of the concrete. The magnitude of the shear reaction depends on the torsional stiffness and the longitudinal and transverse stiffness of all the adjacent units, low stiffness resulting in low load sharing. The precast units are assumed to be cracked longitudinally in the bottom flange, but shear friction generated by transverse restraints in the floor plate ensures integrity at the ultimate state. The deflected profile of the total

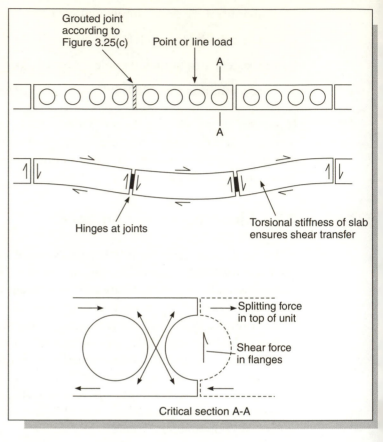

Figure 4.10: Mechanism for lateral load distribution in hollow core floors.

floor slab is computed using finite strips and differential analysis. The cross-section of each floor element is considered as a solid rectangular element and the circular (or oval) voids are ignored. As the result is unsafe, reduction factors of about 1.25 are applied to the shear reactions.[1] Interesting results and further analysis may be found in *The Structural Engineer*, *ACI Journal* and *PCI Journal*.[2,3,4]

Standard edge profiles have evolved to ensure an adequate transfer of horizontal and vertical shear between adjacent units. The main function of the joint is to prevent relative displacements between units. In hollow core units these objectives are achieved using structural grade in situ concrete (C25 minimum) compacted by a small diameter poker in dampened joints. The edges of the slab are profiled to ensure that an adequate shear key of in situ concrete (6 to 10 mm size aggregate), rather than grout, is formed between adjacent units. The manufacturing process is not sympathetic to providing projecting reinforcement across the joint. The capacity of the shear key between the units is sufficient to prevent the adjacent slabs from differential movement. Despite a slight roughening of the surfaces during

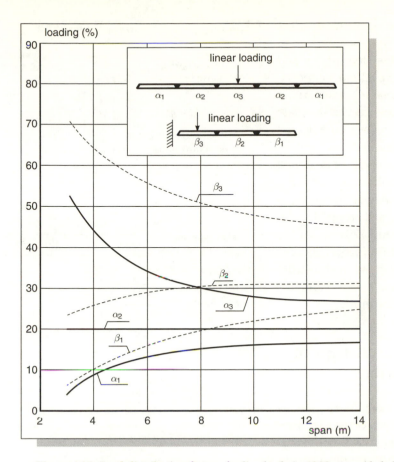

Figure 4.11: Load distribution factors for line loads in 1200 mm wide hollow core floor slabs.

the manufacturing process where indentations of up to 2 mm are present, the surface is classified in BS8110, Part 1, Clause 5.3.7 as 'smooth' or 'normal', as opposed to being 'roughened'. The design ultimate horizontal shear stress is 0.23 N/mm². Vertical shear capacity is based on single castellated joint design with minimum root indentation 40 mm × 10 mm deep.

The transverse moments and shear forces may be distributed (in accordance with BS8110, Part 1, Clause 5.2.2.2) over an effective width equal to the total width of three 1.2 m wide precast units, or one quarter of the span either side of the loaded area. The equivalent uniformly distributed loading on each slab unit may thus be computed. This is a conservative approach as data given in FIP Recommendations[5] show that for spans exceeding 4 m, up to five units are effective, given by α factors in Figure 4.11. The data also show that for edge elements, e.g. adjacent to a large void or free edge, only two slabs contribute significantly in carrying the load.

Welded connections between adjacent double-tee units, or between the units and a supporting member, are shown in Figure 4.12. Electrodes of grade E43 are used to form short continuous fillet welds between fully anchored mild steel plates (stainless steel plates and electrodes may be specified in special circumstances). A small saw cut is made at the ends of the cast-in plate to act as a stress reliever to the heated plate during welding.

Double-tee units are either designed compositely with a structural topping, in which case the flange thickness is 50–75 mm, or are self topped with thicker flanges around 120 mm. In the former vertical and horizontal shear is transferred entirely in the in situ structural topping using a design value for shear stress of $0.45\,\text{N/mm}^2$.

4.3   Structural design of individual units

More than 90 per cent of all precast concrete used in flooring is prestressed, the remainder being statically reinforced. Slabs are designed in

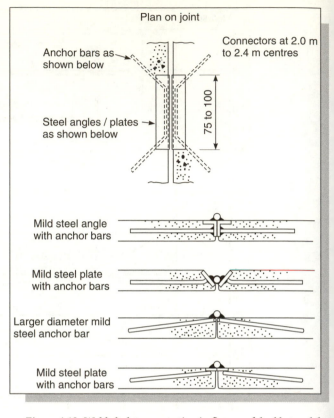

Figure 4.12: Welded plate connection in flanges of double-tee slabs.

accordance with national codes of practice together with other selected literature which deals with special circumstances.[5–10] It is necessary to check all possible failure modes shown diagrammatically in Figure 4.13. These are, from short to long spans respectively:

- bearing capacity

- shear capacity

- flexural capacity

- deflection limits

- handling restriction (imposed by manufacturer).

Standardized cross-sections and reinforcement quantities are designed to cater for all combinations of floor loading and spans. Section sizes are selected at incremental depths, usually 50 mm, and a set of reinforcement patterns are

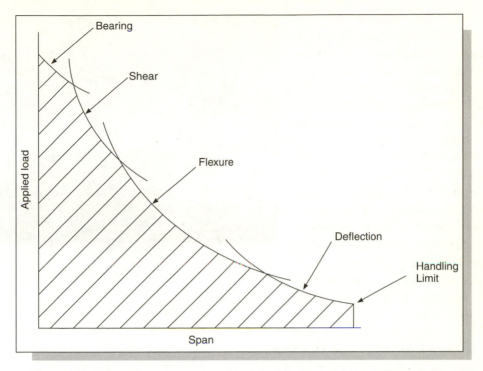

Figure 4.13: Schematic representation of load vs span characteristics in flexural elements.

selected. For example, in the unit shown in Figure 4.2, there are five voids and six webs where reinforcement may be placed. Possible combinations of strand patterns are:

- 6 no. 10.9 mm strands, total area $= 6 \times 71 = 426 \, \text{mm}^2$

- 4 no. 10.9 mm, plus 2 no. 12.5 mm strands, total area $= 4 \times 71 + 2 \times 94 = 472 \, \text{mm}^2$

- 6 no. 12.5 mm strands, total area $= 6 \times 94 = 564 \, \text{mm}^2$.

Moment resistance, shear force resistance and flexural stiffness, i.e. deflection limits, are first calculated and then compared with design requirements. Designers usually have 2 or 3 options of different depths and reinforcements to choose from – the economical one being the shallowest and most heavily reinforced unit, although unacceptable deflections may rule this one out. The additional advantage is that the depth of the 'structural floor zone' is kept to a minimum.

4.3.1 Flexural capacity

The flexural behaviour of precast prestressed concrete is no different to any other type of prestressed concrete. In fact improved quality control of factory cast concrete may actually improve things, and certainly helps to explain the excellent correlation between test results and theory found in precast units. The flexural behaviour of reinforced precast is certainly no different to cast in situ work, all other things being equal. Thus, it is only necessary to discuss further the parameters, both material and geometric, unique to precast concrete.

The major difference in behaviour in precast units is due more to the complex geometry found in voided units such as hollow core and bubble units which have rapidly reducing web thickness

Figure 4.14: Flexural cracking in hollow core slab.

near to the neutral axis (NA). Subjected to a bending moment M, the concrete in the tension face will crack when tensile stress there exceeds the modulus of rupture, i.e. $M/Z_b > f_{ct}$, where Z_b is the section modulus at the tension face, and $f_{ct} = 0.37\sqrt{f_{cu}}$ (although actual values are closer to $0.75\sqrt{f_{cu}}$). After cracking, tension stiffening of the concrete (due to the elasticity of the reinforcement) allows reduced tensile stress in this region, but when the tensile stress reaches the narrow part of the web, cracks extend rapidly through the section and the flexural stiffness of the section reduces to a far greater extent than in a rectangular section. Figure 4.14 shows this behaviour in a flexural test carried out on a 200 mm deep hollow core unit. The serviceability limiting state must be checked to prevent this type of behaviour.

A second reason why the service condition is calculated is that the ratio of the ultimate moment of resistance M_{ur} to the serviceability moment of resistance M_{sr} is usually about 1.7 to 1.8. Thus, with the use of the present load factors (1.35–1.40 for dead and 1.50–1.60 for superimposed), the serviceability condition will always be critical. Finally, the problem of cracking in the unreinforced zones is particularly important with regard to the uncracked shear resistance. It is therefore necessary to ensure that tensile stresses are not exceeded.

4.3.2   Serviceability limit state of flexure

M_{sr} is calculated by limiting the flexural compressive and tensile stresses in the concrete both in the factory transfer and handling condition and in service. Figure 4.15 shows the stress conditions at these stages for applied sagging moments – the diagrams may be inverted for cantilever units subject to hogging moments. Reference should be made to standard texts[11] for a full explanation.

The compressive stress is limited to $0.33f_{cu}$. It is rarely critical in slabs other than the temporary condition in the prestressed solid plank units. The limiting flexural tensile stress f_{ct} depends on whether flexural cracking is allowed or not – usually a durability, viz. exposure, condition. The choice is either

- Class 1, zero tension;

- Class 2, $f_{ct} < 0.45\sqrt{f_{cu}}$ or $3.5\,\text{N/mm}^2$, whichever is the smaller, but no visible cracking.

Most designers specify Class 2, but occasionally Class 1 if the service deflection is excessive.

To optimize the design it is clear from Figure 4.15, that the limiting stresses at transfer should be equally critical with the limiting service stress, and that the top and bottom surface stresses should attain maximum values simultaneously. In practice this is impossible in a symmetrical rectangular section such as a hollow core unit, but can be better achieved in a double-tee section. Also, the balance

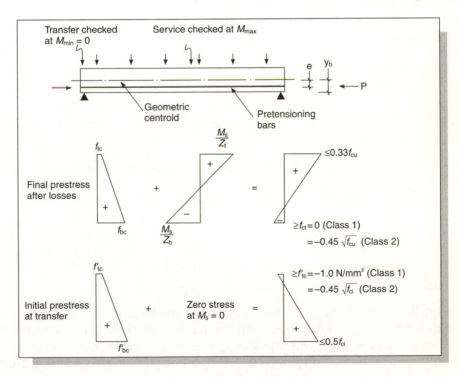

Figure 4.15: Principles of serviceability stress limitations for prestressed elements.

between the limiting concrete stresses at transfer and in service is dictated by the maturity of concrete and the need to de-tension the reinforcement within 12 to 18 hours after casting.

The transfer stress, expressed in the usual manner as the characteristic cube strength f_{ci}, is a function of the final concrete strength f_{cu}. For $f_{cu} = 60 \text{ N/mm}^2$ (the typical strength) f_{ci} should be 38–40 N/mm^2. For $f_{cu} = 50 \text{ N/mm}^2$ it is $f_{ci} = 35 \text{ N/mm}^2$. Use of rapid hardening cements, semi-dry mixes and humid indoor curing conditions are conducive to early strength gain.

Figure 4.16: Production factory for hollow core floor units.

A typical hollow core slab production factory is shown in Figures 2.7 and 4.16. Steel reinforcement, of total area A_{ps}, is stretched between jacking equipment at either end of long steel beds, about 100 m long, after which concrete is cast around the bars. The bars are positioned eccentrically relative to the centroid of the section to produce the desired pretensioning stresses shown in Figure 4.15. The initial prestress (which is set by the manufacturer) is around 70–75 per cent of the ultimate strength $f_{pu} = 1750$ to 1820 N/mm^2. The many different types of reinforcement available simplify to either 10.9 and/or 12.5 mm diameter 7-wire helical strand, or 5 or 7 mm diameter crimped wire. Table 2.3 lists the properties of these.

The reinforcement cannot sustain the initial stress for the following reasons:

1 During tensioning the reinforcement relaxes, and would otherwise creep further under duress, to between 95 and 97.5 per cent of its initial stress – it loses 5 or 2.5 per cent of its stress for Class 1 and Class 2 categories, respectively. A 1000-hour relaxation test value is provided by manufacturers (or as given in BS5896). Codes of practice add safety margins to this value, BS8110 value being 1.2. Thus the relaxation loss is 3–6 per cent.

2 After the concrete has hardened around the reinforcement and the bars are released from the jacking equipment, the force in the bars is transferred to the concrete by bond. The concrete shortens elastically – this may be calculated knowing Young's modulus of the concrete at this point in time transfer. This is called 'elastic shortening' and because the reinforcement is obliged to shorten

the same amount as the concrete has, the stress in it reduces too, by about 5 per cent. Losses 1 and 2 are called 'transfer losses'.

3   Desiccation of the concrete follows to cause a long-term shrinkage loss. This is the product of the shrinkage per unit length (taken as 300×10^{-6} for indoor manufacture and exposure) and modulus of elasticity of the tendons (taken as $E_{ps} = 200 \, \text{kN/mm}^2$, although $195 \, \text{kN/mm}^2$ is more applicable to helical strand which has a slight tendency to unwind when stretched). This gives a shrinkage loss of between 57.5 and $60 \, \text{N/mm}^2$, about 5 per cent.

4   Finally creep strains are allowed for using a specific creep strain (i.e. creep per unit length per unit of stress) of 1.8 for indoor curing and loading at 90 days in the United Kingdom. Creep affects the reinforcement in the same manner as elastic shortening because its effect is measured at the centroid of the bars. Hence, the creep loss is taken as 1.8 times the elastic shortening loss, about 9 per cent.

Total losses range from about 19 to 26 per cent for minimum to maximum levels of prestress. The design effective prestress in the tendons after all losses is given by f_{pe}.

To calculate M_{sr}, the section is considered uncracked and the net cross-sectional area A and second moment of area I are used to compute maximum fibre stresses f_{bc} and f_{tc} in the bottom and top of the section. The section is subjected to a final prestressing force $P_f = f_{pe}A_{ps}$ acting at an eccentricity e from the geometrical NA. Using the usual notation M_{sr} is given for Class 2 permissible tension by the lesser of:

$$M_{sr} = \left(f_{bc} + 0.45\sqrt{f_{cu}} \right) Z_b \qquad \qquad 4.1$$

or

$$M_{sr} = (f_{tc} + 0.33 f_{cu}) Z_t \qquad \qquad 4.2$$

where

$$f_{bc} = P_f\left(\frac{1}{A} + \frac{e}{Z_b} \right) \qquad \qquad 4.3$$

$$f_{tc} = P_f\left(\frac{1}{A} - \frac{e}{Z_t} \right) \qquad \qquad 4.4$$

Double-tee slabs present a special case. Because of its cross-section the centroid of the unit lies close to the top flange, and therefore the section modulus Z_t to the top fibre is very large, typcally three times Z_b. Consequently, as the top fibre does not give a limiting value to M_{sr} the influence of f_{cu} is very small, as given in Eq. 4.1. As the controlling influence in Eq. 4.1 is f_{bc}, the stress at transfer becomes very important. It is therefore necessary with double-tee units to try to achieve the maximum possible transfer stress, say $f_{ci} \approx 40 \, \text{N/mm}^2$.

Example 4.1

Calculate M_{sr} for the 203 mm deep Class 2 prestressed hollow core unit shown in Figure 4.17. The initial prestressing force may be taken as 70 per cent of characteristic strength of the 'standard' 7-wire helical strand. Manufacturer's data gives relaxation as 2.5 per cent. Geometric and material data given by the manufacturer are as follows:

Area $= 135 \times 10^3 \, \text{mm}^2$; $I = 678 \times 10^6 \, \text{mm}^4$; $y_t = 99 \, \text{mm}$; $f_{cu} = 50 \, \text{N/mm}^2$; $E_c = 30 \, \text{kN/mm}^2$; $f_{ci} = 35 \, \text{N/mm}^2$; $E_{ci} = 27 \, \text{kN/mm}^2$; $f_{pu} = 1750 \, \text{N/mm}^2$; $E_{ps} = 195 \, \text{kN/mm}^2$; $A_{ps} = 94.2 \, \text{mm}^2$ per strand; cover to 12.5 mm diameter strand $= 40 \, \text{mm}$. Is the critical fibre stress at the top or bottom of the unit?

Solution

$$\text{Section properties} \quad Z_b = (678 \times 10^6)/(203 - 99) = 6.519 \times 10^6 \, \text{mm}^3$$
$$Z_t = 678 \times 10^6 / 99 = 6.848 \times 10^6 \, \text{mm}^3$$
$$e = 203 - 40 - 6.25 - 99 = 57.7 \, \text{mm}$$
$$\text{Initial prestress in tendons} \quad f_{pi} = 0.7 \times 1750 = 1225 \, \text{N/mm}^2$$
$$\text{Initial prestressing force} \quad P_i = 1225 \times 7 \times 94.2 \times 10^{-3} = 807.8 \, \text{kN}$$

Initial prestress in bottom, top and at level of strands: Eqs 4.3 and 4.4:

$$f_{bc}' = 807.8 \times 10^3 \left(\frac{1}{135 \times 10^3} + \frac{57.7}{6.519 \times 10^6} \right) = +13.14 \, \text{N/mm}^2 \text{ (compression)} < +0.5 f_{ci}$$

$$f_{tc}' = 807.8 \times 10^3 \left(\frac{1}{135 \times 10^3} - \frac{57.7}{6.848 \times 10^6} \right) = -0.83 \, \text{N/mm}^2 \text{ (tension)} < -0.45 \sqrt{f_{ci}}$$

Thus, the transfer conditions are satisfactory without recourse to check the initial losses.

The prestress at level of the centroid of the strands $f_{cc}' = +9.96 \, \text{N/mm}^2$ (compression). Then,

Elastic loss $= 9.96 \times 195/27 = 71.9 \, \text{N/mm}^2$ equal to $100 \times 71.9/1225 = 5.87\%$ loss. Creep loss $= 1.8 \times 5.87 = 10.56\%$ loss

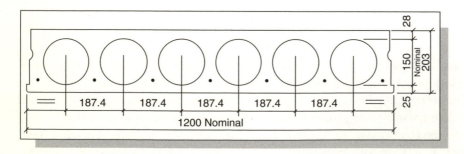

Figure 4.17: Details to Example 4.1.

Shrinkage loss $= 300 \times 10^{-6} \times 195 \times 10^3 = 57.5 \, \text{N/mm}^2$

equal to $100 \times 57.5/1225 = 4.69\%$ loss.

Relaxation loss $= 1.2 \times 2.5 = 3.0\%$ loss.

Total losses $= 24.12\%$, i.e. the residual amount is 0.7588 of the initial prestress values above.

Final prestress in bottom and top

$$f_{bc} = 0.7588 \times (+13.14) = +9.97 \, \text{N/mm}^2 \, (\text{compression})$$

$$f_{tc} = 0.7588 \times (-0.83) = -0.63 \, \text{N/mm}^2 \, (\text{tension})$$

Then, at the bottom fibre, M_{sr} is limited by a tensile stress limit of $0.45\sqrt{50} = 3.2 \, \text{N/mm}^2$

$$M_{sr} = (9.97 + 3.2) \times 6.519 \times 10^6 \times 10^{-6} = \underline{85.8 \, \text{kNm}}.$$

At the bottom fibre, M_{sr} is limited by a compressive stress limit of $0.33 f_{cu} = 16.5 \, \text{N/mm}^2$

$$M_{sr} = (0.63 + 16.5) \times 6.848 \times 10^6 \times 10^{-6} = \underline{117.3 \, \text{kNm}} > 85.8 \, \text{kNm}.$$

The bottom fibre is critical.

Example 4.2
Find the required compressive f_{cu} in Example 4.1 that would equate the service moment based on the top and bottom limiting service stress conditions, thus optimizing the strength of concrete.

Solution
Solve $M_{sr} = (f_{bc} + 0.45\sqrt{f_{cu}})Z_b = (f_{tc} + 0.33 f_{cu})Z_t$. Thus, $(9.97 + 0.45\sqrt{f_{cu}}) \, 6.519 \times 10^6 = (0.63 + 0.33 f_{cu}) \, 6.848 \times 10^6$. Hence $f_{cu} = \underline{34.5 \, \text{N/mm}^2}$.

The result is less than the transfer strength suggesting an impractical solution. (This result further demonstrates that increasing f_{cu} to say $60 \, \text{N/mm}^2$ would have little effect on the value of M_{sr}.) It is therefore necessary to modify the section properties Z_b and Z_t to obtain comparability, as follows.

Example 4.3
Find the required values of Z_b and Z_t in Example 4.1 necessary to equate the value of M_{sr} obtained from limiting stresses. Calculate the new value of M_{sr}. Study the cross-section and check whether the new values of Z_b and Z_t can be achieved practically.

Solution

Solve $M_{sr} = (f_{bc} + 0.45\sqrt{f_{cu}})Z_b = (f_{tc} + 0.33f_{cu})Z_t$. Thus $(9.97 + 3.2)Z_b = (0.63 + 16.5)Z_t$. Then $Z_b/Z_t = 17.13/13.17 = 1.3$, i.e. $y_t/y_b = 1.3$ also $y_b + y_t = 203$ mm.

Solving $y_b = 88.3$ mm and $y_t = 114.7$ mm then $Z_b = \underline{7.678 \times 10^6\,\text{mm}^3}$ and $Z_t = \underline{5.911 \times 10^6\,\text{mm}^3}$ and $M_{sr} = 13.17 \times 7.678 \times 10^6 \times 10^6 = \underline{101.1\,\text{kNm}}. > 85.8$ kNm in Example 4.1.

To achieve this condition, the geometric centroid must be lowered by $114.7 - 99.0 = 15.7$ mm. To achieve this the voids must be repositioned or modified in shape. It is not possible to raise the position of the circular voids by this distance by making the top cover to the cores $28.0 - 15.7 = 12.3$ mm. It would therefore be necessary to change the shape of the voids to non-circular – this may not be welcomed by the manufacturer.

4.3.3 Ultimate limit state of flexure

In calculating the ultimate resistance, material partial safety factors should be applied as per usual, viz. 1.05 for steel and 1.5 for concrete in flexure. The ultimate flexural resistance M_{ur} when using bonded tendons is limited by the following:

1   ultimate compressive strength of concrete, $0.45f_{cu}$;

2   the design tensile stress in the tendons, f_{pb}.

The depth of the (strain responsive) NA X is obtained by considering the equilibrium of the section. The tensile strength of the steel depends on the net prestress f_{pe} in the tendons after all losses and initial prestress levels have been considered. In most hollow core production the ratio $f_{pe}/f_{pu} = 0.50$ to 0.55. Values for X/d and f_{pb} may be obtained from strain compatibility, but as the strands are all located at the same effective depth then BS8110, Part 1, Table 4.4 offers simplified data. This table is reproduced here in Table 4.3. If the strands are located at different levels, as in the case of double-tees, reference should be made to standard theory at ultimate strain (see Kong & Evans[11], Section 9.5).

The ultimate moment of resistance of a rectangular section containing bonded tendons, all of which are located in the tension zone at an effective depth d, is given as:

$$M_{ur} = f_{pb}A_{ps}(d - 0.45X) \qquad\qquad 4.5$$

If the compressive stress block is not rectangular, as in the case of hollow core slabs where $X >$ cover to cores, the depth to the neutral axis must be found by geometrical or arithmetic means.

Table 4.3: Design stress in tendons and depth to neutral axis in prestressed sections (BS8110, Part 1, Table 4.4)

$\dfrac{f_{pu}A_{ps}}{f_{cu}bd}$	Design stress in tendons as a proportion of the design strength, $f_{pb}/0.95f_{pu}$			Ratio of depth of neutral axis to that of the centroid of the tendons in the tension zone, x/d		
	f_{pe}/f_{pu} 0.6	0.5	0.4	f_{pe}/f_{pu} 0.6	0.5	0.4
0.05	1.00	1.00	1.00	0.12	0.12	0.12
0.10	1.00	1.00	1.00	0.23	0.23	0.23
0.15	0.95	0.92	0.89	0.33	0.32	0.31
0.20	0.87	0.84	0.82	0.41	0.40	0.38
0.25	0.82	0.79	0.76	0.48	0.46	0.45
0.30	0.78	0.75	0.72	0.55	0.53	0.51
0.35	0.75	0.72	0.70	0.62	0.59	0.57
0.40	0.73	0.70	0.66	0.69	0.66	0.62
0.45	0.71	0.68	0.62	0.75	0.72	0.66
0.50	0.70	0.65	0.59	0.82	0.76	0.69

Example 4.4

Calculate M_{ur} for the section used in Example 4.1. Is the unit critical at the service or ultimate limit state? Manufacturer's data gives the breadth of the top of the hollow core unit as $b = 1168$ mm.

Solution

$$d = 203 - 46.25 = 156.7 \text{ mm}$$

$$f_{pu}A_{ps}/f_{cu}bd = (1750 \times 659.4)/(50 \times 1168 \times 156.75) = 0.126$$

Also, from Example 4.1, $f_{pe}/f_{pu} = 0.7 \times 0.7588 = 0.531$.

From Table 4.3, $f_{pb}/0.95f_{pu} = 0.966$ (by linear extrapolation). Then the ultimate force in the strands $F_s = 0.95 \times 0.966 \times 1750 \times 659.4 \times 10^{-3} = 1059$ kN. Also the force in the concrete $F_c = 0.45f_{cu}b\,0.9X = 1059$ kN. Then by first iteration $X = 44.8$ mm > 28 mm. But the neutral axis lies beneath the top of the circular cores. This necessitates iteration to find $X = 57$ mm (see Figure 4.18).

The distance to the centroid of the compression block $d_n = 23.9$ mm. Then $M_{sr} = 1059 \times 10^3 \times (156.7 - 23.9) \times 10^{-6} = \underline{140.7 \text{ kNm}}$.

To check whether the unit is critical at ultimate, the ratio $M_{ur}/M_{sr} = 140.7/85.8 = 1.64$. This ratio is greater than the maximum possible ratio of the design ultimate moment to design service moment, i.e. 1.60 using BS8110 load factors. Thus the unit cannot be critical at ultimate.

4.3.4 Deflection

Deflection calculations are always carried out for prestressed members – it is not sufficient to check span-effective depth ratios as in a reinforced section. This is

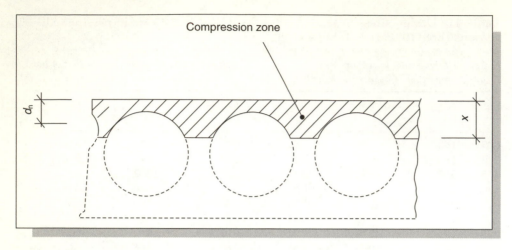

Figure 4.18: Compressive stress zone if neutral axis lies below top flange.

because the strength-to-stiffness ratio of a prestressed section is considerably greater than in a reinforced section. The effects of strand relaxation, creep etc. have greater effects as the degree of prestress increases. The general method of curvature-area may be adopted in prestressed design. For non-deflected strands the curvature diagram is rectangular.

Net deflection is found by superposition of upward cambers due to pretensioning and downward gravity loads. Calculations are based on a flexurally uncracked stiffness $E_c I$ using the transfer value E_{ci} for initial camber due to prestress and the final value E_c and appropriate creep factor for long-term deflections. Precamber deflection comprises of three parts:

1 short term value due to prestressing force P_i' after initial elastic, strand relaxation and shrinkage losses, plus;

2 long term value, due to the prestressing force after all losses P_f; and

3 self weight deflection.

Upward (negative sign) mid-span camber δ is calculated using the following:

$$\delta = -\frac{P_i' e L^2}{8 E_{ci} I} - \frac{\phi P_f e L^2}{8 E_c I} + \frac{5 w_0 L^4 (1 + \phi)}{384 E_c I} \qquad 4.6$$

where ϕ is a creep coefficient for the time interval, and w_0 is the unit uniformly distributed self weight.

In-service long-term deflections are calculated in the usual manner taking into consideration the support conditions, loading arrangement and creep. Service loads are used. Deflections are limited to span/500 or 20 mm where brittle finishes are to be applied, or span/350 or 20 mm for non-brittle finishes – the latter limit of 20 mm is usually critical for spans of more than 8 m. The net deflection (imposed minus precamber) should be less than span/1500.

Example 4.5

Calculate the long-term deflection for the hollow core unit in Example 4.1. The hollow core unit is to be used to carry imposed dead and live loads of $2 \, \text{kN/m}^2$ and $5 \, \text{kN/m}^2$ respectively over a simply supported span of 6.0 m. The finishes are non-brittle. The self weight of the unit is 3.24 kN/m. Use a creep coefficient of 1.8.

Solution

From Example 4.1, initial and final losses are 8.87 per cent and 24.12 per cent, respectively. Then $P'_i = 736 \, \text{kN}$ and $P_f = 613.0 \, \text{kN}$.

The upward camber is:

$$\delta = -\frac{736 \times 57.7 \times 6000^2}{8 \times 27 \times 678 \times 10^6} - \frac{1.8 \times 613 \times 57.7 \times 6000^2}{8 \times 30 \times 678 \times 10^6}$$

$$+\frac{5 \times 3.24 \times 10^{-3} \times 6000^4 \times 2.8}{384 \times 30 \times 678 \times 10^6} = -17 \, \text{mm}$$

Imposed deflection (positive sign) due to a total imposed load of $1.2 \times 7.0 = 8.4 \, \text{kN/m}$ (per 1.2 m wide unit) is:

$$\Delta = +\frac{5wL^4(1 + \phi)}{384E_c I} = +\frac{5 \times 8.4 \times 10^{-3} \times 6000^4 \times 2.8}{384 \times 30 \times 678 \times 10^6}$$

$$= +19.5 \, \text{mm} > L/350 > 17.14 \, \text{mm for non-brittle finishes.}$$

Net deflection $= +2.5 \, \text{mm} < L/1500 < 4.0 \, \text{mm}$.

4.3.5 *Shear capacity*

Calculating the shear capacity of a precast concrete section is no different to any other type of reinforced or prestressed section, providing that localized forces and stresses at end connections are dealt with (see Sections 8.6 and 9.2) and do not interrupt the distribution of shear. Shear capacity is determined only at the ultimate state – unlike flexure there are no limiting service conditions for shear.

However there is a paradox here because design equations derive from an elastic analysis.

Ultimate shear capacity is the sum of the various actions of the concrete (=aggregate interlock), longitudinal bars (=dowel action) and shear reinforcement (=stirrup action). However, many reinforced and prestressed units, such as hollow core units and prestressed planks, have no shear reinforcement due to their manufacturing methods, the only reinforcement being longitudinal rebars or pretensioning strands. The shear capacity of these units therefore depends on the shear resistance of the concrete in combination with dowel action alone.

The calculation of shear resistance carries numerous partial safety factors, i.e. on tensile strength, on the strength due to aggregate interlock, and geometric simplifications, i.e. the equivalent shear area (see Eq. 4.7). There are good reasons for this. Experimental tests have found that shear resistance can vary widely for nominally the same section. Variations in the order of 20 per cent are quite common, especially where there are no shear stirrups and the shear resistance relies on the tensile splitting strength of the concrete. A lot depends on compaction of concrete in the webs and their breadth. In some units, such as hollow core units, small variations in the breadth of webs can lead to premature failure of the entire section. The failure mode is also quite sudden. The Figure 4.19 shows a shear failure in a deep hollow core unit – note how failure has occurred due solely to one large crack through the webs. The position and inclination of this crack has a major influence on the shear load the section can sustain.

In prestressed sections shear capacity is calculated for two conditions: (1) the uncracked section; and (2) the cracked section in flexure (Figure 4.20). The uncracked ultimate shear resistance V_{co} is greater than the ultimate cracked value V_{cr} because the full section properties are considered and a small amount of diagonal tension in the concrete is permitted.

4.3.5.1 Shear capacity in the uncracked region, V_{co}

The term uncracked refers to flexural cracking, and where this exists V_{co} must not be used. The ultimate shear capacity is given as:

$$V_{co} = 0.67b_v h \sqrt{f_t^2 + 0.8f_{cp}f_t} \qquad\qquad 4.7$$

where $f_t = 0.24\sqrt{f_{cu}}$ and f_{cp} is the compressive stress at the centroidal axis due to prestress after all losses.

Although this expression is derived using a rectangular section and not the actual flanged section appropriate to non-rectangular sections such as hollow core units, the difference is accepted as being around 10 per cent, and always on the conservative side (see Example 4.6). The term $0.67b_v h$ may be replaced by the

general term Ib_v/Ay' so that Eq. 4.7 is modified to:

$$V_{co} = \frac{1b_v}{Ay'} \sqrt{f_t^2 + 0.8f_{cp}f_t} \quad 4.7a$$

where y' is the distance from the centroid of the section to the centroid of the area A above the plane considered.

For flanged sections b_v is taken as the web width (=narrower part). In flanged sections where the centroidal axis is in the flange Eq. 4.7 should be applied at the junction of the web-flange, i.e. f_{cp} is calculated there.

Figure 4.19: Shear failure in prestressed hollow core slab.

Design rules recognize the fact that the critical shear plane may occur in the prestress development zone where f_{cp} is not fully developed. It is known that prestressing forces develop parabolically and therefore a reduced value f_{cpx} is used. It may be shown that f_{cpx} is given as:

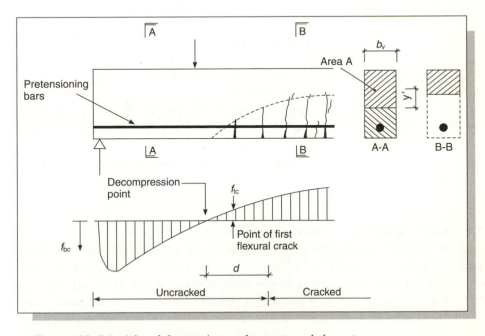

Figure 4.20: Principles of shear resistance for prestressed elements.

$$f_{cpx} = \frac{x}{l_p}\left(2 - \frac{x}{l_p}\right)f_{cp} \qquad\qquad 4.8$$

where l_p is the development length (=greater of transmission length or depth of section) and x is from the end of the unit measured at 45° to the inner bearing edge. The transmission length may be taken as $l_t = K_t\phi/\sqrt{f_{ci}}$, where the coefficient $K_t = 240$ for standard or super strand, 360 for drawn strand, 400 for crimped wire, and 600 for plain or indented wire, and ϕ is the tendon diameter.

Example 4.6.

Calculate the uncracked shear capacity for the section in Example 4.1 using: (a) the Eq. 4.7 using the rectangular area term $0.67\ b_v\ h$; and (b) ditto but replacing $0.67\ b_v\ h$ with the term $Ib_v/A y'$.

The bearing length may be taken as 100 mm.

Solution

From Example 4.1, the prestress after losses at the centroidal axis $f_{cp} = 4.54\,\text{N/mm}^2$

The critical shear point is at 100 (=bearing) + 104 (=height to centroidal axis) = 204 mm

Development length = greater of 203 mm or $240 \times 12.5/\sqrt{35} = 507$ mm. Then $x/l_p = 204/507 = 0.402$, then $f_{cpx} = 2.92\,\text{N/mm}^2$.

Total breadth at centroidal axis $b_v = 1168 - (6 \times 150) = 268$ mm

$$f_t = 0.24\sqrt{50} = 1.7\,\text{N/mm}^2$$

(a) $V_{co} = 0.67 \times 268 \times 203 \times \sqrt{1.7^2 + 0.8 \times 2.92 \times 1.7} \times 10^{-3} = \underline{95.5\,\text{kN}}$

(b) Area above centroidal axis $A = 67\,500\,\text{mm}^2$

Distance to centroid of area above centroidal axis may be found by geometry, $y' = 64.5$ mm. Then $V_{co} = 678 \times 10^6 \times 268\sqrt{1.7^2 + 0.8 \times 2.92 \times 1.7} \times 10^{-3}/67\,500 \times 64.5 = \underline{109.3\,\text{kN}}$

(There is 14 per cent difference in the two values, however the code of practice gives a conservative value.)

4.3.5.2   Shear capacity in the flexurally cracked region V_{cr}

The shear capacity, as calculated by V_{co}, extends to a point one effective depth beyond the point at which the net flexural stress in the tension zone becomes zero, i.e. the decompression point where an applied moment $M_o = Z_b 0.8 f_{bc}$ (f_{bc} is after losses; 0.8 is a factor that allows for possible variations in prestress levels in this critical region) (see Figure 4.20). In the flexurally cracked region shear resistance is a function of both the concrete and dowel action, as in reinforced concrete, plus a contribution of the level of compressive stress causing shear friction in the compressive zone. V_{cr} is calculated using the following (semi-empirical) equation:

$$V_{cr} = \left(1 - 0.55\frac{f_{pe}}{f_{pu}}\right)v_c b_v d + M_o\frac{V}{M} \qquad 4.9$$

where V and M are the values for the **design** ultimate shear force and bending moment at the section considered.

The derivation of Eq. 4.9 is long and one of the problems in using Eq. 4.9 is that the decompression moment M_o and the shear force and bending moment values V and M must be known before V_{cr} may be computed. Now M_o may be computed as above, but V and M are different for each load case and span. It is therefore assumed that at a critical section the design shear force V cannot exceed V_{cr}, and so $V = V_{cr}$. Similarly, the ultimate design moment M cannot exceed $\gamma_f M_s$, and therefore $M = M_u$. Thus, a unique minimum value for V_{cr} exists as:

$$V_{cr,min} = \frac{\left(1 - 0.55\frac{f_{pe}}{f_{pu}}\right)v_c b_v d}{1 - \dfrac{0.8Z_b f_{bc}}{M_u}} \qquad 4.10$$

Example 4.7
Calculate the minimum value for V_{cr} for the slab in Example 4.1.

Solution
As explained above, no knowledge of the actual values for V and M are necessary if a minimum value for V_{cr} is required. Also, the distance to the commencement of the flexurally cracked region need not be known.

$$100\, A_s/b_v d = (100 \times 659.4)/(268 \times 156.7) = 1.57$$

from BS8110, Part 1, Table 3.9, $v_c = 0.79 \times 1.57^{1/3} \times (400/156.7)^{1/4} \times (40/25)^{1/3}/ 1.25 = 1.08\,N/mm^2$.

$$M_o = 6.519 \times 10^6 \times 0.8 \times 9.97 \times 10^{-6} = \underline{52.0\,kNm}$$

M_u from Example 4.4 $= 140.7\,kNm$

$$f_{pe}/f_{pu} = 0.7 \times 0.7588 = 0.531$$

Then, from Eq. 4.9,

$$V_{cr,min} = [(1 - 0.55 \times 0.531) \times 1.08 \times 268 \times 156.7 \times 10^{-3}] + [52.0 \times V_{cr,min}/140.7]$$
$$= \underline{51.0\,kN}.$$

4.3.6 Bearing capacity

In this context 'bearing' capacity refers to the contact bearing pressure at the bearing ledge. We make this distinction as certain literature refers to bearing capacity in calculating shear resistance. It is first necessary to distinguish between 'isolated' components and 'non-isolated' because the bearing allowances differ. Non-isolated components are connected to other components with a secondary means of support. Isolated components rely entirely on their own bearing for total support. Where wide slab units, such as hollow core unit and composite plank, are dry bedded the full width of the unit is not to be taken as the bearing length to allow for uneven bearings where daylight gaps are sometimes seen in the contact plane.

Ultimate bearing capacity is given as:

$$F_b = f_b l_b l_w \qquad\qquad 4.11$$

where the bearing length (perpendicular to floor span) l_b is taken as the least of: (i) the actual bearing length; (ii) one-half of (i) plus 100 mm; and (iii) 600 mm. Conditions (ii) and (iii) are empirical values based on the acceptable degree of flatness possible in members exceeding about 1.0 m in the critical dimension. The net bearing width (parallel to span) l_w, as defined in Figure 4.21, is equal to the nominal bearing width minus the spalling tolerances additive in the supporting and supported members, given in Table 4.4, minus the allowances for constructional inaccuracies. The minimum net bearing width is 40 mm for non-isolated components (or 60 mm for isolated), and the minimum nominal bearing width is therefore 60–75 mm depending on the conditions as follows.

The design ultimate bearing stress, which is based on the cube crushing strength of the weakest of the two, or three, component materials (excluding masonry support), is as follows (BS8110, Clause 5.2.3.4) for:

1   dry bearings on concrete: $0.4f_{cu}$;

2   wet bedded bearings on concrete or mortar: $0.6f_{cu}$ (bearing medium); and

3   elastomeric bearing (called flexible padding): between $0.4f_{cu}$ and $0.6f_{cu}$; use $0.5f_{cu}$ or f_c (bearing material).

The ineffective bearing width, i.e. allowances for spalling, are given in Table 4.4 (reproduced from BS8110, Part 1, Tables 5.1 and 5.2.)

Table 4.4: Ineffective bearing distances (from BS8110)

At the supporting member

Method of Support	Ineffective bearing widths (mm)
Steel	0
Concrete grade ≥C30	15
Concrete grade <C30 and masonry	25
Reinforced concrete less than 300 mm deep at outer edge	Nominal cover to reinforcement at outer edge*
Reinforced concrete where loop reinforcement at outer edge exceeds 12-mm diameter	Nominal cover plus inner radius of bent bar*

At the supported member

Reinforcement at bearing of supported member	Ineffective bearing widths (mm)
Tendons or straight bars exposed at End of member	0
Straight bars, or horizontal or vertical loops not exceeding 12-mm diameter	10, or end cover, whichever is the greater*
Loops exceeding 12-mm diameter	Nominal end cover plus inner radius of bent bar*

Note: *Chamfers occurring in the above zones may be discounted.

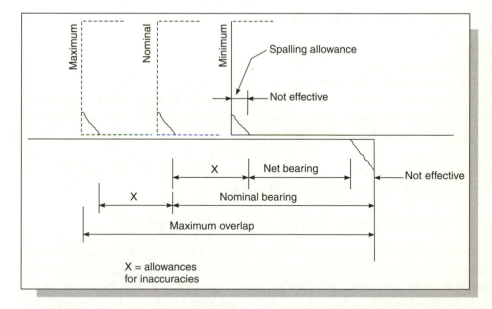

Figure 4.21: Bearing widths.

The depth of the double-tee section is often reduced at the ends of the units to facilitate a recessed bearing on to the supporting beam or wall (see Figure 4.5). The concrete is heavily reinforced at this point because it is subjected to shear, bending and bearing. A truss analogy, or strut and tie method of analysis is used to calculate the forces. Figure 4.22 shows a typical end shear cage in which the vertical bars are provided for the ultimate shear resistance, and the horizontal U-bars are for the lateral bursting resistance. (The black box in Figure 4.22 makes the half joint.)

Example 4.8

Calculate the bearing capacity of the hollow core unit in Example 4.1. The unit has an actual 75 mm dry bearing width onto a reinforced concrete beam where the cover to the bars in the bearing ledge is 30 mm and the depth of

Figure 4.22: Shear and bursting reinforcement at ends of double-tee units.

the bearing ledge is 250 mm. (Note the strands in the hollow core unit are exposed at their ends.) Assume that the unit has secondary support.

Solution

Ultimate bearing stress $f_b = 0.4 \times 50 = 20\,\text{N/mm}^2$

Net bearing width for non-isolated component $l_w = 75$ actual $- 30$ cover in beam $- 0$ ineffective hollow core unit bearing distance $= 45\,\text{mm} > 40\,\text{mm}$ minimum.

Bearing length $l_b =$ least of (i) 1200 mm; (ii) 700 mm; (iii) 600 mm. Use 600 mm. Then $F_b = 20 \times 45 \times 600 \times 10^{-3} = \underline{540\,\text{kN}}$.

4.4 Design of composite floors

4.4.1 Precast floors with composite toppings

The structural capacity of a precast floor unit may be increased by adding a layer of structural reinforced concrete to the top of the unit. Providing that the topping concrete is fully anchored and bonded to the precast unit the two concretes – precast and cast in situ, may be designed as monolithic. The section properties of the precast unit plus the topping are used to determine the structural performance of the composite floor. A composite floor may be made using any type of precast unit, but clearly there is more to be gained from using voided prestressed units, such as hollow core unit, double-tee, which are lightweight and therefore cheaper to transport and erect than solid reinforced concrete units. Figure 4.23a shows the details for the most common types of composite floors. Figure 4.23b gives an indication of the enhanced load capacity of prestressed double-tee floors where a 75 mm thick topping is used. Note the reduced (in fact negative in one case!) increase in performance at large spans where the increase in self-weight counters any increase in structural area.

The minimum thickness of the topping should not be less than 40 mm (50 mm is more practical). There is no limit to the maximum thickness, although 75–100 mm is a practical limit. When calculating the average depth of the topping allowances for camber should be made – allowing span/300 will suffice. The grade of in situ concrete is usually C25 or C30, but there is no reason why higher strength cannot be used except that the increased strength of the composite floor resulting from the higher grade will not justify the additional costs of materials and quality control. See Table 2.1 for the concrete data. The topping *must* be reinforced, but, as explained later, there needs to be tie steel only at the interface between the precast and in situ topping if the design dictates. Mesh reinforcement of minimum area 0.13% × concrete area is the preferred choice – see Section 2.2 for the data.

The main benefit from composite action is in increased bending resistance and flexural stiffness – shear and bearing resistance is barely increased. There are however a number of other reasons why a structural topping may be specified, such as:

- to improve vibration, thermal and acoustic performance of the floor;

- to provide horizontal diaphragm action (see Chapter 7);

- to provide horizontal stability ties across floors; and

- to provide a continuous and monolithic floor finish (e.g. where brittle finishes are applied).

Composite floor design is carried out in *two* stages, before and after the in situ topping becomes structural. (In prestressed concrete the transfer stress condition must also be satisfied.) Therefore, the precast floor unit must carry its own weight plus the self weight of the wet in situ concrete (plus a construction traffic allowance of 1.5 kN/m^2). The composite floor (=precast + hardened topping) carries imposed loads. In the final analysis, the stresses and forces resulting from the two cases (minus the construction traffic allowance which is temporary) are additive. In calculating deflections, the effects of the relative shrinkage of the topping to that of the precast unit must be added to those resulting from loads (and prestressing if applicable).

Figure 4.23b shows the increased bending load capacity for double-tee floors achieved using a 75 mm thick structural topping. Note that the benefit from this

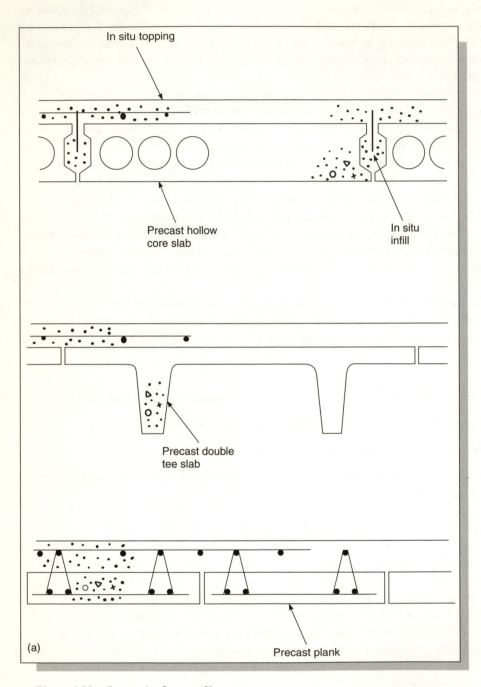

Figure 4.23a: Composite floor profiles.

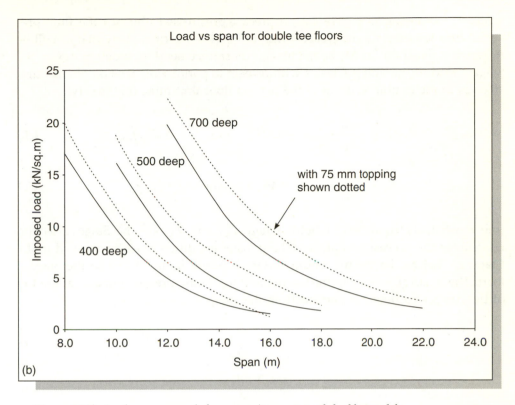

Figure 4.23b: Load vs span graph for composite prestressed double-tee slabs.

decreases as (i) the span increases, viz. self weight of in situ topping nullifies the additional capacity; and (ii) the unit depth increases, viz. the section modulus of the composite section is proportionately less.

The following design procedures are used.

4.4.2 Flexural analysis for composite prestressed concrete elements

4.4.2.1 Serviceability state

Permissible service stresses are checked at two stages of loading – Stages 1 and 2 before and after hardening of the in situ topping as follows:

- Stage 1 for the self weight of the precast slab plus the self weight of the in situ concrete topping, plus an allowance for temporary construction traffic of up to $1.5 \, \text{kN/m}^2$. The section properties of the precast unit alone are used.

- Stage 2 for superimposed loading. The section properties of the composite section are used.

Stage 1. Referring to Figure 4.24, a precast prestressed unit has the final pre-stress after losses of f_{bc} and f_{tc} according to Eqs 4.1–4.4. (It is likely that f_{tc} will be negative = tension.) Let M_1 be the maximum service bending moment due to the Stage 1 load. If the unit is Class 2 with respect to permissible tension, the flexural stresses in the bottom and top of the precast floor unit must first satisfy:

$$f_{bl} = -\frac{M_1}{Z_{bl}} + f_{bc} < +\mathbf{0.33f_{cu}} \quad \text{and} \quad > -0.45\sqrt{f_{cu}} \qquad\qquad 4.13$$

$$f_{tl} = +\frac{M_1}{Z_{tl}} + f_{tc} > -\mathbf{0.45\sqrt{f_{cu}}} \quad \text{and} \quad < +0.33f_{cu} \qquad\qquad 4.14$$

(the most likely condition in **bold**) where Z_{b1} and Z_{t1} are the Stage 1 section moduli for the precast unit alone. (The transfer condition must be checked first.)
Stage 2. Let M_2 be the maximum service bending moment due to the Stage 2 load. The flexural stresses in the bottom and top of the precast unit and at the top of the composite floor are derived from the composite section as:

$$f_{b2} = -\frac{M_2}{Z_{b2}} \qquad\qquad 4.15$$

$$f_{t2} = +\frac{M_1}{Z_{t2}} \qquad\qquad 4.16$$

$$f'_{t2} = +\frac{M_1}{Z'_{t2}} \qquad\qquad 4.17$$

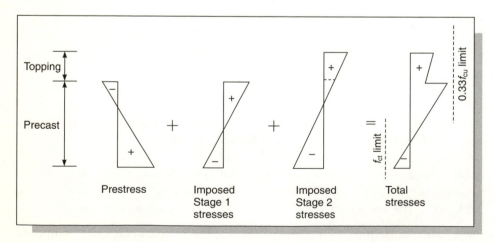

Figure 4.24: Principles of serviceability stress limitations for composite prestressed elements.

where Z_{b2}, Z_{t2} and Z'_{t2} are the Stage 2 section moduli for the composite section at the bottom of the precast unit, the top of the precast unit and the top of the topping, respectively.

To calculate these values, the second moment of area I of the composite section must be determined using the modular ratio approach. This is to allow for the different Young's modulus of the precast and in situ concrete. The breadth of the in situ topping is reduced in the proportion to the Young's moduli, such that the effective breadth of the topping is:

$$b_{\text{eff}} = b E_{c,\text{in situ}} / E_{c,\text{precast}} \qquad 4.18$$

Note that in this case b = full breadth of the precast unit (not the manufactured top breadth).

Then, the stresses from prestressing and Stage 1 loads are added to those from Stage 2 loads, and must satisfy:

$$f_b = f_{bc} - \frac{M_1}{Z_{b1}} - \frac{M_2}{Z_{b2}} > -0.45\sqrt{f_{cu}} \qquad 4.19$$

$$f_t = f_{tc} + \frac{M_1}{Z_{t1}} + \frac{M_2}{Z_{t2}} < 0.33 f_{cu} \qquad 4.20$$

$$f'_t = +\frac{M_2}{Z'_{t2}} < 0.33 f_{cu,\text{in situ}} \qquad 4.21$$

Equation 4.21 is rarely critical and does not affect the design of the precast unit.

The critical situation nearly always occurs in the bottom of the slab because the new position of the neutral axis is often close to the top of the precast unit. Equation 4.19 can be written as:

$$M_{s2} = M_2 > \left(0.45\sqrt{f_{cu}} + f_{bc}\right) Z_{b2} - \left[M_1 \left(\frac{Z_{b2}}{Z_{b1}}\right)\right] \qquad 4.22$$

However, the situation at the top of the precast unit should also be checked for completeness, as:

$$M_{s2} = M_2 > (0.33 f_{cu} - f_{tc}) Z_{t2} - \left[M_1 \left(\frac{Z_{t2}}{Z_{t1}}\right)\right] \qquad 4.23$$

The construction traffic loading need only be considered as part of M_1 when checking that the service stresses do not exceed $0.33 f_{cu}$ and $-0.45\sqrt{f_{cu}}$. It does not have to be included in the *final* calculation in Eq. 4.22 because it is not a permanent load. (Nor does it affect the ultimate limit state.)

It is not possible to prepare in advance standard calculation for composite floor slabs. This is because the final stresses in Eqs 4.19 and 4.20 depend on the respective magnitudes of the Stage 1 and Stage 2 loads and moments, i.e. the same precast concrete unit may have different load bearing capacity when used in different conditions. The following examples will show this.

Example 4.9

Calculate the Stage 2 service bending moment that is available if the hollow core unit in Example 4.1 has a 50 mm minimum thickness structural topping. The floor is simply supported over an effective span of: (a) 4.0 m; (b) 8.0 m

The precamber of the hollow core unit may be assumed as span/300 without loss of accuracy. Use f_{cu} for the topping $= 30\,\text{N/mm}^2$. Self weight of concrete $= 24\,\text{kN/m}^3$.

What is the maximum imposed loading for each span?

Solution

Young's modulus for topping $= 26\,\text{kN/mm}^2$
Effective breadth of topping $= 1200 \times 26/30 = 1040\,\text{mm}$
Total depth of composite section $= 253\,\text{mm}$
Depth to neutral axis of composite section $y_{t2} = (1040 \times 50 \times 25) + (135\,000 \times (50 + 99))/187\,000 = 114.5\,\text{mm}$
Second moment of area of composite section $I_2 = 1266\,\text{mm}^4$
Then $Z_{b2} = 1266 \times 10^6/(253 - 114.5) = 9.14 \times 10^6\,\text{mm}^3$
and $Z_{t2} = 1266 \times 10^6/(114.5 - 50) = 19.63 \times 10^6\,\text{mm}^3$ (at top of hollow core unit)
Then $Z_{b2}/Z_{b1} = 9.14/6.519 = 1.40$, and $Z_{t2}/Z_{t1} = 19.63/6.848 = 2.86$

Solution (a)

Precamber $= 4000/300 = 13\,\text{mm}$
Maximum depth of topping at supports $= 50 + 13 = 63\,\text{mm}$
Average depth of topping $= (50 + 63)/2 = 57\,\text{mm}$
Self weight of topping $= 0.057 \times 24 \times 1.2 = 1.64\,\text{kN/m}$ run for 1.2 m wide unit
Self weight of hollow core unit $= 3.24\,\text{kN/m}$ run
Stage 1 moment $M_1 = (3.24 + 1.64) \times 4.0^2/8 = 9.76\,\text{kNm}$

$$M_2 = (3.2 + 9.97) \times 9.14 - 9.76 \times 1.40 = 106.7\,\text{kNm} \qquad (\textit{using Eq. 4.22})$$
$$M_2 = (16.5 - (-0.63)) \times 19.63 - 9.76 \times 2.86$$
$$= 308.3\,\text{kNm (clearly not critical!)} \qquad (\textit{using Eq. 4.23})$$

The allowable imposed load $= 8 \times 106.7/4.0^2 = \underline{53.4\,\text{kN/m}}$
(Note that the *total* $M_s = 9.76 + 106.7 = 116.5\,\text{kNm} > 85.8\,\text{kNm}$ for the basic unit in Example 4.1.)

Solution (b)

Precamber $= 8000/300 = 26\,\text{mm}$

Average depth of topping $= (50 + 76)/2 = 63\,\text{mm}$
Self weight of topping $= 0.063 \times 24 \times 1.2 = 1.81\,\text{kN/m}$ run for 1.2 m wide unit
Stage 1 moment

$$M_1 = (3.24 + 1.81) \times 8.0^2/8 = 40.4\,\text{kNm}$$
$$M_2 = (3.2 + 9.97) \times 9.14 - 40.4 \times 1.40 = \underline{63.8\,\text{kNm}} \qquad \textit{(using Eq. 4.22)}$$
$$M_2 = (16.5 - (-0.63)) \times 19.63 - 40.4 \times 2.86 = 220.7\,\text{kNm}$$
$$\text{(clearly not critical!)} \qquad \textit{(using Eq. 4.23)}$$

The allowable imposed load $= 8 \times 63.8/8.0^2 = \underline{7.98\,\text{kN/m}}$
(Note that the *total* $M_s = 40.4 + 63.8 = 104.2\,\text{kNm}$ is less than the total moment in case (a). This is because the Stage 1 moment, which causes greater stresses than an equivalent Stage 2 moment, is greater than in case (a).)

4.4.2.2 Ultimate limit state

The design at ultimate limit state is also a two-stage process, with the flexural stresses resulting from the self weight of the precast element plus any wet in situ concrete being carried by the precast unit alone. The lever arm is the same as in a non-composite design, i.e. d. The method is to calculate the area of steel, A_{ps1}, required in Stage 1, and to add the area, A_{ps2}, required in Stage 2 using an increased lever arm (see Figure 4.25a). The effect of the structural topping is to increase the lever arm to the steel reinforcement by an amount equal to the thickness of the topping h_s, proving the depth to the neutral axis is less than h_s (see Figure 4.25b. In Stage 2, the effective breadth of the topping $b_{eff} = b \times f'_{cu}$ (in situ/f_{cu} (precast), where b is the full breadth of the precast unit, not the manufactured breadth.

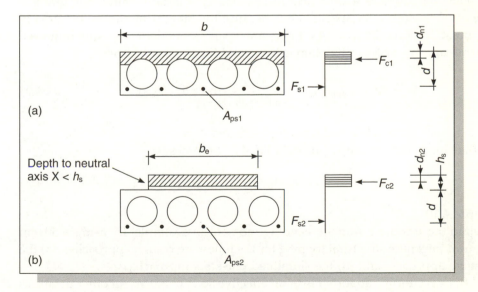

Figure 4.25: Principles of ultimate strength for composite prestressed elements.

Most design engineers choose not to separate the design into two stages, using the composite section properties alone. This is obviously less conservative, but the differences are quite small as will be shown in Example 4.10.

Adopting a two-stage approach (Figure 4.25a), equilibrium in the section due to Stage 1 stresses is:

$$f_{pb}A_{ps1} = 0.45f_{cu}(\text{precast})\, b\, 0.9\, X_1 \qquad\qquad 4.24$$

but $d_{n1} = 0.45X_1$
then

$$d_{n1} = f_{pb}A_{ps1}/0.9f_{cu}b \qquad\qquad 4.25$$

Then

$$M_{u1} = f_{pb}A_{ps1}(d - d_{n1}) \qquad\qquad 4.26$$

From which d_n and A_{ps1} may be determined.

At Stage 2, the area of steel to resist M_{u2} is $A_{ps2} = A_{ps} - A_{ps1}$. But to allow for f_{pb}/f_{pu} being less than 0.95, f_{pb} and d_{n2} are obtained from Table 4.3 for specific levels of prestress and strength ratio $f_{pu}A_{ps2}/f_{cu}b_{eff}(d + h_s)$. Then:

$$M_{u2} = f_{pb}A_{ps2}(d + h_s - d_{n2}) \qquad\qquad 4.27$$

It is seen that there is a difficulty in calculating standard values for ultimate moment of resistance for specific units because the Stage 1 moments and area of steel must be first known. As these are span dependent the superimposed moment capacity M_{u2} is a function of span and Stage 1 loads. Then:

$$M_{u2} = f_{pb}\left[A_{ps} - \frac{M_{U1}}{f_{pb}(d - d_{n1})}\right](d + h_s - d_{n2}) \qquad\qquad 4.28$$

Where d_{n1} has been calculated from Eqs 4.25 and 4.26.

In the simplified one step approach, Eq. 4.27 becomes:

$$M_u = f_{pb}A_{ps}(d + h_s - d_n) \qquad\qquad 4.29$$

Example 4.10
Calculate the imposed ultimate bending moment in Example 4.1 using a 50 mm minimum thickness structural topping for the following design approaches: (a) the two-stage approach; and (b) the simplified one-step approach.

The floor is simply supported over an effective span of 8.0 m. All other details as Example 4.9.

What is the maximum ultimate imposed loading. Is the composite slab critical at service or at ultimate?

Solution
Effective depth in hollow core unit $d_{n1} = 156.7\,\text{mm}$
Effective depth in composite section $d_{n2} = 156.7 + 50 = 206.7\,\text{mm}$
Effective breadth of topping $b_{\text{eff}} = 1200 \times 30/50 = 720\,\text{mm}$

(a) Two stage approach
Stage 1 moment $M_{u1} = 1.4 \times 40.4 = 56.6\,\text{kNm}$
Using Eqs 4.24 to 4.26.

$$M_{u1} = 56.6 \times 10^6 = (0.95 \times 1750 \times A_{ps1} \times 156.7) - \frac{(0.95 \times 1750 \times A_{ps1})^2}{0.9 \times 50 \times 1168}$$

Hence, $A_{ps1} = 227\,\text{mm}^2$ and $d_{n1} = 7.2\,\text{mm}$. Then $A_{ps2} = 659 - 227 = 432\,\text{mm}^2$

$$\frac{f_{pu}A_{ps}}{f_{cu}b_{\text{eff}}(d + h_s)} = \frac{1750 \times 432}{50 \times 720 \times 206.7} = 0.10$$

From Table 4.3, $f_{pb}/0.95f_{pu} = 1.0$

$$d_{n2} = 22.2\,\text{mm} \qquad (\text{using Eq. 4.25})$$

Then $M_{u2} = 1.0 \times 0.95 \times 1750 \times 432 \times (206.7 - 22.2) \times 10^6 = \underline{132.5\,\text{kNm}}$
Ultimate load $= 8 \times 132.5/8.0^2 = \underline{16.5\,\text{kN/m run}}$
    Refer to Example 4.9. Ratio of ultimate/service imposed load $= 16.5/7.98 = 2.07$. As this value is greater than 1.6 the composite slab is critical at the service limit state.

(b) One step approach

$$f_{pu}A_{ps}/f_{cu}b_{\text{eff}}(d + h_s) = 1750 \times 659/50 \times 720 \times 206.7 = 0.155$$

From Example 4.1, $f_{pe}/f_{pu} = 0.531$
From Table 4.3, $f_{pb}/0.95f_{pu} = 0.93$, $X/d = 0.324$, hence $X = 67\,\text{mm}$ and $d_n = 30\,\text{mm}$.
Then $M_u = 0.93 \times 0.95 \times 1750 \times 659 \times (206.7 - 30) \times 10^6 = \underline{180.0\,\text{kNm}}$
    Total ultimate load $= 8 \times 180/8.0^2 = \underline{22.5\,\text{kN/m run}}$
    Subtract self weight of hollow core unit and topping $= 1.4 \times 5.05 = 7.1\,\text{kN/m}$, leaving imposed load $= 22.5 - 7.1 = 15.4\,\text{kN/m}$. This is some $1.2\,\text{kN/m}$ less than in the two stage approach. This is because the bars attain only $0.93 \times 0.95f_y$, whereas in the two stage design they both achieve the full $0.95f_y$.

4.4.3   *Propping*

Propping is a technique which is used to increase the service moment capacity by reversing the Stage 1 stresses particularly at mid-span. This is achieved by placing a rigidly founded support, 'Acrow prop' or similar, in the desired place whilst the in situ concrete topping is hardening (see Figure 4.26). To ensure that the props are always effective, many contractors prefer to use two props rather than one – just in case the foundation to one of the props is 'soft'. However, the following analysis will consider a single mid-span propped floor slab. The reader can easily extend the same analysis to multiple props.

Figure 4.26: Propping of composite plank floor.

The benefit derives from the fact that the Stage 1 moments due to the weight of the wet concrete topping are determined over a continuous double span, each of $L/2$. When the props are removed the prop reaction R creates a new moment which is carried by the composite section. Finally, the superimposed loads are added as shown in Figure 4.27.

Under the action of the prop, the hogging moment is $-wL^2/32$ (or $-0.031\ 25wL^2$), where w = self weight of the wet in situ topping (allowing for precamber of the slab) and L = effective span of slab. The prop reaction is $R = 0.625\ wL$, such that the additional moment at Stage 2 following the removal of the prop is $M = +0.156\ 25wL^2$. Equation 4.22 is therefore modified as:

$$M_{s2} = M_2 > (0.45\sqrt{f_{cu}} + f_{bc})Z_{b2} - \left[M_1\left(\frac{Z_{b2}}{Z_{b1}}\right)\right] - 0.156\ 25wL^2 \qquad 4.30$$

The economical and practical benefits of propping wide slabs such as hollow core unit and double-tees should be carefully considered. Propping can be quite expensive and may slow down site erection rates.

Example 4.11
Repeat Example 4.9b with the hollow core unit propped at mid-span. The span is 8.0 m.

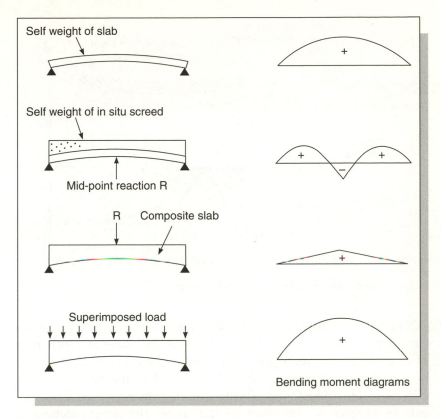

Figure 4.27: Bending moments resulting from propping.

Solution

Stage 1 moment due to self weight of hollow core unit $M_1 = 3.24 \times 8.0^2/8 = +25.9\,$kNm

Negative moment due to propping $M_{\mathrm{prop}} = -0.031\,25 \times 1.81 \times 8.0^2 = -3.6\,$kNm

Net Stage 1 moment $= 25.9 - 3.6 = 22.3\,$kNm (compared with $40.4\,$kNm in Example 4.9(b))

Prop reaction moment $= +0.156\,25 \times 1.81 \times 8.0^2 = +18.1\,$kNm

$$M_2 = (3.2 + 9.97) \times 9.14 - 22.3 \times 1.4 - 18.1$$

$\qquad = \underline{71.0\,\text{kNm}}$ (compared with $63.8\,$kNm in Example 4.9(b))   (*using Eq.* 4.30)

Imposed load $= 8 \times 71.0/8.0^2 = 8.9\,$kN/m.

4.4.4 *Interface shear stress in composite slabs*

Under the action of vertical flexural shear, the horizontal interface between the precast unit and in situ topping will be subjected to a horizontal shear force,

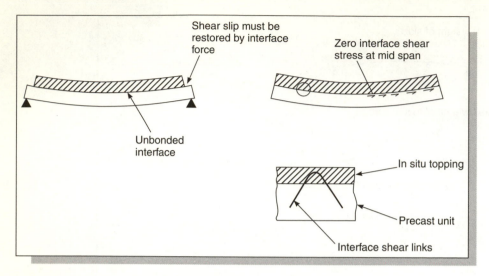

Figure 4.28: Interface shear stress and forces in composite elements.

(Figure 4.28). Often termed 'shear flow', because it is measured in force per linear length, this shear derives from the equilibrium of the vertical shear at a section. It is the result of imposed loads present only after the in situ concrete topping has hardened. The distribution of interface shear is identical to the imposed shear force distribution and must therefore be checked at all critical sections. Interface shear need only be checked for the ultimate limit state. The design method is based on experimental evidence, and will ensure that serviceability conditions are satisfied.

Providing that the in situ topping is fully bonded to the precast unit, full interaction is assumed, i.e. there is no relative slippage between the two concretes. The horizontal shear force F_v at the interface is equal to the total force in the in situ topping due to imposed loads. It is therefore necessary to have carried out a calculation for the ultimate limit state in flexure and to have determined the depth to the neutral axis X before this is attempted. (This is a benefit from having carried out a two-stage approach to ultimate flexure.)

If the neutral axis is below the interface, $X > h_s$, then

$$F_v = 0.45 f_{cu} b_{eff} h_s \qquad\qquad 4.31$$

If the neutral axis is above the interface, $X < h_s$, then

$$F_v = 0.45 f_{cu} b_{eff} 0.9X \qquad\qquad 4.32$$

The force F_v only acts at the point of maximum bending moment – elsewhere it is less than this and may even change sign in a continuous floor. Therefore, the distance over which this force is distributed along the span of the floor is taken as the distance from the maximum to the zero or minimum moment.

The average ultimate shear stresses at the interface may be calculated as:

$$v_{ave} = \frac{F_V}{bL_z} \qquad 4.33$$

where v_{ave} = the average shear stress at the cross-section of the interface
considered at the ultimate limit state

b = the transverse width of the interface

L_z = distance between the points of minimum and maximum bending moment.

The average stress is then distributed in accordance with the magnitude of the vertical shear at any section, to give the design shear stress v_h. Thus, for uniformly distributed superimposed loading (self weight does not create interface stress) the maximum stress $v_h = 2v_{ave}$. For a point load at mid-span $v_h = v_{ave}$ and so on.

If v_h is greater than the limiting values given in Table 4.5 (reproduced from BS8110, Part 1, Table 5.5) all the horizontal force should be carried by reinforcement (per 1 m run) projecting from the precast unit into the structural topping. The amount of steel required is:

$$A_f = \frac{1000bv_h}{0.95f_y} \qquad 4.34$$

but not less than 0.15 per cent of the contact area. The reinforcement should be adequately anchored on both sides of the interface. If loops are used, as shown in Figure 4.29, the clear space beneath the bend should be at least 5 mm + size of

Table 4.5: Design ultimate horizontal shear stress at interface (N/mm^2)

Precast unit	Surface type	Grade of in situ concrete		
		C25	C30	C40+
Without links	As cast or as extruded	0.40	0.55	0.65
	Brushed, screeded or rough tamped	0.60	0.65	0.75
	Washed to remove laitance, or treated with retarding agent and cleaned	0.70	0.75	0.80
Nominal links	As cast or as extruded	1.2	1.8	2.0
Projecting into in situ concrete	Brushed, screeded or rough tamped	1.8	2.0	2.2
	Washed to remove laitance, or treated with retarding agent and cleaned	2.1	2.2	2.5

aggregate. The spacing of links should not be too large, 1.2–1.5 m being typical for hollow core slabs. If v_h is less than values in Table 4.5, no interface shear reinforcement is required, although some contractors choose to place R10 or R12 loops (as shown in Figure 4.29) at about 1.2 m intervals. The loops should pass over the top of the bars in the structural topping and be concreted into the joints between the precast units.

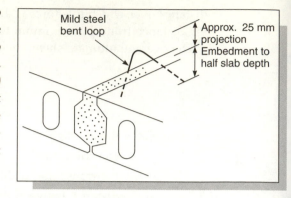

Figure 4.29: Interface shear links (or loops) in composite hollow core floors.

Example 4.12

Calculate the shear reinforcement necessary to satisfy the ultimate horizontal shear force at the precast – in situ interface in Example 4.10. The top surface of the hollow core unit is 'as extruded' finish. Use HT reinforcement $f_y = 460\,\text{N/mm}^2$.

Solution

From Example 4.10(a), $X_2 = (22.2/0.45) = 49.3\,\text{mm} < h_s = 50\,\text{mm}$

$$F_v = 0.45 \times 50 \times 720 \times 0.9 \times 49.3 \times 10^{-3} = 719\,\text{kN} \qquad (using\ Eq.\ 4.32)$$

The distance L_z = half the span = 4000 mm
    The interface breadth = 1200 mm (not the effective breadth of 720 mm)

$$v_{ave} = (719 \times 10^3)/(1200 \times 4000) = 0.15\,\text{N/mm}^2 \qquad (using\ Eq.\ 4.33)$$

If the imposed loading is uniformly distributed, $v_h = 2 \times 0.15 = 0.3\,\text{N/mm}^2 < 0.55\,\text{N/mm}^2$ from Table 4.5.

4.5   Composite plank floor

Shallow precast slabs, hence the name 'planks', are laid between supports and are used as permanent formwork for an in situ concrete topping. The precast plank is between 65 and 100 mm thick, depending on the span, and the depth of the complete floor is between 150 and 200 mm. Steel bars, wires or tendons placed in the precast plank units act as the flexural sagging reinforcement, and a light steel mesh (e.g. A142, A193) in the in situ concrete acts as hogging reinforcement. The diagonal bars in the lattices provide shear resistance during the construction stage. The planks can also be pretensioned using longitudinal wires only. Lattices are positioned next to the prestressing wires.

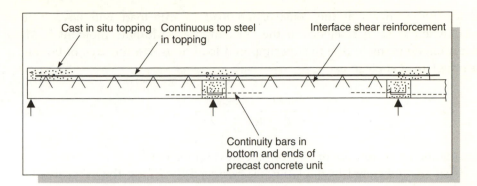

Figure 4.30: Continuity in composite plank floor.

The composite floor slab may be designed as either simply supported or continuous as shown in Figure 4.30. Note that there should also be continuity bars at the bottom of a continuous slab at the support to ensure rigidity in the compression zone and to cater for reversals of bending moments due to creep, shrinkage, temperature effects etc. when no (or small) superimposed load is present.

In reinforced planks (i.e. non-prestressed) deflections are catered for by checking that the actual span/effective depth ratio is within the allowable limit (same as in any r.c. design). Only in exceptional circumstances where a deflection violation using span/depth ratios may occur would the actual deflection be calculated. In prestressed planks, the same procedures as in Section 4.3.4 are adopted.

In the temporary stage, the precast plank is simply supported. The unit may be designed so that unpropped spans of upto 4 m are possible, usually by increasing the number of lattices to increase shear stiffness, but not necessarily increasing the number of bottom bars. The top bar is in compression, but is firmly restrained both vertically and horizontally by the inclined bars making the lattice. The unit is most critical when the self weight of the wet in situ concrete is added to the self weight of the precast plank. An allowance for construction traffic of upto $1.5\,\text{kN/m}^2$ is added to the temporary loading when calculating the sizes of bars required. In the permanent situation, the hardened in situ concrete provides the compressive resistance. The flexural sagging resistance at mid-span is governed by the strength of the bottom reinforcing bars, as in an ordinary under-reinforced situation.

If the maximum bending moment in the temporary condition is M_1 (inclusive of the construction traffic) then the area of top steel in the lattice is given as:

$$A'_s = \frac{M_1}{z_1 0.95 f_y}$$
4.35

where z_1 is the centre-to-centre vertical distance between the bars in the lattice. The area of the bottom steel is specified after the full service load is considered, but without the effects of the construction load as this load will have been removed when in service. Hence, if the net temporary ultimate moment is M'_1 and the ultimate moment due to superimposed loading is M_2, the area of bottom steel in the lattice is given as:

$$A_{s1} = \frac{M'_1}{z_1 0.95 f_y} + \frac{M_2}{z_2 0.95 f_y} \qquad 4.36$$

where z_2 is the lever arm obtained from the flexural design.

The shear reinforcement in the lattices is designed by taking the vertical component of the axial force in the inclined bars as the only shear resistance against the temporary shear force V_1. The area of the lattice's 'shear links' is given as:

$$A_{sv} = \frac{\sqrt{2} V_1}{0.95 f_{yv}} \qquad 4.37$$

Two diagonal lattice bars are used to provide the shear reinforcement. They are also used to transfer shear forces due to superimposed loads V_2 in the precast-in situ interface. Thus, if the design horizontal shear stress v_h given as:

$$v_h = \frac{V_2}{bd} \qquad 4.38$$

is greater than the limiting value in Table 4.5 the interface shear steel should carry the interface shear force $F = v_h b$ per unit length of the interface as:

$$A_h = \frac{v_h b}{0.95 f_y} \qquad 4.39$$

But not less than 0.15 per cent of the contact area.

Example 4.13
Design a composite reinforced concrete plank floor to carry a characteristic superimposed live load of $5 \, kN/m^2$ over a simply supported effective span of 3.5 m. No propping is allowed.

Use concrete $f_{cu} = 40 \, N/mm^2$ for the precast, $f_{cu} = 25 \, N/mm^2$ for the in situ. Use HT bars in the precast plank, and a square wire mesh in the in situ topping, using $f_y = 460 \, N/mm^2$. Cover to all reinforcement = 25 mm.

Check the design at both the temporary and permanent stages for flexural and vertical shear only.

Solution
Deflection control. BS8110, Part 1, Tables 3.10 and 3.11.

Basic span/$d = 20$. Modification factor (for an initial estimate $M/bd^2 = 1.0$) is 1.38. Thus $d = 3500/20 \times 1.38 = 127$ mm

$$h = 127 + 25 + \text{say } 8 = 160 \text{ mm}$$

Consider 1 m width of floor, using 50 mm deep precast with 110 mm deep in situ topping.

Loading on lattice in temporary condition

	kN/m^2
Self weight of 50 mm deep precast unit $= 0.050 \times 24$	$= 1.20$
Self weight of 110 mm deep wet concrete $= 0.110 \times 24$	$= 2.64$
Construction traffic allowance, in this case say 1.0 kN/m^2	$= 1.00$
Total	$= 4.84$

$$M_{u1} = 1.4 \times 4.84 \times 3.5^2/8 \quad = 10.37 \text{ kNm/m}$$
$$V_{u1} = 1.4 \times 4.84 \times 3.5/2 \quad = 11.86 \text{ kN/m}$$

Assume lattice top and bottom bar size $= 16$ mm
Lever arm $= 160 - (25 + 8) - (25 + 8) = 94$ mm
Force in top and bottom bars $= 10.37 \times 10^3/94 = 110.3$ kN/m run

$A_s = (110.3 \times 10^3)/(0.95 \times 460) = 252 \text{ mm}^2/\text{m} \times 0.6$ m centres $= 152 \text{ mm}^2$ per lattice.

Use 1 no. T 16 top bar (201) in lattices at 600 mm centres.
Bottom bars will be specified after full service loads considered. Subtract the effects of the construction load when calculating the force in bottom steel (as this load will have been removed when in service). Hence $A_s = (3.84/4.84) \times 252 = 200 \text{ mm}^2/\text{m}$.

Shear per lattice $= 0.6 \times 11.86 = 7.12$ kN.

Lattice bars at $45°$ inclination, thus force in diagonal bar $= 7.12/\sin 45° = 10.0$ kN.

$$A_s = (10 \times 10^3)/(0.95 \times 250) = 42 \text{ mm}^2/2 \text{ no. bars} = 21 \text{ mm}^2.$$

Try double R 6 diagonal lattice bars (28) inclined at 45° but check minimum interface requirement.
The height of the lattice $= 160 - 25$ top cover $- 25$ bottom cover $= 110$ mm. Therefore, distance between diagonal lattice bars $= 110$ mm.

Then, area of diagonal lattice bars per interface area crossing interface $= (28 \times \sqrt{2} \times 2)/(600 \times 110) \times 100 = 0.12\%$.

This is less than the minimum value of 0.15 per cent, therefore increase diagonal bars to R8.
Use double R8 diagonal lattice bars (50) inclined at 45°.

Service loading

When the in situ concrete has hardened it is effectively stress free because the deflections have all occurred whilst the concrete was wet. Therefore, the only stresses in the in situ topping derives from superimposed load $=5.00\,\text{kN/m}^2$

$$M_{u2} = 1.6 \times 5.00 \times 3.5^2/8 \quad = 12.25\,\text{kNm/m}$$
$$V_{u1} = 1.6 \times 5.00 \times 3.5/2 \quad = 14.0\,\text{kN/m}$$

Flexural design

$$f_{cu} = 25\,\text{N/mm}^2, \, b = 1000\,\text{mm}, \, d = 160 - 33 = 127\,\text{mm}$$
$$K = (12.25 \times 10^6)/(25 \times 1000 \times 127^2) = 0.03 < 0.156$$

Then $z/d = 0.95$, and the area of the bottom steel is $A_s = (12.25 \times 10^6)/$ $(0.95 \times 127 \times 0.95 \times 460) = 232\,\text{mm}^2/\text{m}$ plus $200\,\text{mm}^2/\text{m}$ from the construction stage $= 432\,\text{mm}^2/\text{m} \times 0.6\,\text{m}$ centres $= 259\,\text{mm}^2$ per lattice.

Use 2 no. T16 bottom bars (402) in lattices at 600 mm centres.

Shear design

$v = (14.0 \times 10^3)/(1000 \times 127) = 0.11\,\text{N/mm}^2 <$ minimum value in BS8110, Part 1, Tables 3.9 and 5.5, therefore no additional reinforcement to the lattice required.

References

1 Van Acker, A., Transversal Load Distribution of Linear Loadings in Hollow Core Floors, FIP Conference, Calgary, Canada, 25–31 August 1984, pp. 27–33.
2 Moss, R. M., Load Testing of Hollow Plank Concrete Floors, *The Structural Engineer*, Vol. 73, No. 10, May 1995, pp. 161–168.
3 Pfeifer, D. W. and Nelson, T. A., Tests to Determine the Lateral Distribution of Vertical Loads in a Long Span Hollow Core Floor Assembly, *PCI Journal*, Vol. 23, No. 6, 1983, pp. 42–57.
4 Stanton, J., Proposed Design Rules for Load Distribution in Precast Concrete Decks, *ACI Structural Journal*, September–October 1987, pp. 371–382.
5 FIP Recommendations: Precast Prestressed Hollow Cored Floors, FIP Commission on Prefabrication, Thomas Telford, London, 1988, 31p.
6 Precast/Prestressed Concrete Institute, PCI Manual for the Design of Hollow Core Slabs, Chicago, USA, 1991, 88p.
7 Concrete Manufacturers' Association, Precast Concrete Floor Slabs Design Manual, Johannesburg, South Africa, 90p.
8 Walraven, J. C. and Mercx, W., The Bearing Capacity of Prestressed Hollow Core Slabs, Heron, 28, No. 3, University of Delft, Netherlands, 1983, 46p.
9 Girhammar, U. A., Design Principles for Simply Supported Prestressed Hollow Core Slabs, *Structural Engineering Review*, Oxford, UK, Vol. 4, No. 4, 1992, pp. 301–316.
10 Pajari, M., Shear Resistance of PHC Slabs Supported on Beams, *Journal of Structural Engineering*, Vol. 124, No. 9, 1998, Part 1: Tests, 1050–1061. Part 2: Analysis, 1062–1073.
11 Kong, F. K. and Evans, R. H., *Reinforced and Prestressed Concrete*, 3rd edition Van Nostrand Reinhold, 1987, 508p.

5 Precast concrete beams

5.1 General introduction

Beams are the main horizontal load carrying members in skeletal structures. They are, by definition, relatively small prismatic sections of large flexural (typically 300–800 kNm) and shear (100–500 kN) capacity. In a precast concrete structure they must at some point in time support the self-weight of the floor slabs alone and should therefore be capable of resisting all of the possible load combinations that precast construction brings – for example, torsion will be present if, in the temporary construction stage, the floor units are all positioned on one side of the beam. This must be allowed for both in the design of the beam and at the end connections to the column.

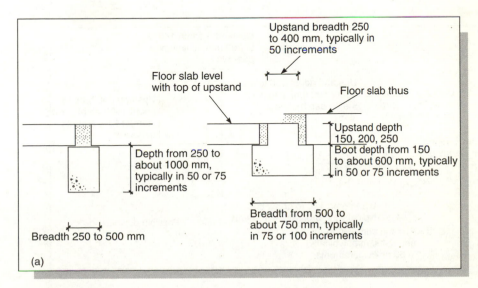

Upstand breadth 250
to 400 mm, typically in
50 increments

Floor slab level
with top of upstand

Floor slab thus

Upstand depth
150, 200, 250
Boot depth from 150
to about 600 mm, typically
in 50 or 75 increments

Depth from 250 to
about 1000 mm,
typically in 50 or 75
increments

Breadth from 500 to
about 750 mm, typically
in 75 or 100 increments

Breadth 250 to 500 mm

(a)

Figure 5.1: Types of beams. (a) Internal rectangular and inverted tee.

Beams fall into two distinct categories: (1) internal; and (2) external. Internal beams are usually symmetrically loaded, i.e. floor slabs are on both sides of the beam, and therefore the beam is symmetrical in cross-section as shown in Figure 5.1a. The limiting design criterion is often minimum depth in order to maximize headroom and reduce the drop beam, or 'downstand' depth defined in this figure. For this reason, internal beams are often pretensioned to maximize their structural

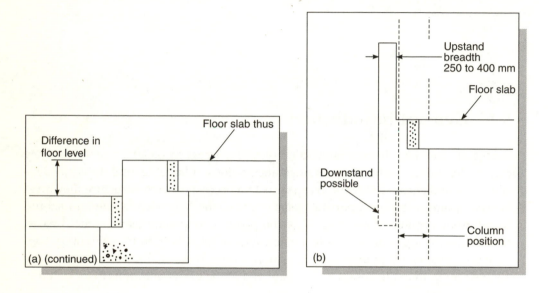

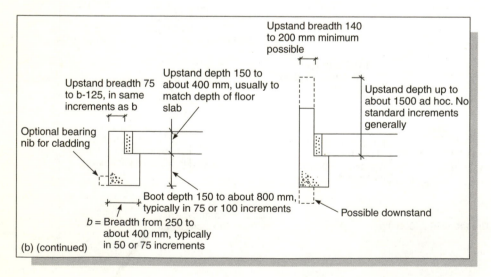

Figure 5.1: Types of beams. (b) Edge L and spandrel beam.

performance. To minimize the downstand, part of the beam may be recessed within the depth of the floor slab, giving rise to the so-called 'inverted-tee' beam. Internal beams may be designed compositely, with the floor slab acting as a compression flange (see Section 5.3).

External (edge) beams are, by nature, asymmetrically loaded. Take for example the deep spandrel beam shown in Figure 5.2, where torsion will result when floor slabs sit on the bearing nib because the line of action of the load is not coincident with the shear centre of the beam. Torsion must therefore be considered in design.

Cross-section may be rectangular, but to avoid having to place formwork at an external edge the cross-section tends to be L-shape, as shown in Figure 5.1b. Beams with tall upstands are known as 'spandrel' beams. They are frequently used around the perimeter of buildings such as in car parks where they form part of the impact barrier. Spandrels are often used to form a *dry envelope* around the perimeter of the building by making a temporary weather shield between successive storey heights. Edge beams are not pretensioned – their non-symmetrical shape is the main reason, but as beam depth is not a limiting factor there is no reason to do so. Edge beams may be designed compositely with the floor slab, but for the same reasons as above, there is often little need to do so.

The L-shape edge beams support non-symmetrical floor loads. The part of the beam supporting the floor is called the 'boot' and the main web is the 'upstand'. There are two types of edge beam, shown in Figure 5.3:

Figure 5.2: Deep spandrel edge beam to support double-tee floor slabs.

Type I, where a wide upstand is part of the structural section (Figure 5.3 right).

Type II, where a narrow upstand provides a permanent formwork to the floor slab, and is considered monolithic with the in situ concrete infill at the ends of the floor slab (Figure 5.3 left).

In type I beams, the minimum width of the upstand should be approximately $b_w = 150-175$ mm. The ledge width is the sum of the nominal slab-bearing length (75 mm), a fixing tolerance (10 mm) and the clear space for in situ infill (50 mm), giving a total dimension of 135 mm. Thus the minimum breadth of a type I beam is

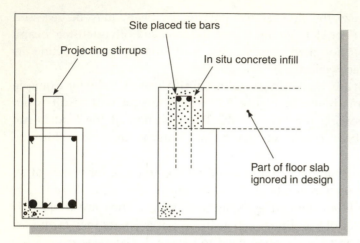
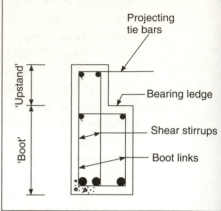

Figure 5.3: L-section edge beams – composite (left) and non-composite (right).

about $b = 300$ mm. The precast upstand width in Type II beams is 75–100 mm, and the minimum breadth is about 250 mm. Minimum depth is often determined by the size of the connector in the end of the beam. The minimum depth would be equal to the depth of the floor slab (h_s) plus the minimum boot depth of 150 mm.

The design of beams is based on ordinary reinforced or prestressed concrete principles for specified loads and support conditions. Support conditions may be simple or continuous. Semi-rigid supports are not generally adopted although some research data do exist. Unlike monolithic r.c. design where the cross-section and reinforcement is designed to satisfy project requirements, in precast concrete design it is the reverse. A predetermined set of standardized beam sections is selected by a manufacturer according to the requirements of most building structures. Flexural and shear reinforcements are computed for the optimum reinforcement quantities appropriate to each size of beam. Standardized designs are prepared in advance for beams that may vary only in depth, breadth and quantity of reinforcement. Simple computer programs or spread-sheets are used to do this. The following sections will illustrate these methods.

Although a designer is able to specify any grade of concrete, in practice a manufacturer would want to restrict this to two grades, one for r.c. work and one for prestressed work. For practical reasons of demoulding and detensioning, grade C40 is used for r.c. beams and C50 for prestressed beams. Similarly, one type of rebar is used, i.e. HT deformed bar of $f_y = 460$ N/mm^2. Although mild steel is perfectly suitable, its cost differential (compared with HT bar) and smooth surface (difficult to make stable cages) make it less attractive. Mild steel is used for projecting dowels which must be hand-bent into position on site. One type of pretensioning bar is used, i.e. 7-wire helical strand of between $f_{pu} = 1750$ and $1820\,N/mm^2$. (Pretensioning wire would not generally be used because of the large force demand in beams.)

Cover to reinforcement must satisfy fire and durability requirements. The usual approach is to fix the cover distance for the various faces of the beam, and to quote these properties as part of the beam specification. External surfaces usually have a cover of 30 or 40 mm, whilst internal (=protected) surfaces have 25 mm cover. The clear distances between bars in tension should satisfy code requirements, e.g. Table 3.28 in BS8110, Part 1. Minimum and maximum quantities of reinforcement should be checked.

5.2   Non-composite reinforced concrete beams

Non-composite construction utilizes the properties of the basic beam alone. For specified cross-section and flexural and shear reinforcement patterns, the following may be calculated:

1   Ultimate moment of resistance;

2   Ultimate shear resistance;

3   Torsional resistance;

4   Bearing ledge resistance; and

5   Flexural stiffness (=deflection limit)

The reinforcement quantities are curtailed according to the distribution of design moments and shear forces. Figure 5.4 shows a typical reinforcement pattern for

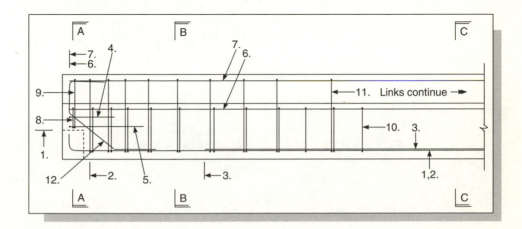

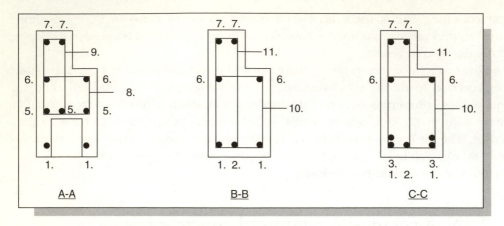

Figure 5.4: Typical reinforcement details in non-composite L beam.

an L beam. This beam will be used to demonstrate the design procedures in the following sections.

5.2.1   Ultimate moment of resistance

Type I beam. These beams may be designed using all types of floor units, i.e. hollow core, double-tee, plank floors. Reinforcement in the top of the boot is ignored. Referring to Figure 5.5, let $\rho = A_s/b\,h$ and assume that the depth to the NA $X > h_s$. Then:

$$T = 0.95 f_y \rho b h \qquad\qquad 5.1$$

$$C_1 = 0.45 f_{cu} b_w h_s \qquad\qquad 5.2$$

$$C_2 = 0.45 f_{cu} b (0.9X - h_s) \qquad\qquad 5.3$$

$$T - C_1 = C_2$$

hence X: check $X < 0.5d$.

$$Z_1 = d - h_s/2 \quad \text{and} \quad Z_2 = (d - 0.45X) - h_s/2 \qquad\qquad 5.4$$

$$M_R = C_1 Z_1 + C_2 Z_2 \qquad\qquad 5.5$$

$$\text{If } C_1 < T, \text{ then } X < h_s, \text{ and } M_R = T(d - 0.45X) \qquad\qquad 5.6$$

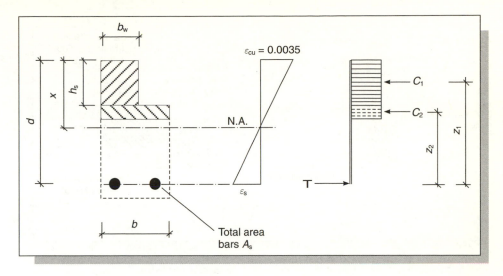

Figure 5.5: Design method for ultimate moment of resistance in L beam.

If $X > 0.5d$, the beam should be doubly reinforced or the value of A_s should be reduced to remain singly reinforced. Adding compression reinforcement to the top of the upstand of the beam is often not practical because of the limited space between the bars. It is better to increase the compressive resistance of the concrete by utilizing cast in situ infill concrete at the ends of the slabs. This gives rise to type II beams.

Type II beam. These beams may be designed using only those floor types that permit the placement of fully confined in situ infill, i.e. hollow core and plank floors. Double-tee units are not permitted. In order to be economic and obtain maximum M_R the strength of the infill concrete should be similar to that used in the beam, however a lower strength, say $f_{cu} = 30\,\text{N/mm}^2$ may be suitable. In Figure 5.6, the effective breadth b_{eff} of the compression zone is equal to the beam breadth minus the slab-bearing length, usually taken as 75 mm, but may be greater. The calculation is as above except that in Eq. 5.2 b_w is replaced with b_{eff} and f_{cu} is the strength of the in situ infill concrete (not the precast beam).

5.2.2 Ultimate shear resistance

The design for shear follows the normal procedures for r.c. sections – that is the ultimate shear capacity is the sum of the concrete resistance (=aggregate interlock + dowel action) plus the ultimate (=yield stress) capacity of the shear stirrups. In L beams the effective breadth of the web b_v used in shear calculations depends on whether the NA of the section occurs in the upstand or in the boot of

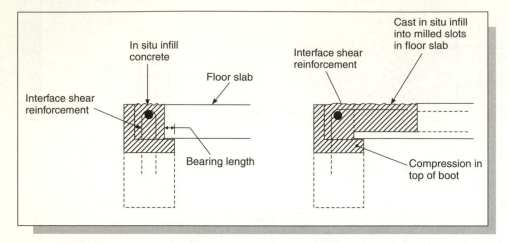

Figure 5.6: Interface shear bars or loops between L beam and floor slab.

the beam as shown in Figure 5.7, where the NA lies in the upstand (Figure 5.7a), then b_v = upstand width*. However, where the NA lies in the boot, the critical section may lie either in the upstand or in the boot, and both cases must be considered as shown in Figure 5.7a,c.

Elastic shear stress distribution function $\tau = VS/Ib$ is used to determine the shear stress at the two critical sections. In Figure 5.7 the critical sections are at: (i) the top level of the boot; and (ii) the NA. In this calculation S is the first moment of area above the critical section, I is the second moment of area of the whole beam (=using transformed section), b is the effective breadth at the critical section and V is the shear force. We are not interested in the actual value of τ, only where it is a maximum. Because V and I are constant, we require the maximum value of the term S/b.

At the level at the top of the boot,

$$S/b = \frac{b_v h_s^2}{2b_v} = 0.5h_s^2 \qquad\qquad 5.7$$

At the NA,

$$S/b = \frac{\dfrac{bX^2}{2} - (b - b_v)h_s\left(X - \dfrac{h_s}{2}\right)}{b} \qquad\qquad 5.8$$

But let $X = 0.5d$ in the limit, then Eq. 5.8 reads

---

*(Notation – the upstand width b_v is used in shear calculations whereas b_w is used for the same parameter in flexural calculations.)

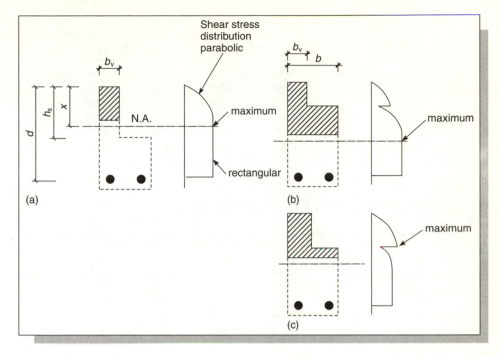

Figure 5.7: Principles of shear stresses in reinforced L beams.

At the NA,

$$S/b = \frac{d^2}{8} - 0.5\left(1 - \frac{b_v}{b}\right)h_s(d - h_s) \qquad\qquad 5.9$$

Given values of b_w, b, h_s, and d, the maximum value of S/b may be found. Equations 5.7 and 5.9 may be further simplified by considering the ratios h_s/d and b_v/b. Table 5.1 gives the values of S/bd^2 for the two cases for typical values of h_s/d and b_v/b. Critical values are given in bold. From these data it is possible to determine where the critical shear section should be taken. Generally it is found that the critical shear section lies at the top of the boot when $h_s > 0.3d$.

Designed shear reinforcement must be placed in the appropriate section. Shear stirrups will be placed in the upstand of the beam as shown in Figure 5.4. Links must also be provided in the boot of the beam to carry the bearing ledge forces, as explained in Section 5.2.3. In the latter, shear stirrups must be provided in the upstand *or* in the boot depending on where the effective breadth is taken. Designed shear stirrups must be additional to those required for torsion of bearing ledge reactions.

Table 5.1a: Values of S/bd^2 at the top of the boot

b_v/b	$h_s/d=0.1$	$h_s/d=0.2$	$h_s/d=0.3$	$h_s/d=0.4$	$h_s/d=0.5$	$h_s/d=0.6$
0.1	0.005	0.020	0.045	0.080	0.125	0.180
0.2	0.005	0.020	0.045	0.080	0.125	0.180
0.3	0.005	0.020	0.045	0.080	0.125	0.180
0.4	0.005	0.020	0.045	0.080	0.125	0.180
0.5	0.005	0.020	0.045	0.080	0.125	0.180
0.6	0.005	0.020	0.045	0.080	0.125	0.180

Table 5.1b: Values of S/bd^2 at the neutral axis

b_v/b	$h_s/d=0.1$	$h_s/d=0.2$	$h_s/d=0.3$	$h_s/d=0.4$	$h_s/d=0.5$	$h_s/d=0.6$
0.1	0.085	0.053	0.031	0.017	0.013	0.017
0.2	0.089	0.061	0.041	0.029	0.025	0.029
0.3	0.094	0.069	0.052	0.041	0.038	0.041
0.4	0.098	0.077	0.062	0.053	0.050	0.053
0.5	0.103	0.085	0.073	0.065	0.063	0.065
0.6	0.107	0.093	0.083	0.077	0.075	0.077

In Figure 5.8 and in BS8110, Table 3.9, the area of designed shear reinforcement is given by $A_{sv} = b_v s_v (v - v_c)/0.95 f_{yv}$. The area of shear reinforcement must be appropriate to the critical shear section; in this case it is assumed this is at the top of the boot. The design shear stress may be given as:

$$v = \frac{0.95 f_{yv} A_{sv}}{b_v S_v} + v_c > (0.4\,\text{N/mm}^2 + v_c) < 0.8\sqrt{f_{cu}} \qquad 5.10$$

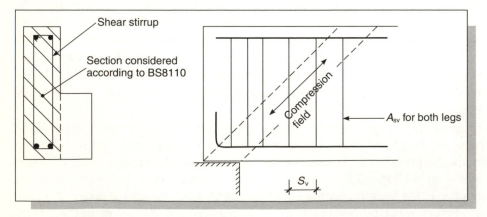

Figure 5.8: Shear design in L beams.

and the design shear capacity is given as:

$$V_R = v b_v d \qquad\qquad 5.11$$

Example 5.1

Calculate the M_R and V_R of a 550 mm deep $\times$ 300 mm wide L beam shown in Figure 5.4. Main steel at mid-span comprises 3 no. T25 bars, reducing to 3 no. T16 near to the supports. Shear stirrups are T10 bars at 100 mm spacing. The upstand is 200 mm deep $\times$ 165 mm wide. Cover to stirrups = 40 mm, use $f_{cu} = 40\,\text{N/mm}^2$ and $f_y = 460\,\text{N/mm}^2$.

Solution

Flexure

At mid-span point, 3 no. T25 bars are present

$d = 550 - 40 - 10 - 12.5 = 488\,\text{mm}$

$\rho = 3 \times 491/300 \times 550 = 0.0089$ (= 0.89% > minimum 0.13% for $b_w/b > 0.4$, and < maximum 4%)

$T = 0.95 \times 460 \times 0.0089 \times 300 \times 550 = 643.7 \times 10^3\,\text{N}$

$C_1 = 0.45 \times 40 \times 165 \times 200 = 594.0 \times 10^3\,\text{N}$ (in the upstand) See Figure 5.5.

$T > C_1$ then $X > h_s$

$C_2 = 0.45 \times 40 \times 300 \times (0.9X - 200) = 4860X - 1080 \times 10^3\,\text{N}$

Then $4860X - 1080 \times 10^3 = 643.7 \times 10^3 - 594.0 \times 10^3$
Therefore $X = 232\,\text{mm}$ and $C_2 = 49.7\,\text{kN}$

$X/d = 0.475 < 0.5$

Lever arm $z_1 = 488 - 100 = 388\,\text{mm}$, and $z_2 = 488 - (0.45 \times 232) - 100 = 284\,\text{mm}$

$$M_R = 594 \times 0.388 + 49.7 \times 0.284 = \underline{244.6\,\text{kNm}}$$

Shear

At the support, 3 no. T16 longitudinal bars ($A_s = 603\,\text{mm}^2$) and T10 stirrups at 100 mm spacing are present. From Table 5.1 and Eqs 5.7 and 5.9 it is found that the critical shear section lies at the top of the boot. Therefore, the effective breadth $b_v = 165\,\text{mm}$ and $d = 550 - 40 - 10 - 8 = 492\,\text{mm}$

$$100A_s/b_v d = 0.743$$

$$f_{cu}/25 = 40/25 = 1.6$$

Then concrete shear stress (BS8110, Table 3.9) $v_c = 0.79 \times 0.743^{1/3} \times 1.6^{1/3}/1.25 = 0.67\,\text{N/mm}^2$

$$\text{Design shear stress } v = \frac{0.95 \times 460 \times 157}{165 \times 100} + 0.67$$

$$= 4.83\,\text{N/mm}^2 < 0.8\sqrt{f_{cu}} \qquad (\textit{using Eq. 5.10})$$

$$V_R = 4.83 \times 165 \times 492 \times 10^{-3} = 392\,\text{kN} \qquad (\textit{using Eq. 5.11})$$

5.2.3 Boot design

The boot of the beam must be reinforced using links around the full perimeter of the boot. If the depth of the boot is less than 300 mm it should be designed as a short cantilever in bending. Otherwise the behaviour is nearer to the strut and tie action. The bending method gives a slightly greater area of tie back steel (about 5–10 per cent).

As with all projecting nibs (a nib is a short-bearing ledge) it is first necessary to preclude a shear failure at the root of the nib. The enhanced shear stress given in BS8110, Part 1, Clause 3.4.5.9 usually takes care of any vertical shear problems. If not, then the depth of the boot should be increased in preference to providing shear reinforcement. A 'shallow' boot is where the lever arm 'a' in Figure 5.9 is greater than $0.6d''$, where d'' is the effective depth to the steel in the top of the boot from the bottom of the beam. Otherwise the nib is classed as 'deep'. The compressive strut in the boot is inclined at θ to the vertical, where $\theta = \tan^{-1} a/x$, where $x = (d'' - c)$ is the centre-to-centre distance of the boot link, and c the edge distance to the centroid of the steel bar in the top of the boot.

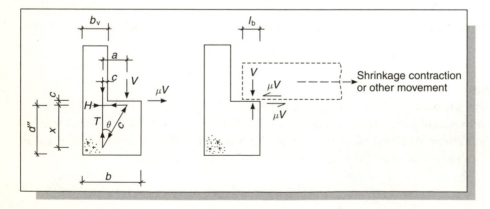

Figure 5.9: Boot reinforcement design in L beams.

If the floor slab is placed in direct contact with a shallow bearing nib, a horizontal force resulting from possible contractions or other movement (e.g. thermal effects) of the floor slab relative to the beam will develop at the interface. Referring to Figure 5.9, if the floor reaction is V per unit length of beam, the horizontal force is μV, where μ is the coefficient of friction between two concrete surfaces taken as 0.7. Thus, the horizontal tie force is:

$$H = V \tan \theta + \frac{x+c}{x} \mu V \qquad 5.12$$

The horizontal bars placed in the top of the boot must satisfy:

$$A_{sh} = \frac{H}{0.95f_y} \qquad 5.13$$

The bars are formed into links, but do not contribute to vertical shear strength of the beam unless the boot is sufficiently deep, where $h_s > 0.3d$ (see Table 5.1). The upstand width is fairly small, typically 150 mm, such that the bars in the top of the boot must extend a full anchorage length in the rear face of the beam. This means that the bars are stressed beyond a point which is more than four diameters from the corner of the bar, and the bend radius must be checked so that the bursting stresses caused by small bend radii are not a problem. The usual practice is to provide T8 or T10 links at a spacing not greater than 155 mm (BS8110, Part 1, Table 3.28).

The compressive strut force is:

$$C = \frac{V}{\cos \theta} \qquad 5.14$$

which must be resisted by a compressive strut in the uncracked part of the nib. The uncracked zone may extend to a point at $0.5d''$ from the bottom of the beam. The limiting compressive strength of the concrete is $0.4f_{cu}$. Thus the strut capacity is:

$$C = 0.2f_{cu}d'' \sin \theta \text{ per unit length of beam} \qquad 5.15$$

The tie force due to μV in Eq. 5.12 is resisted by compressure struts in the upstand of the beam. Assuming strut action takes place at 45° (because the upstand is deep in relation to its breadth) the tie force $T = 0.5\mu V/\tan 45°$. The total tie force is given by

$$T = V + 0.5\mu V/\tan 45° = V(1 + 0.5\mu) \qquad 5.16$$

If the floor slab is fully tied to the beam using reinforced in situ strips capable of generating the frictional force μV, then this force may be ignored in the above design, in which case $T = V$. The area of one leg of vertical stirrups is:

$$A_{sv} = \frac{T}{0.95f_y} \qquad\qquad 5.17$$

This steel must be in addition to any design shear requirement.

In a deep boot, the floor slab reactions would be carried directly into the web of the beam by diagonal strut action assuming $\theta = 45°$. If the level of the bearing surface is above the NA the only steel required would be the horizontal steel A_{sh}.

In fact the design of all the above reinforcement should be carried in two stages, before and after in situ concrete has been added to the ends of the slab. This is because the in situ concrete increases the bearing length to the full ledge width, and hence reduces the lever arm a. Before the in situ concrete is added the lever arm is $a = c + (b - b_v) - l_b/2$, and the slab reaction is due to the self-weight of the slab plus the in situ concrete infill. Afterwards $a = c + (b - b_v)/2$, and the slab reaction is due to superimposed dead and live loading.

Example 5.2
Calculate the minimum bearing ledge capacity of the boot of the L beam in Example 5.1 and Figure 5.4. Assume that the line of action of the force is at the mid-point of the bearing ledge, i.e. at $135/2 = 67$ mm from the edge. Use T8 boot links at 150 mm spacing. $f_y = 460$ N/mm^2. Minimum upstand links are T10 at 300 mm spacing. Cover to bearing ledge $= 25$ mm, otherwise 40 mm.

Solution

$$d'' = 350 - 25 - 4 = 321 \text{ mm}$$
$$x = 321 - 40 = 281 \text{ mm}$$
$$a = 135/2 + 25 + 5 = 97 \text{ mm}$$
$$\theta = \tan^{-1} 97/281 = 19°$$

Horizontal tie steel capacity $H = 0.95 \times 460 \times 50 \times 1000 \times 10^{-3}/150 = 145.6$ kN/m run. Changing the subject,

$$V = \frac{145.6}{\tan 19° + 0.7 \times 306/281} = 132.0 \text{ kN/m run} \qquad (\textit{using Eq. 5.12})$$

and changing the subject,

$$V = 0.2 \times 40 \times 321 \times 1000 \times \sin 19° \cos 19° \times 10^{-3} = 790 \text{ kN/m}$$

run (clearly not critical)   (*using Eqs 5.14 and 5.15*)

Vertical force in the stirrups in the upstand $= 0.95 \times 460 \times 78.5 \times 1000 \times 10^{-3}/300 = 114.3 \, \text{kN/m}$ run. Changing the subject,

$$V = \frac{114.3}{1 + (0.5 \times 0.7)} = 84.7 \, \text{kN/m run} \qquad \textit{(using Eq. 5.16)}$$

Thus, the minimum bearing ledge capacity is 84.7 kN/m run.

5.3 Composite reinforced beams

Precast reinforced beams may act compositely with certain types of floor slabs, such as hollow core and plank units, by the introduction of appropriate interface shear mechanisms and cast in situ concrete infill. Typical details are shown in Figure 5.10. It is usual, but not obligatory, for only internal beams to be designed compositely as there is rarely a need to enhance the strength of edge beams in this way.

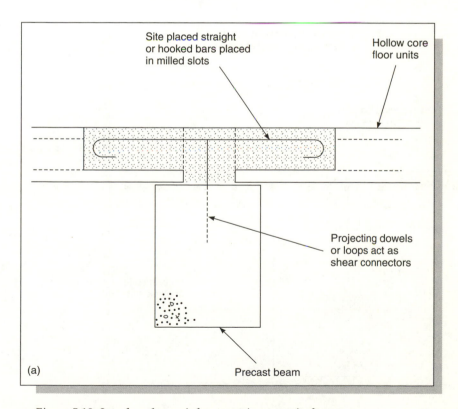

Figure 5.10: Interface shear reinforcement in composite beams.

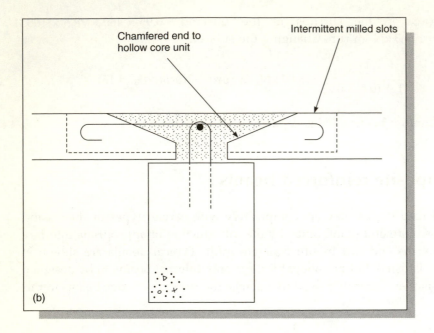

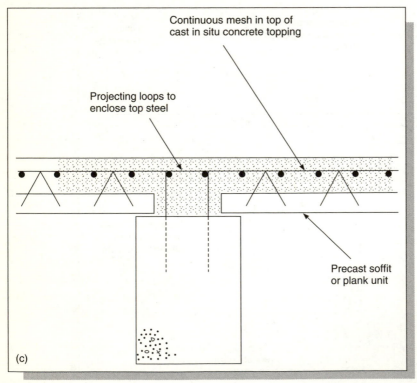

Figure 5.10 (continued): Interface shear reinforcement in composite beams.

The characteristic cube strength of the in situ infill concrete is either 25 or 30 N/mm². The main benefit in using a composite beam is to increase the flexural strength and stiffness (=reduce deflections). These must be carefully considered against the additional cost and design responsibility, particularly if the floor slab is designed by other parties. For these reasons composite r.c. beams are not often used – unlike composite prestressed beams which have greater benefits (see Section 5.5).

It is necessary to reinforce the cast in situ concrete such that it will develop full design strength $0.45f_{cu}$. In the case of hollow core units (Figure 5.10a), the milled slots over the top of all the cores are removed for a distance of approximately 500 mm to receive site placed tie steel. The tie steel may be loose bars or bent bars projecting from the beam. It is usually HT grade 460. The tie steel also serves several other functions, including diaphragm action (Section 7.4) and stability tie steel (Section 10.4), but in this context a steel area of 0.2 per cent of the transverse area has been found by experimentation to be adequate,[1] e.g. T12 bars at 300 mm spacing for floor depths upto 200 mm. Flooring manufacturers should be consulted over the practicalities of opening cores as the end of the floor unit may become unstable where many cores are opened.

The ends of the hollow core unit may be chamfered with sloping ends, as shown in Figure 5.10b to benefit the placement of in situ infill. The chamfer is usually around 250 mm long – manufacturers will provide exact details. In this case not all of the hollow cores will be opened as slots. Experimental results show that opened cores at 300 mm centres are sufficient.

Full interaction between the in situ concrete in the cores and the precast hollow core unit is assumed. The full depth of the slab is used in design. The effective breadth of the flange is taken as equal to the actual length of the filled cores, and this is equal to one bond length for the transverse steel that is placed in the core, i.e. 40 diameters for HT bar in grade C30 infill. If the length of the slot becomes excessive, say greater than 600 mm, the ends of the transverse bar are hooked. The bars should be placed at mid-depth of the slot – however due to lazy site practice they tend to rest on the bottom of the core at less than 50 mm from the bottom (the effect of this is not known).

In the case of solid plank flooring (Figure 5.10c), transverse tie steel will automatically be present as part of the topping/floor design, but as before a minimum area ratio of 0.2 per cent is recommended. The positions of the transverse bars is more accurate than in the case of hollow core units, with the top cover being about 50 mm. Full interaction between the precast plank and the in situ topping is assumed. The full depth of the slab and an effective flange breadth of 1/10 of the simply supported span of the beam, L_z, are used in design.

Composite beams are not designed for vertical shear. However, interface shear calculations for shear forces due to imposed loads should be made according to Section 4.4.4. Note that the contact breadth may be small, typically 150 mm in 300 mm wide beams, resulting in large interface shear stress. Interface reinforcement

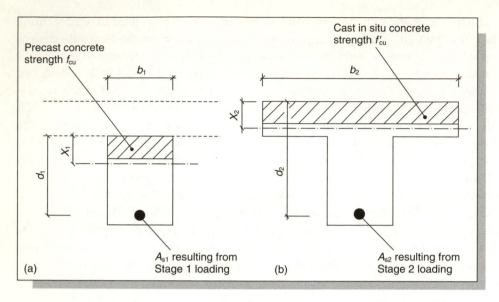

Figure 5.11: Ultimate state design in composite beams.

is always used (irrespective of the value of v_h) in the form of loops or dowels, as shown in Figure 5.10.

5.3.1 Design in flexure of composite reinforced concrete beams

Flexural design is carried out in two stages and the resultant effects are added at the ultimate limit state. At Stage 1 the self-weight of the beam, precast floor units and the cast in situ infill/topping is carried by the precast beam alone. The area of reinforcement required for this is called A_{s1}. At Stage 2 after the in situ has developed full strength, the imposed dead (=services, partitions, ceiling, etc.) and live loads are carried by the composite beam. The area of steel here is A_{s2}. Then the total steel area $A_s = A_{s1} + A_{s2}$ (Figure 5.11).

There is a fundamental difficulty in this approach as that part of the concrete in the top of the precast beam may be called on to resist compressive stress at both Stage 1 and Stage 2. To solve this in a rigorous manner, the strain history of the beam should be followed and the resulting stresses found. In this case an idealized 'rectangular stress block' approach is not appropriate. However, it is found that in most cases the depth of the compressive stress block at Stage 2, given in Figure 5.11b as X_2, is less than the depth of the slab.

Referring to Figure 5.11a, let the ultimate design moment due to self weight be M_1. Then if the strength of the precast concrete is f_{cu}:

$$(BS8110, \ clause \ 3.4.4.4) \ K_1 = \frac{M_1}{f_{cu}b_1d_1^2} \qquad 5.18$$

from which the lever arm z_1, NA depth X_1 are found according to BS8110 rectangular stress block approach. Check $X_1 < 0.5d_1$. Check $z_1 < 0.95d_1$. The area of steel is:

$$A_{s1} = \frac{M_1}{z_1 0.95f_y} \qquad 5.19$$

Let the ultimate design moment due to imposed dead and live loads be M_2. Then if the strength of the infill/topping concrete is f'_{cu}:

$$K_2 = \frac{M_2}{f'_{cu}b_2d_2^2} \qquad 5.20$$

where $d_2 = d_1 + h_s$. The lever arm z_2, NA depth X_2 are found according to BS8110 rectangular stress block approach. Check $X_2 < h_s < 0.5d_2$. The area of steel is:

$$A_{s2} = \frac{M_2}{z_2 0.95f_y} \qquad 5.21$$

then $A_s = A_{s1} + A_{s2}$

The remainder of the precast beam will be reinforced according to Section 5.2.

5.3.2 *Deflections in composite r.c. beams*

In everyday design, deflections are controlled by the procedure of limiting span-to-effective depth ratio, a procedure which is entirely satisfactory for most loading conditions in singular sections. The method involves equating beam curvature and strain distributions with a limiting deflection of span/250. This cannot be adopted in a composite beam because there are deflections due to Stage 1 loads that (obviously) respond to a completely different flexural stiffness E_cI, than those due to Stage 2 loads. Also, the Young's modulus of concrete E_c changes with time due to creep such that Stage 1 deflections take place as the floor slabs are being positioned at between 7 and 28 days after casting typically, whilst Stage 2 deflections take place over many years, 30 years being the design 'period'. It is often forgotten that Stage 1 loads act over the long term and must be considered in the Stage 2 calculation.

The effects of the relative shrinkage of the precast beam to the in situ concrete must also be considered in deflection calculations, particularly in the case of plank floors where the volume of wet concrete is large. A value of relative shrinkage

strain of $\epsilon_{sh} = 100\,\mu\epsilon$ is adopted. This is
not necessary when using hollow core
units, as the shrinkage of these units is
small and undisturbed by the in situ infill
placed into individual opened cores.

The design method adopts the *area-
moment* and *partially cracked section*
method. Reference should be made to
standard texts for details (e.g. Ref. 4.11).
For uniformly distributed loading acting
on a beam of effective length L, referring
to Figure 5.12 the mid-span deflection is
given as:

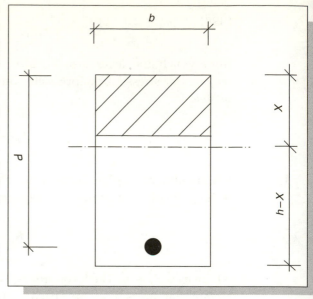

$$\delta = \frac{L^2 M_{net}}{9.6\ E_c I_c} < \frac{L}{250} \qquad 5.22$$

where

$$M_{net} = M_s - \frac{b(h-X)^3 f_{ct}}{3(d-X)} \qquad 5.23$$

Figure 5.12: Definition of terms for deflection check in composite beams.

and M_s is the service moment at mid-span.
The value of E_c is appropriate to the load-
ing stage, i.e. at Stage 1 it is E_c, and at stage 2 it is $E_c/(1+\phi)$ where ϕ is the creep
coefficient. The value of I_c is for the full flexurally cracked section. The permissible
tension f_{ct} is 1.0 and 0.55 N/mm^2 for short- and long-term effects, respectively.
The NA depth is calculated using the transformed area method (see Ref. 4.11) as:

$$X = d\left(\sqrt{\alpha^2 \rho^2 + 2\alpha\rho} - \alpha\rho\right) \qquad 5.24$$

where $\alpha = E_{steel}/E_{concrete}$, with $E_{steel} = 200\,kN/mm^2$ and $\rho = A_s/bd$, with each
parameter being used appropriately at each loading stage. The second moment
of area is:

$$I_c = bX^3/3 + \alpha A_s(d-X)^2 \qquad 5.25$$

Example 5.3

The composite beam shown in Figure 5.13 is simply supported over a span of
6.0 m. It carries 200 mm deep hollow cored slabs which span 6.0 m and have a self
weight (including in situ infill) of 3.0 kN/m^2. The imposed floor loading is 5.0
kN/m^2 live and 1.0 kN/m^2 dead. Calculate the area of reinforcement required
to satisfy the ultimate bending moment, and the mid-span deflection at service.
Use $f_{cu} = 40\,N/mm^2$ and $E_c = 28\,kN/mm^2$ for the precast beam, and $f'_{cu} = 25\,N/mm^2$ and $E_c = 25\,kN/mm^2$ for the in situ infill, $f_y = 460\,N/mm^2$. Assume a
creep coefficient $\phi = 1.8$.

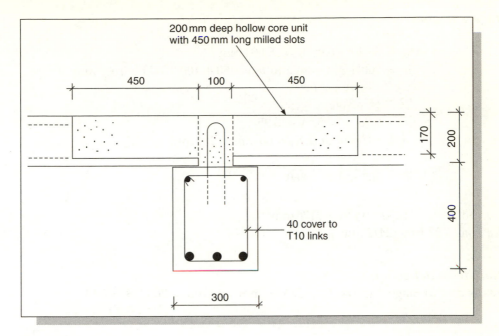

Figure 5.13: Detail to Example 5.3.

Solution

Loading	Service kN/m	γ_f	Ultimate kN/m
Self-weight of floor slab $= 3\,\text{kN/m}^2 \times 6\,\text{m}$	18.00		
Self-weight of beam $= 0.4 \times 0.3 \times 24$	2.88		
Total Stage 1	**20.88**	×1.4	**29.23**
Live $5\,\text{kN/m}^2 \times 6\,\text{m}$	30.00	×1.6	48.00
Dead $1\,\text{kN/m}^2 \times 6\,\text{m}$	6.00	×1.4	8.40
Total Stage 2	**36.00**		**56.40**

Stage 1 flexure

$$M_{u1} = 29.23 \times 6.0^2/8 = 131.5\,\text{kNm}$$
$$d_1 = 400 - 40 - 10 - 16\ (\text{say}) = 334\,\text{mm}, \quad \text{and}\ b_1 = 300\,\text{mm}$$
$$K_1 = \frac{131.5 \times 10^6}{40 \times 300 \times 334^2} = 0.098$$
$$z_1/d_1 = 0.875 < 0.95$$
$$z_1 = 292\,\text{mm and } X_1 = 93\,\text{mm} < 0.5d_1$$
$$\text{Then } A_{s1} = \frac{131.5 \times 10^6}{292 \times 0.95 \times 460} = 1030\,\text{mm}^2$$

Stage 2 flexure

$$M_{u2} = 56.40 \times 6.0^2/8 = 253.8 \, \text{kNm}$$
$$d_2 = 200 + 334 \, \text{mm, and } b_2 = 450 + 100 + 450 = 1000 \, \text{mm}$$
$$K_2 = \frac{253.8 \times 10^6}{25 \times 1000 \times 534^2} = 0.036$$
$$z_2/d_2 = 0.96 > 0.95, \quad \text{use } 0.95$$
$$z_2 = 507 \, \text{mm and } X_2 = 60 \, \text{mm} < 200 \, \text{mm} < 0.5d_2$$
$$\text{Then } A_{s2} = \frac{253.8 \times 10^6}{507 \times 0.95 \times 460} = 1146 \, \text{mm}^2$$

Then total $A_s = 1030 + 1146 = 2176 \, \text{mm}^2$
Use 3 no. T32 bars (2412 mm^2).

Short-term deflection
This is due to Stage 1 loads. $E_c = 28 \, \text{kN/mm}^2$, then $\alpha = 200/28 = 7.14$.

$$\rho = 2412/(300 \times 334) = 0.024$$

$$\alpha\rho = 0.172$$

Then, Eq. 5.24, $X_1 = 0.44 \, d_1 = 146 \, \text{mm}$

$$I_{c1} = 300 \times 146^3/3 + 7.14 \times 2412 \times (334 - 146)^2$$
$$= 920 \times 10^6 \, \text{mm}^4 \quad (using \, Eq. \, 5.25)$$
$$M_{net,1} = 20.88 \times 6.0^2/8 - \frac{300 \times (400 - 146)^3 \times 1.0 \times 10^{-6}}{3 \times (334 - 146)} = 94.0 - 8.7$$
$$= 85.3 \, \text{kNm} \quad (using \, Eq. \, 5.23)$$
$$\delta_1 = \frac{6000^2 \times 85.3 \times 10^6}{9.6 \times 28\,000 \times 920 \times 10^6} = 12.4 \, \text{mm} \quad (using \, Eq. \, 5.22)$$

Long-term deflection
This is due to Stage 1 loads, which continue to act, and Stage 2 loads. $E_c = 28/(1 + 1.8) = 10 \, \text{kN/mm}^2$, then $\alpha = 200/10 = 20.0$.
Effective breadth of flange $= 1000 \times 25/28 = 893 \, \text{mm}$

$$\rho = 2412/(893 \times 534) = 0.005$$

$$\alpha\rho = 0.1$$

Then, Eq. 5.24, $X_2 = 0.36d_2 = 191 \, \text{mm} < 200 \, \text{mm}$ depth of hollow core unit

$$I_{c2} = 893 \times 191^3/3 + 20.0 \times 2412 \times (534 - 191)^2$$

$$= 7750 \times 10^6 \, \text{mm}^4 \quad (using \; Eq. \; 5.25)$$

$$M_{\text{net},2} = 56.88 \times 6.0^2/8 - \frac{893 \times (600 - 191)^3 \times 0.55 \times 10^{-6}}{3 \times (534 - 191)} = 256.0 - 32.6$$

$$= 223.4 \, \text{kNm} \quad (using \; Eq. \; 5.23)$$

$$\delta_2 = \frac{6000^2 \times 223.4 \times 10^6}{9.6 \times 10\,000 \times 7750 \times 10^6} = 10.8 \, \text{mm} \quad (using \; Eq. \; 5.22)$$

Total deflection $= 12.4 + 10.8 = 23.2 \, \text{mm}$

Span$/250 = 6000/250 = 24 \, \text{mm}$, deflection OK.

5.4 Non-composite prestressed beams

The design of prestressed beams is less versatile than reinforced beams because tendons positions are restricted to a predetermined pattern by an array of holes in the jacking heads, which is usually a permanent fixture at a precasting works. Figure 5.14 shows a full array of possible tendon positions in an inverted-tee beam, and an example of a typical tendon layout. Note the symmetry. The tendons are placed at all the corners; a 40 to 50 mm centroidal cover distance being used in most cases.

The minimum breadth of the beam is a function of the type of floor slab to be used. The breadth is equal to twice the recess width, plus the upstand width. The same reasoning as for the L beam is used if floor ties are intended to be placed within the recess and concealed in the depth of hollow cored floor slabs. The minimum recess width for this condition is 100–125 mm. If the ties are to be located elsewhere the recess width may be 90–100 mm. The minimum upstand breadth is 250–300 mm. Beam depths depend on three factors:

1   The flexural and shear capacities;

2   The size of the end connector; and

3   The depth of the boot required to carry the floor loads.

5.4.1 Flexural design

The design procedure is identical to the design of prestressed floor units given in Section 4.3.1 with the additional consideration of satisfying transfer, as well as

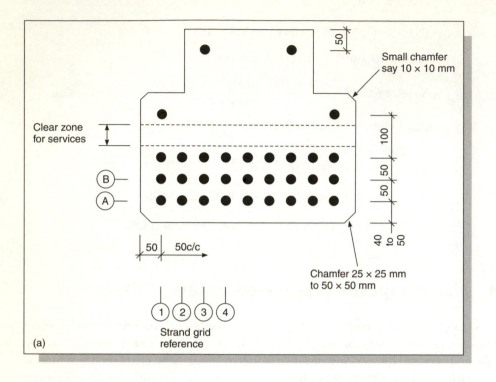

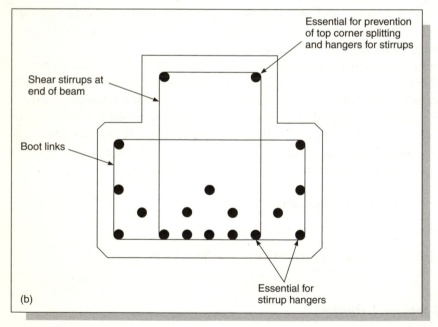

Figure 5.14: (a) Prestressing strand array in inverted-tee beams; (b) Typical strand and link arrangement in inverted-tee beam.

working, stress conditions. This is because strands can be either deflected or debonded. There is much more freedom in selecting the strand pattern than in floor units as the design of the beam can be optimized (=economy of strands) by choosing a pattern that will simultaneously satisfy transfer at the ends of the beam and working loads at the point of maximum imposed bending moment.

It is desirable that the permissible stress at transfer $0.5f_{ci}$ after the initial losses (due to elastic shortening) is made equal (or as close as possible) to the working stress $0.33f_{cu}$ after all losses. In most beam designs the initial and final losses are about 8 and 25 per cent, respectively, meaning that the ratio f_{ci}/f_{cu} should be at least 0.55. In fact the transfer strength for grade C60 concrete is at least $40\,\text{N/mm}^2$, and therefore *transfer* stresses will (nearly) always govern for parallel, unbonded tendons. To overcome this problem it is desirable to debond a small number of strands, say four in a typical situation. It is also wise to actually restrict the top fibre stress to something less than $0.45\sqrt{f_{ci}}$, say half this value (i.e. $0.225\sqrt{f_{ci}}$), whilst accepting that there will be a small loss in moment capacity. (This factor of safety is based on the experience gained from transporting and handling highly stressed prestressed beams, and the need to avoid flexural cracking for the sake of durability.)

There is a further refinement to the design of prestressed beams borne out of practical experience as follows. The actual initial pretensioning force which takes place at the moment of detensioning will not include the instantaneous elastic shortening loss, defined here as ξ, and calculated according to Section 4.3.2. Therefore, the actual pretensioning force at transfer is $P_i(1-\xi)$, where $P_i = \eta A_{ps}f_{pu}$. Therefore, the stress conditions at transfer are:

$$f_{bci} = P_i(1-\xi)\left(\frac{1}{A} + \frac{e}{Z_b}\right) < +0.5f_{ci} \quad \text{at the bottom, and} \qquad 5.26$$

$$f_{tci} = P_i(1-\xi)\left(\frac{1}{A} - \frac{e}{Z_t}\right) > -0.225\sqrt{f_{ci}} \quad \text{at the top} \qquad 5.27$$

Manipulation of the simultaneous Eqs. 5.26 and 5.27 will give optimum values for the initial prestressing force P_i and the eccentricity e, as follows:

$$P_i = \frac{A}{2(1-\xi)}\left[(0.5f_{ci} - 0.225\sqrt{f_{ci}}) + \left((0.5f_{ci} + 0.225\sqrt{f_{ci}})\frac{1-\alpha}{1+\alpha}\right)\right] \qquad 5.28$$

where $\alpha = Z_t/Z_b$ and ξ = initial prestress loss due to elastic shortening at the level of the centroid of the strands (expressed as a decimal fraction). The number of tendons required is:

$$N = P_i/\eta A_{ps}f_{pu} \qquad 5.29$$

where η = degree of prestress usually taken as 0.7. The eccentricity e is given by:

$$e = \frac{0.5 f_{ci} + 0.225\sqrt{f_{ci}}}{P_i \beta} \qquad 5.30$$

where $\beta = \left(\dfrac{1}{Z_b} - \dfrac{1}{Z_t}\right)$, and P_i is based on the value obtained from Eq. 5.28.

Given P_i and e, the actual bottom and top fibre stresses may be calculated from Eq. 5.26 and 5.27 from which the actual (as opposed to the guessed at value in Eq. 5.26) value of ξ is found – if this differs by more than 2–3 per cent, iteration should take place. The elastic shortening loss is given by:

$$\xi = \frac{f_{cci} E_s}{E_{ci} \eta f_{pu}} \qquad 5.31$$

where

$$f_{cci} = P_i \left(\frac{1}{A} + \frac{e^2}{I}\right) \qquad 5.32$$

Total losses may be determined as given in Section 4.3.2. The final bottom and top stresses f_{bc} and f_{tc} after losses may be calculated in the usual manner. The service moment of resistance M_{sr} is given by the lesser of:

$$M_{sr} = (f_{bc} + 0.45\sqrt{f_{cu}})Z_b \qquad 5.33$$

or

$$M_{sr} = (f_{tc} + 0.33 f_{cu})Z_t \qquad 5.34$$

Tables 5.2–5.5 give the minimum value of M_{sr} for a range of typical sizes for prestressed inverted-tee beams.

Example 5.4

Calculate the initial prestressing requirements for an inverted-tee beam shown in Figure 5.15. The beam is Class 2 according to permissible tension. Use $f_{cu} = 60 \text{N/mm}^2$, $E_c = 32 \text{kN/mm}^2$, $f_{ci} = 40 \text{N/mm}^2$, $E_{ci} = 28 \text{kN/mm}^2$, strand diameter = 12.5 mm, $A_{ps} = 94.2 \text{mm}^2$, $f_{pu} = 1750 \text{kN/mm}^2$, $\eta = 0.7$, $E_s = 195 \text{kN/mm}^2$, and 2.5 per cent strand relaxation. Assume $\xi = 8$ per cent initial losses.

Solution
Geometric data:

$$A = 307.5 \times 10^3 \text{mm}^2, \quad y_b = 269.8 \text{mm}, \quad Z_b = 30.215 \times 10^6 \text{mm}^3,$$

$$Z_t = 24.685 \times 10^6 \text{mm}^3$$

$$\alpha = 0.817, \quad \beta = 7.36 \times 10^{-8} \text{mm}^{-1}$$

Prestressing force per strand $= 0.7 \times 94.2 \times 1750 \times 10^{-3} = 115.5 \text{kN}$

Table 5.2: Service and ultimate moments of resistance for 500 mm wide grade C50 prestressed inverted-tee beams

Soffit breadth (mm)	Boot depth (mm)	Upstand depth (mm)	Upstand breadth (mm)	Number of strands	Initial prestressing force (kN)	Eccentricity (mm)	Service moment of resistance (kNm)	Ultimate moment of resistance (kNm)
500	200	150	300	11	1269.3	58.0	124.4	242.2
500	300	150	300	15	1730.9	71.6	210.0	412.8
500	400	150	300	18	2077.1	91.1	328.6	625.5
500	500	150	300	22	2538.7	106.4	468.9	881.9
500	600	150	300	26	3000.3	122.1	636.5	1181.8
500	200	200	300	12	1384.7	69.1	161.8	292.9
500	300	200	300	16	1846.3	81.7	254.9	488.2
500	400	200	300	20	2307.9	95.5	373.4	713.5
500	500	200	300	24	2769.5	110.1	518.4	981.8
500	600	200	300	27	3115.7	129.9	699.6	1292.0
500	200	250	300	14	1615.5	75.0	199.6	356.1
500	300	250	300	17	1961.7	92.4	306.4	574.1
500	400	250	300	21	2423.3	105.4	432.0	811.8
500	500	250	300	25	2884.9	119.3	583.4	1092.1
500	600	250	300	29	3346.5	133.9	761.3	1415.6

All sections have $f_{cu} = 50\,N/mm^2$; $f_{ci} = 30\,N/mm^2$; $f_{pu} = 1750\,N/mm^2$; $A_{ps} = 94.2\,mm^2$ per strand.

Table 5.3: Service and ultimate moments of resistance for 600 mm wide grade C50 prestressed inverted-tee beams

Soffit breadth (mm)	Boot depth (mm)	Upstand depth (mm)	Upstand breadth (mm)	Number of strands	Initial prestressing force (kN)	Eccentricity (mm)	Service moment of resistance (kNm)	Ultimate moment of resistance (kNm)
600	200	150	350	13	1500.1	58.0	146.9	279.3
600	300	150	350	18	2077.1	70.6	247.4	477.1
600	400	150	350	22	2538.7	88.4	385.5	725.4
600	500	150	350	27	3115.7	102.9	550.2	1026.6
600	600	150	350	31	3577.2	121.7	756.2	1378.7
600	200	200	350	15	1730.9	65.3	186.9	345.6
600	300	200	350	19	2192.5	81.3	300.7	565.5
600	400	200	350	24	2769.5	94.2	439.8	828.7
600	500	200	350	28	3231.1	111.9	618.2	1142.2
600	600	200	350	33	3808.0	126.2	821.5	1509.1
600	200	250	350	16	1846.3	77.5	238.2	414.7
600	300	250	350	21	2423.3	88.4	355.8	667.1
600	400	250	350	26	3000.3	100.7	502.0	945.0
600	500	250	350	30	3461.9	117.7	687.3	1272.7
600	600	250	350	34	3923.4	135.4	906.0	1651.6

All sections have $f_{cu} = 50\,N/mm^2$; $f_{ci} = 30\,N/mm^2$; $f_{pu} = 1750\,N/mm^2$; $A_{ps} = 94.2\,mm^2$ per strand.

Table 5.4: Service and ultimate moments of resistance for 500 mm wide grade C60 prestressed inverted-tee beams

Soffit breadth (mm)	Boot depth (mm)	Upstand depth (mm)	Upstand breadth (mm)	Number of strands	Initial prestressing force (kN)	Eccentricity (mm)	Service moment of resistance (kNm)	Ultimate moment of resistance (kNm)
500	200	150	300	13	1500.1	55.8	146.5	275.3
500	300	150	300	17	1961.7	71.9	251.2	469.2
500	400	150	300	21	2423.3	88.9	384.7	712.8
500	500	150	300	25	2884.9	106.5	543.0	1006.6
500	600	150	300	29	3346.5	124.5	730.5	1351.0
500	200	200	300	14	1615.5	67.4	191.4	334.9
500	300	200	300	19	2192.5	78.2	299.7	558.0
500	400	200	300	23	2654.1	94.4	444.3	815.8
500	500	200	300	27	3115.7	111.3	621.8	1123.4
500	600	200	300	31	3577.2	128.8	820.7	1481.2
500	200	250	300	16	1846.3	74.6	238.1	407.2
500	300	250	300	20	2307.9	89.4	361.3	658.2
500	400	250	300	24	2769.5	104.9	514.9	930.2
500	500	250	300	29	3346.5	117.0	692.0	1253.8
500	600	250	300	33	3808.0	133.9	909.4	1625.5

All sections have $f_{cu} = 60\,\text{N/mm}^2$; $f_{ci} = 40\,\text{N/mm}^2$; $f_{pu} = 1750\,\text{N/mm}^2$; $A_{ps} = 94.2\,\text{mm}^2$ per strand.

Table 5.5: Service and ultimate moments of resistance for 600 mm wide grade C60 prestressed inverted-tee beams

Soffit breadth (mm)	Boot depth (mm)	Upstand depth (mm)	Upstand breadth (mm)	Number of strands	Initial prestressing force (kN)	Eccentricity (mm)	Service moment of resistance (kNm)	Ultimate moment of resistance (kNm)
600	200	150	350	15	1730.9	57.2	174.4	318.9
600	300	150	350	21	2423.3	68.9	292.9	545.5
600	400	150	350	26	3000.3	85.1	454.2	831.0
600	500	150	350	31	3577.2	102.0	653.4	1177.2
600	600	150	350	35	4038.8	122.6	872.8	1581.6
600	200	200	350	17	1961.7	65.6	223.5	394.6
600	300	200	350	22	2538.7	79.9	356.7	648.3
600	400	200	350	28	3231.1	91.8	520.6	951.6
600	500	200	350	33	3808.0	108.0	729.1	1313.6
600	600	200	350	38	4385.0	124.6	977.3	1736.1
600	200	250	350	19	2192.5	74.2	280.0	478.7
600	300	250	350	24	2769.5	88.0	424.1	767.1
600	400	250	350	29	3346.5	102.7	604.1	1085.7
600	500	250	350	35	4038.8	114.8	813.5	1465.9
600	600	250	350	40	4615.8	130.9	1069.1	1904.4

All sections have $E_c = 60\,\text{N/mm}^2$; $E_{ci} = 40\,\text{N/mm}^2$; $f_{pu} = 1750\,\text{N/mm}^2$; $A_{ps} = 94.2\,\text{mm}^2$ per strand.

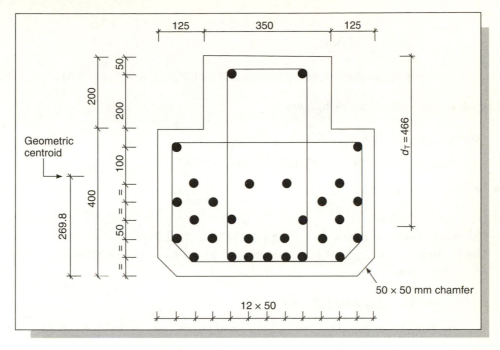

Figure 5.15: Detail to Example 5.4.

Limiting bottom and top stresses at transfer are $f_{bci} < +20.0\,\text{N/mm}^2$ and $f_{tci} > - < 0.225\sqrt{40} = -1.42\,\text{N/mm}^2$

$$P_i = \frac{307.5 \times 10^3}{2 \times (1 - 0.08)} \times \left[18.58 + \left(21.42 \times \frac{0.183}{1.817}\right)\right] = 3465.6\,\text{kN} \qquad (\text{using Eq. 5.28})$$

$$e = \frac{21.42}{0.92 \times 3465.6 \times 10^3 \times 7.36 \times 10^{-8}}$$
$$= 91\,\text{mm} \text{ (to nearest mm)} \qquad (\text{using Eq. 5.30})$$

$$N = 3465.6 \times 10^3/115.5 = 30.005 \qquad (\text{using Eq. 5.29})$$

but rounded down to 29 to prevent the possibility of overstress.

The actual prestressing force $P_i = 29 \times 115.5 = 3349.5\,\text{kN}$, and the first estimate after initial losses $= 0.92\,P_i = 3081.5\,\text{kN}$. The maximum fibre stresses at transfer are:

$$f_{bci} = \frac{3081.5}{307.5} + \frac{3081.5 \times 91}{30\,215} = 10.02 + 9.28 = +19.30\,\text{N/mm}^2$$
$$< +20.0\,\text{N/mm}^2 \qquad (\text{using Eq. 5.26})$$

$$f_{tci} = \frac{3081.5}{307.5} - \frac{3081.5 \times 91}{24\,685} = 10.02 - 11.36 = -1.34\,\text{N/mm}^2$$
$$> -1.42\,\text{N/mm}^2 \qquad (\text{using Eq. 5.27})$$

$f_{cci} = 14.36 \, \text{N/mm}^2$ before the initial loss   (*using Eq. 5.32*)

$$\xi = \frac{14.36 \times 195\,000}{28\,000 \times 0.7 \times 1750}$$

$= 0.0815$ (original assumption of 0.08 is OK)   (*using Eq. 5.31*)

The other final losses are as follows:

- creep $= 1.8 \times 8.15\% = 14.67\%$

- shrinkage $= 300 \times 10^{-6} \times 195/0.7 \times 1750 = 4.77\%$

- relaxation $= 1.2 \times 2.5 = 3.0\%$

Final losses $= 30.6\%$.

The final prestress force $= (1 - 0.306) \times 3349.5 = 2324.5 \, \text{kN}$.

The final working fibre stresses are $f_{bc} = +14.56 \, \text{N/mm}^2$ and $f_{tc} = -1.01 \, \text{N/mm}^2$.

$$M_{sr} = (14.56 + 3.5) \times 30.215 \times 10^6 \times 10^{-6}$$

$= 545.7 \, \text{kNm based on the bottom fibre stress, or}$ (*using Eq. 5.33*)

$$M_{sr} = (1.01 + 19.8) \times 24.685 \times 10^6 \times 10^{-6}$$

$= 513.7 \, \text{kNm based on the top fibre stress}$ (*using Eq. 5.34*)

Therefore the critical value is $M_{sr} = 513.7 \, \text{kNm}$

The strands are arranged such that the distance Y to their centroid from the bottom of the beam is as near as possible to $Y = y_b - e$. If the number of strands is N, the sum of the first moment of area of the strands from the bottom of the beam must be NY. Figure 5.15 shows a suggested strand arrangement (there are of course several possible arrangements) obtained from the following table:

No. strands in each row I	Distance Y_i from bottom (mm)	$\Sigma N_i \, Y_i$ (mm)
2	550	1100
2	350	700
4	250	1000
4	200	800
4	150	600
6	100	600
7	50	350
$\Sigma = 29$		$\Sigma = 5150$

Then $Y = 5150/29 = 177.6 \, \text{mm}$

$e = y_b - Y = 269.8 - 177.6 = 92.2 \, \text{mm} \approx 91 \, \text{mm required}$.

5.4.2 Ultimate flexural design

Ultimate limit state flexural design of prestressed beams follows the procedures adopted for prestressed floor units given in Section 4.3.3, but with one major addition. Because the position of many of the strands in beams is nearer to the NA than is the case in floor units, the strains in each of the strands should be calculated to determine whether they attain their yield value. In many cases it is possible that no strand will attain its yield value and therefore the assumption that $f_{pb} = 0.95\, f_{pu}$ is not valid at all – in other cases some of the strands will attain $0.95 f_{pu}$. Because of having to satisfy the fairly stringent serviceability stress limits, most prestressed rectangular and inverted-tee beams are over reinforced at ultimate – this is why composite prestressed beams are efficient in enabling most of the strands to attain yield value.

The basic analytical procedure is given in standard texts (e.g. Ref. 4.11), but a simplified method is presented here. It is assumed that the stress distribution (after losses) and a strand pattern to satisfy the service condition are known.

Firstly, the strands in the top of the beam and at the top of the boot are ignored. This leaves the number of strands as N_T. The first estimate for the depth to the NA X and the ultimate stress f_{pb} is found by using the Table 4.4 in BS8110. Then, knowing X the strain in the strands at the next row down (below the top of the boot) is found. In Figure 5.16, if the distance from the top of the beam to the strands is g, the ultimate strain in the strand at g is given by:

$$\varepsilon_g = \frac{f_{pe}}{E_s} + \frac{g - X}{X} 0.0035 + \frac{f_g}{E_c} \qquad\qquad 5.35$$

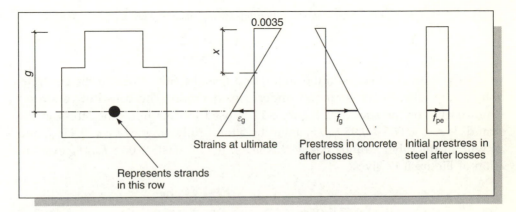

Figure 5.16: Strains in tendons at ultimate limit state.

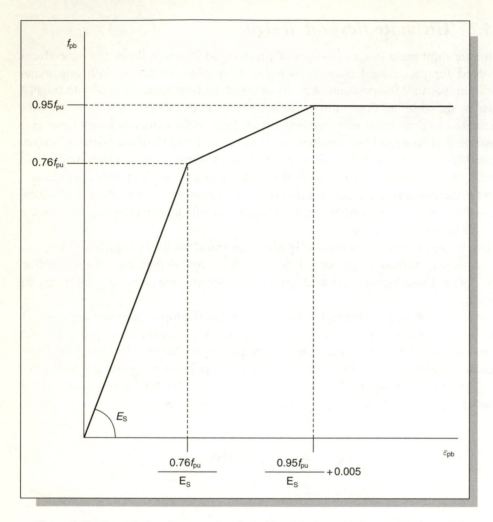

Figure 5.17: Constitutive stress–strain relationship for pretensioning strand (from BS8110).

where f_{pe} is the final stress in the strand and f_g is the final stress in the concrete, both at the level of the strand(s) under consideration. The constitutive stress–strain curve for the strand in Figure 5.17 is used to determine the stress in the strand. If $\varepsilon_g > 0.005 + 0.95\, f_{pu}/E_s$ then the strand fully stresses $f_{pb} = 0.95 f_{pu}$, and so will all the other strands below this level. If $\varepsilon_g < 0.005 + 0.95\, f_{pu}/E_s$ examination of Figure 5.17 gives:

$$f_{pb} = 0.76 f_{pu} + \frac{0.19 f_{pu} E_s \varepsilon_g - 0.144 f_{pu}^2}{0.19 f_{pu} + 0.005 E_s} \qquad 5.36$$

Knowing the values of f_{pb} force equilibrium gives:

$$F_c = 0.45\, f_{cu}b\, 0.9X \text{ in the concrete} \qquad\qquad 5.37$$

$$F_s = \Sigma f_{pb}A_{ps} \text{ for each row of strands below the top of the boot} \qquad 5.38$$

Hence X is found and resubstituted into Eq. 5.35 for iteration. The analysis is clearly long and tedious and involves having to calculate f_{pb} at each level. An approximate method is to consider that the average strain exists at the centroid of the strands in the tension zone, i.e. $g = d_T$ where d_T is the effective depth to these strands. Equations 5.35–5.38 are still valid. The ultimate moment of resistance M_{ur} is given as:

$$M_{ur} = \Sigma f_{pb}A_{ps}(d_T - 0.45X) \qquad\qquad 5.39$$

Note that in inverted-tee beams the compression zone comprises the upstand, of breadth b_w and depth h_s, plus the part of the boot of breadth b. Equations 5.37 and 5.39 are modified to:

$$F_c = 0.45f_{cu}b_w h_s + 0.45f_{cu}b\,(0.9X - h_s) \qquad\qquad 5.37a$$

$$M_{ur} = \Sigma f_{pb}A_{ps}(d_T - d_n) \qquad\qquad 5.39a$$

where d_n is the centroid of the inverted-T shape compressive zone. Tables 5.2–5.5 give the minimum value of M_{ur} for a range of typical sizes for prestressed inverted-tee beams.

Example 5.5

Calculate M_{ur} for the beam in Example 5.4 and Figure 5.15. The chosen strand pattern is shown in Figure 5.15b.

Solution

Effective depth to strands in tension zone

$$d_T = \frac{(7 \times 550) + (6 \times 500) + (4 \times 450) + (4 \times 400) + (4 \times 350)}{25} = 466\,\text{mm}$$

Initially compression breadth $b = 350\,\text{mm}$

First estimate of X and f_{pb} from BS8110, Part 1, Table 4.4

$$\frac{f_{pu}\,A_{ps}}{f_{cu}\,bd} = \frac{1750 \times 25 \times 94.2}{60 \times 350 \times 466} = 0.421$$

Now $f_{pe}/f_{pu} = 0.7 \times 0.694 = 0.486$

Then $X/d \approx 0.68$ and $f_{pb} \approx 0.69 \times 0.95 \times 1750 = 1147\,\text{N/mm}^2$

$$F_s = 25 \times 94.2 \times 1147 = 2701 \times 10^3\,\text{N} \qquad \text{(using Eq. 5.38)}$$

$$F_c = (0.45 \times 60 \times 350 \times 200) + [0.45 \times 60 \times 600 \times (0.9X - 200)]$$

$$= 14\,580X - 1350 \times 10^3\,\text{N} \qquad \text{(using Eq. 5.37a)}$$

Hence $X = 278\,\text{mm}$

The strain at the centroidal level $g = 466$ mm

$$\varepsilon_g = \frac{0.486 \times 1750}{195\,000} + \frac{466 - 278}{278} \times 0.0035 + \frac{11.08}{32\,000} = 0.007 \qquad \text{(using Eq. 5.35)}$$

$$f_{\text{pb},g} = 1330 + \frac{64.84 \times 10^6 \times 0.007 - 441\,000}{1307.5} = 1340\,\text{N/mm}^2 \qquad \text{(using Eq. 5.36)}$$

(A second iteration might be made at this point leading to $X = 309$ mm, but the effect on the final answer will be small.)

The depth of the compressive stress block $= 0.9 \times 278 = 250$ mm. The depth to its centroid $d_n = 137.5$ mm

$$M_{\text{ur}} \approx 25 \times 94.2 \times 1340 \times (466 - 137.5) \times 10^{-6} = 1036.6\,\text{kNm} \qquad \text{(using Eq. 5.39a)}$$

(Note: the ratio $M_{\text{ur}}/M_{\text{sr}} = 2.02$. This shows that the ultimate moment will not be critical as the maximum possible ratio between ultimate and service loads is 1.6 (assuming live load only). Therefore the approximations made in the calculation of M_{ur} are unlikely to be important.)

5.4.3   Shear in prestressed beams

Ultimate shear design of prestressed beams follows the procedures adopted for prestressed floor units given in Section 4.3.5 in which the uncracked V_{co} and the flexurally cracked V_{cr} shear resistances are calculated. In many cases due to the high degree of prestress in beams and the large spans over which they operate, designed shear reinforcement is often quite small and the nominal area suffices. However, shear reinforcement, in the form of inclined bars or links, is required at the ends of the beam in the vicinity of the connections, and although the design does not require it, shear reinforcement is added at progressively greater spacing. Figure 5.18 shows an example of where shear stirrups placed in the end cage are progressively spaced further apart until the nominal area of stirrups is reached.

In the calculation for V_{co} inverted-tee beams, the shear should be considered at both the centroidal axis and at the intersection of the upstand and boot. No general guidance can be given as to which of the two positions is critical because the shear resistance is a function of geometry and prestress. The term $0.67b_v h$ (in Eq. 4.7) should be replaced by Ib_v/Ay'. In rectangular beams only the centroidal axis is considered and the term is $0.67b_v h$ satisfactory.

Shear reinforcement in the flexurally cracked region is also rarely necessary because the ultimate shear resistance of prestressed members is a function of the ultimate flexural requirements. The shear span (M_u/V_u) for most beams in precast structures is $L/4$. Thus, the shear force at the position of flexural decompression rarely exceeds $0.5V_{\text{cr}}$ – the value given in BS8110 deemed not to necessitate shear

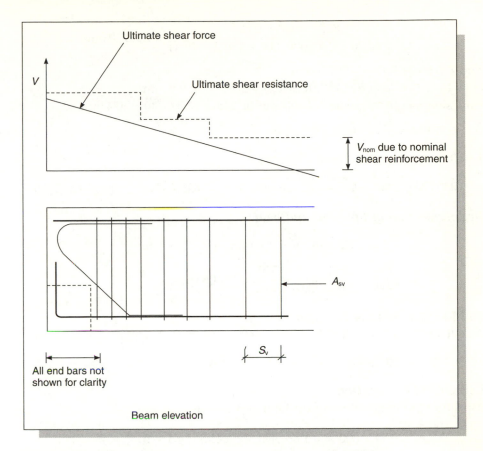

Figure 5.18: Shear reinforcement according to shear stress distribution.

reinforcement. As with prestressed floor units (Section 4.3.5) a minimum value of V_{cr} is calculated on the assumption that at the critical section $M_u = M_{ur}$ and $V_u = V_{cr}$ (see Eq. 4.10).

Example 5.6

Calculate V_{co} for the inverted-tee beam in Example 5.4 using the following for the effective shear area: (a) the true cross-section Ib_v/Ay'; (b) the BS8110 term $0.67b_vh$.

The bearing length may be taken as 150 mm.

Solution

(a) Using the true cross-section at the centroidal axis

$$b_v = 600 \, \text{mm}$$

$$I = y_b Z_b = 8152 \times 10^6 \, \text{mm}^4$$

$$y_t = 600 - 269.8 = 330.2 \, \text{mm}$$

$$\frac{Ib_v}{Ay'} = \frac{8152 \times 10^6 \times 600}{(350 \times 200 \times 230.2) + (600 \times 130.2 \times 65.1)} = 230\,720\,\text{mm}^2$$

f_{cp}(from Example 5.4) = $7.55\,\text{N/mm}^2$
Distance to critical plane = 150 (bearing) + 269.8 = 419.8 mm
Transmission length = greater of 600 mm or $240 \times 12.5/\sqrt{40} = 474$ mm

$$f_{cpx} = 6.87\,\text{N/mm}^2 \qquad (using\ Eq.\ 4.8)$$

$$f_t = 0.24\sqrt{f_{cu}} = 1.86\,\text{N/mm}^2$$

$$V_{co} = 230\,720\sqrt{1.86^2 + 0.8 \times 6.87 \times 1.86} \times 10^{-3} = 853.4\,\text{kN} \qquad (using\ Eq.\ 4.7a)$$

At the intersection of upstand and boot

$$b_v = 350\,\text{mm}, \quad y = 200\,\text{mm}, \quad y' = 230.2\,\text{mm}$$

$$\frac{Ib_v}{Ay'} = \frac{8152 \times 10^6 \times 350}{350 \times 200 \times 230.2} = 177\,063\,\text{mm}^2$$

f_{cp}(from Example 5.4) = $4.18\,\text{N/mm}^2$
Distance to critical plane = 150 (bearing) + 400 = 550 mm. Then $f_{cpx} = 4.15\,\text{N/mm}^2$

$$V_{co} = 177\,063\sqrt{1.86^2 + 0.8 \times 4.15 \times 1.86} \times 10^{-3} = 549.6\,\text{kN}$$

(b) Using BS8110 equation
By inspection the intersection of upstand and boot is critical

$$V_{co} = 0.67 \times 350 \times 600 \times \sqrt{1.86^2 + 0.8 \times 4.15 \times 1.86} \times 10^{-3} = 436.7\,\text{kN}$$

The least of these three values would be used in design.

5.5   Composite prestressed beam design

Composite action in prestressed beams is achieved in exactly the same manner as described in Section 5.3 for reinforced concrete beams. However, there is potential for greater enhancement in the flexural capacity of prestressed composite beams than with composite reinforced beams and composite slabs because of the greatly increased section modulus at the top of the beam. This is because the NA in the composite section is near to the top of the beam, and as seen in Example 5.5 the need to render the beam flexurally 'balanced' or 'under reinforced' is important. Shear design is not carried out using the composite properties, although (if wished) the term for h in Eq. 4.7 may be taken as the total depth of construction. Deflections are significantly reduced in composite beams due to the greater I values in the composite section. Composite action is not considered where double-tee floor slabs are used.

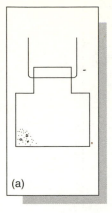

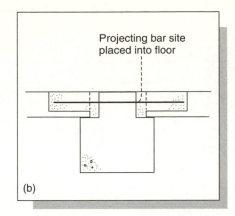

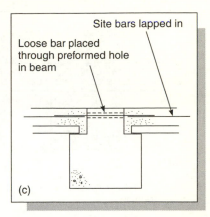

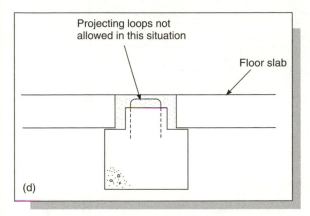

Figure 5.19: Procedures for placing interface shear reinforcement and tie steel in upstands.

The precast-in situ interface must be reinforced, even if only nominal steel at an area of 0.15 per cent times the interface area is used. If the interface is horizontal, projecting loops or dowels are provided as shown in Figure 5.10. If the interface is vertical, as is the case with inverted-tee beams shown in Figure 5.19, projecting bars cast into the beam are positioned (on site by hand) into the opened cores of hollow core units, Figure 5.19a–b, or into the topping over plank floors, Figure 5.19c. It is not satisfactory to provide loops in the top of inverted-tee beams as shown in Figure 5.19d – it is not possible to generate the necessary compressive force in the small quantity of in situ concrete in this region.

5.5.1 Flexural design

As with composite slab design, Section 4.4, service and ultimate stresses are checked at two stages of loading (or three if a structural topping is added to the floor) and superimposed elastically. If a structural topping is used the loading conditions are as follows:

Stage	Loading	Section properties based on
1	Self-weight of beam and dry floor slab	Precast beam
2	As Stage 1 plus self-weight of topping only	As Stage 1 plus in situ near beam
3	Superimposed	As Stage 2 plus topping

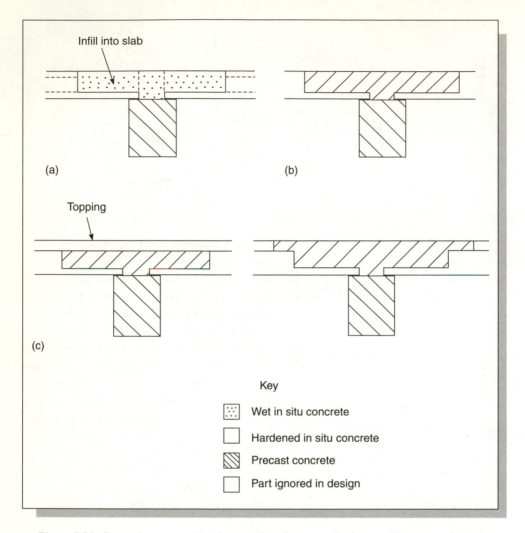

Figure 5.20: Stages in cross-sectional properties of a composite beam with structural topping.

The essential features of the design procedure are given in Figure 5.20. Stage I stresses exist in the precast beam only (Figure 5.20a). These are the result of prestress and relaxation (if the beam is prestressed), self weight of the beam, precast floor units and wet concrete. Stage 2 stresses exist in the precast beam and in situ infill (Figure 5.20b), and are due to the self weight of the topping (Figure 5.20c). Stage 3 stresses in the composite beam, Figure 5.20d are in addition to the above, and are the result of superimposed, services and partition loading, differential shrinkage and the total creep relaxation after hardening of the in situ concrete. Equations 4.13 to 4.23 are appropriate, except that if a structural topping is included Eq. 4.19 is modified to:

$$f_b = f_{bc} - \frac{M_1}{Z_{b1}} - \frac{M_2}{Z_{b2}} - \frac{M_3}{Z_{b3}} > -0.45\sqrt{f_{cu}} \qquad 5.40$$

where M_3 and Z_{b3} is the moment and section modulus for Stage 3.

Ultimate moment is calculated as per Section 4.4.2.2, except that if a structural topping is included the moments of resistance due to Stages 2 and 3 are added together (the difference compared with separate calculation for the two stages is negligible). Equation 4.27 is now deleted, and the subscripts 2 become 2,3 (=Stages 2 and 3 combined). Then,

$$M_{u2,3} = f_{pb}A_{ps2,3}(d + h_t + h_s - d_{n2,3}) \qquad 5.41$$

where $M_{2,3}$ is the moment for Stages 2 and 3 added, and h_t and h_s are the minimum depths of topping and slab, respectively.

The effective breadth of the flange (based on the full depth of the in situ concrete) is as given in Section 4.4.2.2. Figure 5.21 shows a composite beam test being carried out on an inverted-tee beam and hollow cored floor slab. It was found that the effective breadth of the floor was at least 1.0 m wide in spite of the fact that the floor section includes many hollow cores. The resulting span-to-depth ratio for this type of construction is around 18.

The breadth of the beam can be made as wide as possible, in some cases 1200 mm as shown in Figure 5.22, where the precast beam is acting essentially as a permanent

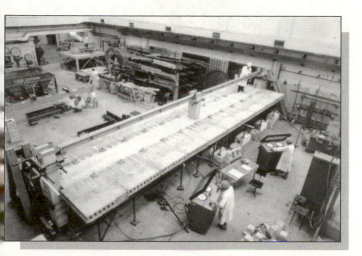

Figure 5.21: Full scale testing of composite beam with hollow core floor slab (courtesy Tarmac Precast, UK).

Figure 5.22: Precast concrete element with rebar girder awaiting composite action with the floor slab.

shuttering to hollow cored slabs. The rebar girder is acting as a bending truss to carry the self weight of the beam and floor slabs until in situ concrete is placed over the top of the beam to form a composite section, as shown in Figure 5.23. To further increase the moment capacity of the beam and reduce the mid-span sagging moment, continuity reinforcement is placed in the top of the rebar girder, and passes through the column, thus producing a hogging moment of resistance equal to the sagging moment. In cases such as these the columns are usually dis-

Figure 5.23: Reinforced cast in situ concrete forms a composite beam with hogging resistance and continuity.

continuous at the floor level, which leaves space for the continuity reinforcement.

Example 5.7

Determine the serviceability prestressing requirements f_{bc} and f_{tc}, and hence P and e, for the beam used in Example 5.4 and in Figure 5.24 for the following:

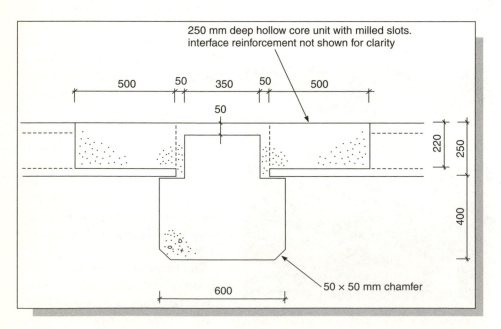

Figure 5.24: Detail to Example 5.7.

(a) the basic precast beam alone; (b) composite action with 250 mm deep hollow core units.

The floor units span 6.0 m and have self weight of 3.5 kN/m^2. (The depth of the bottom flange of the hollow core unit is 30 mm.) The beam is simply supported over a span of 9.0 m. The imposed dead and live loads on the floor are 1.0 and 5.0 kN/m^2, respectively. Use the same concrete beam and pretensioning data as in Example 5.4, $f'_{cu} = 25 \, \text{N/mm}^2$ and $E'_c = 25 \, \text{kN/mm}^2$ for the in situ infill. Assume final prestressing losses of 25 per cent.

Solution

(a) Non composite beam

Self weight of beam $= 7.38 \, \text{kN/m}$
Total load from floor $= (3.5 + 1.0 + 5.0) \times 6.0 = 57.0 \, \text{kN/m}$
Service moment $= (57.0 + 7.38) \times 9.0^2 / 8 = 652 \, \text{kNm}$
Changing the subject,

$$f_{bc} = \frac{652}{30.215} - 3.5 < +18.1 \, \text{N/mm}^2 \text{ after losses} \qquad (\textit{using Eq. 4.1})$$

Changing the subject,

$$f_{tc} = 19.8 - \frac{652}{24.685} > -6.61 \, \text{N/mm}^2 \text{ after losses} < -0.45\sqrt{f_{cu}} \qquad (\textit{using Eq. 4.2})$$

Clearly these are impossible to achieve within the defined parameters.

(b) Composite with hollow core unit

Actual breadth of flange $= 550 + 550 = 1100 \, \text{mm}$
Effective breadth $= 1100 \times 25/32 = 860 \, \text{mm}$
Depth of in situ flange $= 250 - 30 = 220 \, \text{mm}$
Composite section properties:

Section	$A \times 10^3$	y_b	$A \, y_b \times 10^6$	$A \, y_i^2 \times 10^6$	$I_{oo} \times 10^6$
Beam	307.5	269.8	82.96	3275	8152
In situ flange	189.2	540.0	102.17	5276	763
Totals	496.7	–	185.13	8551	8915

$$y_{b2} = 373 \, \text{mm}, \quad y_{t2} = 600 - 373 = 227 \, \text{mm}$$

$I_2 = 17\,470 \times 10^6 \, \text{mm}^4$, $Z_{b2} = 46.83 \times 10^6 \, \text{mm}^3$, $Z_{t2} = 76.96 \times 10^6 \, \text{mm}^3$ at top of beam (note the large increase compared with Z_{t1})

Stage 1 load $= 7.38 + (3.5 \times 6.0) = 28.38 \, \text{kN/m}$

$$M_1 = 287.3 \, \text{kNm}$$

Stage 2 load $= (1.0 + 5.0) \times 6.0 = 36.0 \, \text{kN/m}$

$$M_2 = 364.5 \, \text{kNm}$$

$$f_{bc} = \frac{287.3}{30.215} + \frac{364.5}{46.83} - 3.5 = +13.8 \, \text{N/mm}^2 \qquad (\text{using Eq. 4.19})$$

$$f_{tc} = 19.8 - \frac{287.3}{24.685} - \frac{364.5}{76.96} = +3.4 \, \text{N/mm}^2 \qquad (\text{using Eq. 4.20})$$

But replacing the limiting stresses there with f_{bc} and f_{tc} and

$$\xi = 0.25 \qquad (\text{using Eq. 5.28})$$

$$P_i = \frac{307.5 \times 10^3}{2(1 - 0.25)} \left[(13.8 - 3.4) + \left((13.8 + 3.4) \frac{1 - 0.817}{1 + 0.817} \right) \right] = 2487 \, \text{kN}$$

$$N = 2487/115.5 = 21 \, \text{strands} \qquad (\text{using Eq. 5.29})$$

$$e = \frac{13.8 + 3.4}{2487 \times 10^3 \times 7.36 \times 10^{-8}} = 94 \, \text{mm} \qquad (\text{using Eq. 5.30})$$

5.6 Propping

The basic concept for propping composite sections before the in situ concrete is added is as given in Section 4.4.3 for floors. The potential for structural enhancement is much greater for prestressed beams than in floors because the stresses due to the self weight of the floor slab, in situ infill and topping are reduced by the use of 2 or 3 props. The props are added after the beam is positioned, and therefore the stresses due to the self weight of the beam are calculated for the unpropped span.

It is worthwhile propping beams of over 6 m to 10 m span with 2 or 3 props, respectively, particularly if a structural topping is being used and the ratio of the geometric properties of the composite section to the basic section exceeds about 1.50. Propping increases beam erection time by about 15–25 per cent, and so the delicate balance between material saving and time lost on site has to be considered. Only the precast manufacturer can give exact information on this.

Let w_0 = self weight of beam, w_1 = self weight of precast floor units plus wet concrete infill and (if present) structural topping, and w_2 = imposed loads (all beam loads, e.g. kN/m). The service moments are (see Section 4.4.3):

$$M_0 = +0.125w_0L^2$$

$$M_1 = -0.03125w_1L^2$$

$$M_2 = +0.156\,25w_1L^2 + 0.125w_2L^2$$

The resulting prestress requirements are:

$$f_{bc} = \frac{(M_{s0} + M_{s1})}{Z_{b1}} + \frac{M_{s2}}{Z_{b2}} - 0.45\sqrt{f_{cu}} \qquad 5.42$$

$$f_{tc} = 0.33\,f_{cu} - \frac{(M_{s0} + M_{s1})}{Z_{t1}} + \frac{M_{s2}}{Z_{t2}} \qquad 5.43$$

5.7   Horizontal interface shear

The integrity of composite construction relies on the long-term continuity between the precast and in situ concrete. Although floor loading in buildings are predominantly static, there may be instances where it is variable and cyclic in nature, resulting in a fluctuation in shear stress levels. For this reason primary composite beams are designed with interface shear links or other similar mechanical fasteners. Shear friction is not recommended for beams.

As with floor units the ultimate interface shear stress v_h is checked for the uncracked section (BS8110, Part 1, Clause 5.4.7.2) against values in Table 5.5 of BS8110. Nominal interface shear links should always be provided. The minimum area of links is 0.15 per cent of the contact area. If v_h is greater than the values given in the design table, typically $1.2\,\text{N/mm}^2$, shear steel is provided according to BS8110, Part 1, Eq. 62 to carry the *entire* shear force.

Example 5.8
Determine the ultimate interface shear stress in the composite inverted-tee beam and floor slab in Example 5.7. Assume that there are 5 hollow cores opened in each 1200 mm hollow core unit. Use $f_{yv} = 460\,\text{N/mm}^2$ for the interface shear reinforcement.

Solution
Ultimate imposed load on the beam after in situ infill has hardened $= (1.0 \times 6.0 \times 1.4) + (5.0 \times 6.0 \times 1.6) = 56.4\,\text{kN/m}$.
Ultimate moment due to imposed load $= 56.4 \times 9.0^2/8 = 571\,\text{kNm}$. There are 21 strands in the beam at an effective depth of 472 mm from the top of the floor. Effective

depth to 17 strands in tension zone (assuming 2 no. strands at 100 mm and 2 no. at 300 mm from top of floor) $d_T = (21 \times 474 - 2 \times 100 - 2 \times 300)/17 = 538$ mm.

$$\frac{f_{pu}A_{ps}}{f_{cu}bd} = \frac{1750 \times 17 \times 94.2}{25 \times 1100 \times 538} = 0.19, \text{ then } X = 0.38d_T$$

$$X = 204.4 \text{ mm} < 220 \text{ mm available}$$

$$d_n = 0.45 \times 204.4 = 92 \text{ mm}$$

Total force in in situ infill $F_v = M_{u2}/(d - d_n) = (571 \times 10^6)/(538 - 92) = 1280$ kN. In this case the transverse interface contact width is equal to twice the depth of the upstand of the beam in contact with the in situ infill $= 2 \times 200 = 400$ mm

$$v_{ave} = \frac{1280 \times 10^3}{400 \times 4500} = 0.71 \text{ N/mm}^2 \qquad (\textit{using Eq. 4.33})$$

The maximum interface stress (for UDL) $v_h = 2 \times 0.71 = 1.42 \text{ N/mm}^2$

The limiting value in BS8110, Table 5.5 for as-cast surface and C25 in situ concrete $= 1.2 \text{ N/mm}^2$

$$A_f = \frac{1000 \times 400 \times 1.42}{0.95 \times 460} = 1300 \text{ mm}^2/\text{m run per 2 faces}$$

$$= 650 \text{ mm}^2/\text{m run per face} \qquad (\textit{using Eq. 4.34})$$

The area of interface shear steel per hollow core unit $= 650 \times 1.2 = 780 \text{ mm}^2$ per unit

Use 1 no. T16 bar into each opened core (1005).

Reference

1 Lam, D., Elliott, K. S. and Nethercot, D. A., Experiments on Composite Steel Beams with Precast Concrete Hollow Core Floor Slabs, *Proc. Inst. of Civil Engineers Structures and Building*, Vol. 140, May 2000, pp. 127–138.

6   Columns and shear walls

6.1   Precast concrete columns

Precast columns are the main vertical load carrying members in skeletal frames. They may also be used as the horizontal load carrying members in sway frames, but their capacity to achieve this in buildings of more than three storeys is very limited. Typically, a precast column is manufactured in grey concrete and is square or rectangular in cross-sections. However, there are many instances where columns of other shapes are used as part of the external architecture, shown for example in Figures 6.1 and 6.2. The 12 m high circular columns in Figure 6.1 were manufactured horizontally in circular moulds, sometimes they are cast vertically in storey height pieces. Generally columns are manufactured in the largest length possible to erect on site. In Europe, this is typically 12 to 18 m, but in the United States of America lengths of 25–30 m are possible. If a column is joined on site the connection is called a 'splice'. Figure 6.3 shows 25 m long columns in eight-storey parking garages in Las Vegas. This is acceptable providing the crane is being fully utilized at this lifting capacity, but often it is not. It therefore makes sense to reduce weight commensurate with the weights of other components, beams and slabs each weighing 4–5 tons.

The maximum length-to-depth ratio for lifting and pitching purposes is about 50:1. Great care must be taken in lifting the column from the mould. Many would say 40:1 is the limit. Such slender columns may be pretensioned axially to about $3 \, \text{N}/\text{mm}^2$, as shown in Figure 6.4, using small diameter helical strand, to prevent permanent damage through flexural cracking. In

Figure 6.1: Circular painted columns at The Shires Retail Centre, Leicester, UK.

Figure 6.2: White concrete columns at Sunbury Business Centre, UK (courtesy British Cement Association, UK).

many cases these cracks are temporary and will close beyond detection by eye once in service, however one must anticipate inclement erection conditions and the possibility of impact loading if perfect handling and fixing conditions cannot be guaranteed. (Another reason why columns are pretensioned is to take advantage of being able to specify grade C60 concrete in design.)

The minimum cross-section is often governed by the size of the beam–column connector, typically 250–300 mm. Maximum dimensions have no theoretical limit, 600 × 1200 mm being the practical maximum breadth and depth.

Most columns are manufactured horizontally, such as those shown in Figure 6.5. However, some single

Figure 6.3: 25 m long single piece columns at MGM car park, Las Vegas (courtesy Andrew T Curd & Partners, USA).

Figure 6.4: Prestressed columns.

Figure 6.5: Circular columns at Colorado Convention Centre, Denver, USA (courtesy PCI Journal, *USA).*

storey columns upto 3 m length with complex profiles are cast vertically. The famous spiral columns shown in Figure 1.16 are good examples, although these were more decorative pillars than structural columns. The moulds are mainly accurate steel faces, where dimensional tolerances of less than ±3 mm and clean unblemished surfaces result. Less expensive and versatile timber moulds allow ad hoc variations to profiles, e.g. channels or chases for M&E installations, but the refurbishment costs may be significant after many uses.

To take advantage of the precasting method, it is possible to use up to $A_s/A_c = 10\%$ high tensile reinforcement (the recommended maximum value). The minimum diameter of bar used is 12 mm, even if design value gives less, to ensure that the cage is sufficiently robust to be handled in the factory, or transported from a subcontract supplier. The maximum size of bar is 40 mm, with not more than two bars grouped at any point. The bars are distributed around the perimeter, primarily at the corners and mid-sides. Links are distributed according to the normal rules of using a diameter not less than 1/4 the size of the main bars and at a spacing of not more than 12 times the main bars. Additional links are provided at the foot and head of the column, adjacent to connectors (see Section 9.4) and around lifting points. Cover to reinforcement is according to the standard durability and fire regulations, but is taken as not less than 30 mm to links for the practical reason of fitting in connectors in such a way that the inevitable high stresses around them are confined by links.

The characteristic cube strength of the concrete is usually 50 N/mm^2, but, because of the early strength required for lifting in the factory, actual characteristic strengths are in the range $60\text{--}70 \text{ N/mm}^2$. Reinforcement is nearly always HT deformed bar, even for the links.

Columns placed into a foundation pocket (see Section 9.4.2) have roughened surfaces in the contact region, typically over a distance of 1.5 times the column breadth measured from the foot of the column. Retarding agents, which slow down cement hydration and hardening for the first 24 hours after mixing, are applied to the mould where roughening is carried out by mechanical scabbling. Different techniques are used to expose the coarse aggregate without causing structural damage to the cement matrix. The trowelled face of the column is roughened simply by making indentations (e.g. by cross hatching with the pointed edge of a trowel) into the concrete approximately 5 mm deep just prior to the concrete stiffening, say 90 minutes after casting.

6.2 Column design

The structural design of precast columns is no different to ordinary reinforced concrete columns, once the different loading history that a precast column experiences during manufacture, transportation and erection, has been catered for. The main design differences at the ultimate limit state are more a function of the type of precast structure and the type of connections than the resulting analysis of the

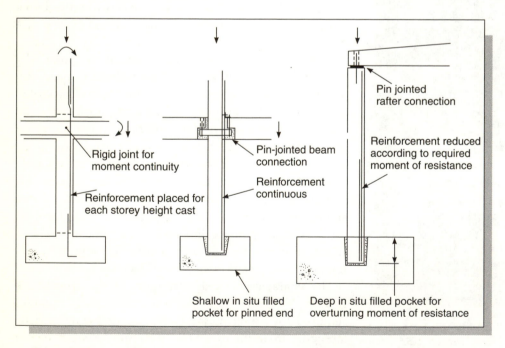

Figure 6.6: Column design philosophy for cast in situ, precast skeletal and precast portal frames.

column section. This is illustrated in Figure 6.6 for cast in situ, precast skeletal and precast portal frames.

Precast column design is carried out in three stages for the following:

1   Factory handling and transportation, where a set of predetermined reinforcement and lifting calculations are made independent of any project requirements.

2   On site erection and temporary stability, where design is carried out to ensure safety during frame erection prior to the final service load requirements of each project.

3   In service, where ultimate limit state calculations are carried out to satisfy the requirements of each project.

Design Stage 1 is at the behest of the manufacturer in which the project engineer should not interfere. This does not mean that the engineer is not informed of the casting and handling procedures. The objective of design Stage 1 is to deliver to site a safe and structurally sound precast component, which means manufacturing and handling and transporting it properly from the mould to the crane hook on site.

Design Stage 2 is carried out by the project engineer in consultation with the erection contractor, which may or may not be under the same management. This may prove to be critical particularly in long, slender columns where on site pitching may govern. Good communications between site and design office are essential during this stage. The maximum axial force and bending moment carried by the column in its temporary condition may be much more onerous than in service. All combinations of the sequence in which the precast frame is attached to the column must be considered in design Stage 2 so that the erection sequence is unhindered.

The ultimate limit state design Stage 3 is carried out according to the usual ultimate limit state procedures, providing that aspects of design particular to the design of the precast structure are catered for. These include localized stress in the region of the connection, special foundation details, etc., all of which are discussed in Chapter 9.

6.2.1   *Column design for factory handling*

Columns are demoulded after a minimum of 15–18 hours to facilitate a 24-hour casting cycle. The concrete is fast cured under steam or similar favourable methods such that the compressive cube strength of the concrete is at least $20 \, \text{N}/\text{mm}^2$. This is obtained by testing two cubes match cured from the same mix. After removing mould sides the column is lifted at 2 points, or 4 points in long columns, at predetermined positions to minimize deflection and tensile strain.

Lifting points are usually at 1/4 to 1/6 of span from either end of the column because the optimum situation for equal sagging and bending moment is where the lifting point is at 0.208L from the ends of a unit length L. Thus 0.2L is used as shown in Figure 6.7a. Very long columns of slender section (length/depth >50 to 55 may require 4 point lifting because of unacceptable deflection in a 2 point lift. Distances between these points are shown in Figure 6.7b. Bending moments and shear forces are calculated on self weight plus a 25 per cent mould suction and impact allowance. Reinforcement is designed using $f_s = 0.95f_y$ together with an ultimate load factor of 1.4. Concrete is taken as $f_c = 20\,N/mm^2$ mean, or $f'_{cu} = 15\,N/mm^2$ characteristic compressive cube strength.

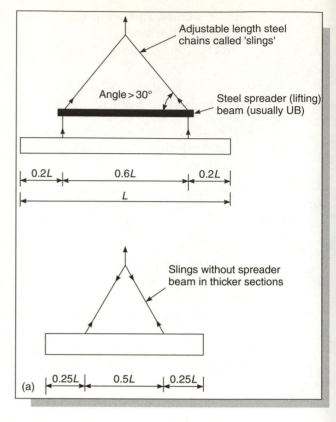

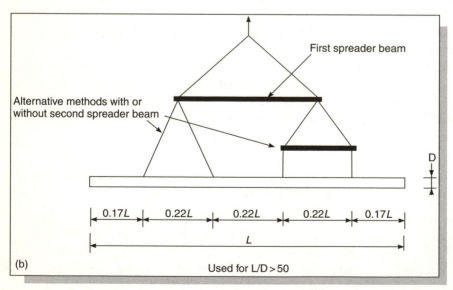

Figure 6.7: Flat handling of columns: (a) Two-point lifting; and (b) four-point lifting.

6.2.2   *Column design for pitching and erection*

Single point pitching is the most practical – it is often difficult enough releasing one sling from the crane hook near to the top of a tall column than two. The minimum bending moment (in prismatic columns of constant cross-section of self weight w per unit length) is found when the pitching point is at $0.3L$ from the end of the column length L, as shown in Figure 6.8. The resulting ultimate moment is given as:

$$M = 0.113wL^2 \qquad\qquad 6.1$$

($\gamma_f = 1.4$ and a 25% impact allowance included)
and the maximum ultimate shear force is:

$$V = 0.74wL \qquad\qquad 6.2$$

Dual pitching using sliding chains may be used for very long columns. The pitching points are placed at $0.16L$ and $0.60L$ from the top end of the column and the maximum ultimate bending moment is:

$$M = 0.023wL^2 \qquad\qquad 6.3$$

i.e. a five-fold reduction in relative moment compared to a single pitch point.

The centre of gravity of non-prismatic columns, e.g. those with protruding arms or other facilities as shown in Figure 6.9, is not coincident with the centre of pitching, and this causes the column to swivel when it leaves the ground. It is important that the pitching point is located at the centre of gravity of columns to avoid tilting during the fixing operation. Although this has almost no effect on the design of columns, verticality in placement ensures that all the reinforcement provided for the temporary condition is equally stressed, e.g. anti-bursting bars at the foot of the column, column splice bars.

Temporary stability of columns is usually achieved by propping against 'Acrow' (or similar) props which, if firm ground conditions allow unrestricted access to all sides, is achieved using

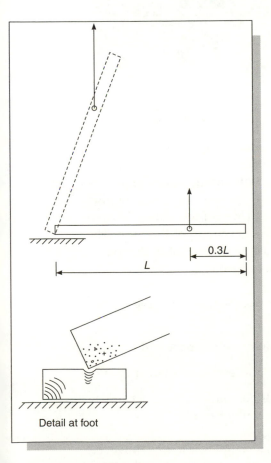

Detail at foot

Figure 6.8: Pitching points for columns.

Figure 6.9: Pitching asymmetrical columns.

Figure 6.10: Stabilizing columns with compression props and guy ropes.

pairs of compression props on opposite sides of the column. Figure 6.10 shows six-storey columns propped in this way. Column loads derive from wind loads acting on the incompleted frame, overturning moments due to spurious inclinations, or 'lack of plumb', and imbalanced loads at beam positions. Propping forces are calculated assuming a 60° rake reacting at a position about 500 mm below the first floor beams, i.e. about 2.5 m above the lower end of the prop. The horizontal reaction resulting from the prop force must be catered for at the foot of the column. Temporary instability conditions are not onerous if:

1   the columns are securely fixed by adequate props, positioned at the correct rake and height;

2   the sequence in which beams are fixed to the columns gives balanced loading;

3   the structural frame is tied as each subsequent floor level is erected; and

4   precautions are taken when stabilizing very tall columns in (forecast) gusty weather, e.g. additional guy ropes at 8–10 m above floor level.

Calculating wind forces acting on skeletal frames is a complex science. However, linear wind pressures acting on isolated one-dimensional elements have been estimated as being equivalent to three times the mean wind pressure (q) acting on the impermeable building.[1] Vortex behaviour causes drag on the tops and soffits of floor slabs (of area A per column), which are present in the frame before the column is structurally tied in. This produces a characteristic horizontal point force at the level of the beams approximately equivalent to:

$$H = 0.05qA \qquad\qquad 6.4$$

Lack of plumb inclinations δ may be taken as:

$$\delta = H/500 \qquad\qquad 6.5$$

where H is the height of the column at the point of measurement from the top of the foundation.

Example 6.1
Calculate the flexural lifting requirement for a 15 m long column, 300 mm × 350 mm in cross-section which is lifted from the mould using only two points at the optimum position. Use $f'_{cu} = 15\,\text{N/mm}^2$, $f_y = 460\,\text{N/mm}^2$ and 30 mm cover to 8 mm diameter links. Self weight of concrete = 24 kN/m^3.

Solution
Lifting points to be at $0.208L$ from ends = 3.12 m from ends
Self weight = $0.350 \times 0.300 \times 24 = 2.52\,\text{kN/m}$ plus 25 per cent suction and impact for a narrow mould = 3.15 kN/m.
Ultimate moment $M = 1.4 \times 3.15 \times 3.12^2/2 = 21.5\,\text{kNm}$

It is not known, at this stage, which orientation the column will be manufactured. Assume depth is minimum. Assume 25-mm diameter bars and 8 mm links.

$$b = 350\,\text{mm}; \ d = 300 - 30 - 8 - 12 \ (\text{say}) = 250\,\text{mm}$$
$$K = (21.5 \times 10^6)/(15 \times 350 \times 250^2) = 0.066; \text{ hence } z/d = 0.92$$
$$A_s = (21.5 \times 10^6)/(0.95 \times 460 \times 0.92 \times 250) = 214\,\text{mm}^2$$

Use 2 no. T 12 each face (226).

Example 6.2
The column in Example 6.1 is pitched on solid ground to the vertical using single pitching point at the optimum height. Calculate the flexural and shear reinforcement required to cater for pitching. Use $f_{cu} = 50\,\text{N/mm}^2$ and $f_y = 460\,\text{N/mm}^2$.

Solution
Again, with no knowledge of the orientation of the column during pitching, assume minimum depth.

$$M = 0.113 \times 2.52 \times 15^2 = 64.1 \, \text{kNm} \qquad (\textit{using Eq. 6.1})$$
$$K = (64.1 \times 10^6)/(50 \times 350 \times 250^2) = 0.059 \text{ (note almost same } K \text{ value as lifting).}$$
$$\text{Hence, } z/d = 0.93$$
$$A_s = (64.1 \times 10^6)/(0.95 \times 460 \times 0.93 \times 250) = 631 \, \text{mm}^2$$

Use 2 no. T 20 each face (628, OK).

$$V = 0.74 \times 3.15 \times 15 = 35.0 \, \text{kN} \qquad (\textit{using Eq. 6.2})$$

Shear stress $v = (35 \times 10^3)/(350 \times 250) = 0.4 \, \text{N/mm}^2$.

Concrete shear stress (BS8110, Part 1, Table 3.9) $v_c = 0.75 \, \text{N/mm}^2$. As $v > 0.5 v_c$ use nominal links $A_{sv}/s_v = 0.59 \, \text{mm}^2/\text{mm}$.

Use T8 links at 170 mm, or T10 at 250 mm spacing.

Practical tip: provide two extra links at 50 mm spacing near to pitching hole.

Example 6.3

The column in Examples 6.1 and 6.2 is erected in the vertical position and secured using 4 props, one per face, connected to the column at a height of 3.0 m above the top of the foundation. The props are inclined at 55° to the horizontal. This is an edge column which supports a beam connected to the 350 mm wide face of the column. The beam carries 40 m² of floor slab resulting in a temporary construction load of 120 kN per floor. The distance from the face of the column to the centre of the beam end reactions is 65 mm. The height to the soffit of the first and second floor beams is 4.0 m and 7.5 m above the foundation, respectively. The characteristic wind pressure for the complete structure may be taken as 0.71 kN/m². Use $\gamma_f = 1.4$ for wind loads.

Determine the maximum ultimate overturning moment in the column, the ultimate horizontal reaction at the foundation, and hence the maximum propping reaction given that the first floor beams are NOT to be fully tied to the column until the second floor level is constructed.

Solution

Figure 6.11 shows the arrangement.

Ultimate wind pressure $= 1.4 \times 0.71 = 1.0 \, \text{kN/m}^2$

Moment due to wind on the sail face 350 mm wide $\times$ 15.0 m high

$$M = 1.0 \times 0.35 \times 15.0^2/2 = 39.4 \, \text{kNm}$$

Wind drag force on the floors and beams $= 0.05 \times 1.0 \times 40 = 2.0 \, \text{kN}$

$$M = (2.0 \times 4.0) + (2.0 \times 7.5) = 23.0 \, \text{kNm}$$

Lack of plumb at 1st floor $= 4000/500 = 8 \, \text{mm}$, and 15 mm at 2nd floor

Eccentricity of temporary beam load at 1st floor $= 300/2 + 65 + 8 = 223 \, \text{mm}$, and 230 mm at 2nd floor

$$M = (120 \times 0.223) + (120 \times 0.230) = 54.4 \, \text{kNm}$$

Total ultimate overturning moment $= 39.4 + 23.0 + 54.4 = \underline{116.8 \, \text{kNm}}$

Lever arm to prop $= 3.0 \, \text{m}$

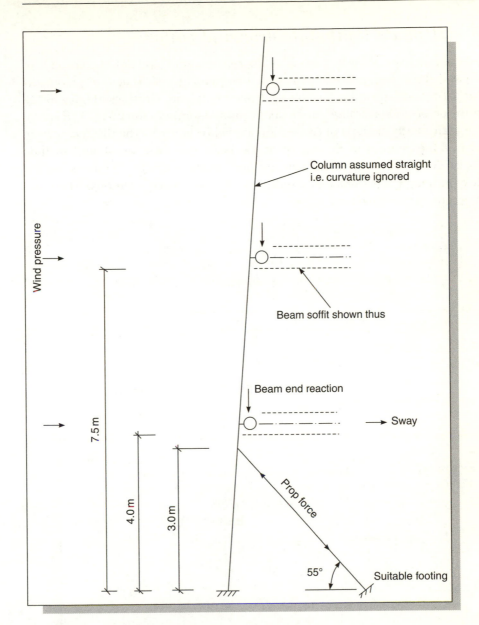

Figure 6.11: Detail to Example 6.3.

Ultimate horizontal force = 116.8/3.0 = <u>38.9 kN</u>
Ultimate propping reaction = 38.9/ cos 55° = 59.9 kN
If $\gamma_f = 1.4$ for wind loads, the characteristic prop reaction = 59.9/1.4 = <u>42.8 kN</u>

6.2.3   *Ultimate limit state design of columns*

Precast columns are subjected to axial forces, bending moments (axial and/or biaxial) and shear. In pinned-jointed structures (see Figure 3.9) bending moments are due to the eccentric load resulting from beam end reactions (Figure 6.12). There are no so-called 'frame' moments in pinned-jointed structures. The eccentricity e varies with the type of connection, corbel or haunch. The distance x from the face of the column to the centre of the beam-end reaction should include tolerances Δ to allow for manufacturing and erection errors. 15 mm is recommended, although most erectors would say this was very generous. The maximum eccentricity is given as:

$$e = \frac{h}{2} + x + \Delta \qquad\qquad 6.6$$

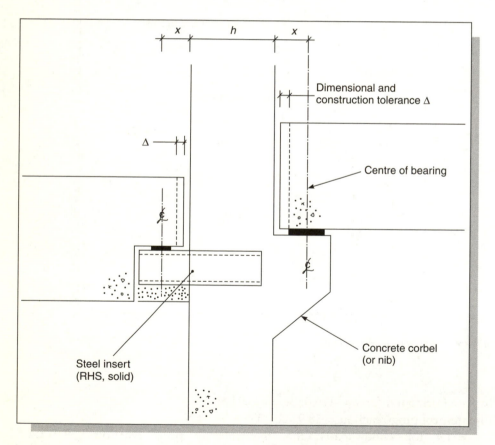

Figure 6.12: Column bending moments due to eccentric beam end reactions.

The bending moment acting at the level of the beams, called a 'nodal' moment (connection position = *node*) is $M = Ve$, where V is the ultimate beam end reaction. However, load combinations where two or more beams meet at a column should be considered to determine the maximum nodal moment. At internal columns supporting a (nearly) symmetrical arrangement of beams the moment is obtained by the summation of moments at each side of the column. Patch loading is used so that the maximum and minimum beam reactions on either side of the column are:

$$V_{max} = 1.4 \times \text{dead load} + 1.6 \times \text{live load}$$

$$V_{min} = 1.0 \times \text{dead load}$$

6.7

The net overturning moment is obtained from the worst possible scenario when the construction tolerance Δ is **added** to the eccentricity of the greater load and **subtracted** from the eccentricity of the smaller. Thus, if the distance from the centroid of the column to the centre of the beam reaction is e, the net column moment is:

$$M_{net} = M_{max} - M_{min} = V_{max}(e + \Delta) - V_{min}(e - \Delta)$$

6.8

The net eccentricity is given as:

$$e_{net} = \frac{M_{net}}{V_{max} + V_{min}}$$

6.9

The analysis may now proceed in the same manner as for the single sided beam. A similar approach is adopted for three-way and four-way beam connections where biaxial bending moments are present in the column.

The resulting bending moments are distributed in the column in proportion to the stiffness EI/h of the column between adjacent floor levels (Figure 6.13). The stiffness factors are $4EI/h$ for a continuous column and $3EI/h$ for a column where the remote end is pinned. The storey height h is the nodal distance, not the clear height between beams of l_o.

In some precast structures there are only beams (except at gable ends or around lifts or stairwells) in one of the two orthogonal directions. In this case, the clear height l_o is taken to the centre depth of the first floor slab based on the ground to first floor height. Subsequent upper storey columns are designed using the same effective length factor as before because the column is structurally continuous in these types of structures. Bending moments due to eccentric loading, horizontal forces and second-order deflection are combined to give the most onerous design condition. The deflection induced bending moments M_{add}

are distributed throughout the structure in proportion to the stiffness of all the columns.

6.2.4   *Design rules in BS8110 for columns in precast structures*

Effective column height according to Part 2, Clause 2.5.
Braced columns: the effective height l_e for columns in framed structures may be taken as the lesser of:

$$l_e = l_o[0.7 + 0.05(\alpha_{c1} + \alpha_{c2})] < l_o \qquad\qquad 6.10$$

$$l_e = l_o(0.85 + 0.05\alpha_{c,\,min}) < l_o \qquad\qquad 6.11$$

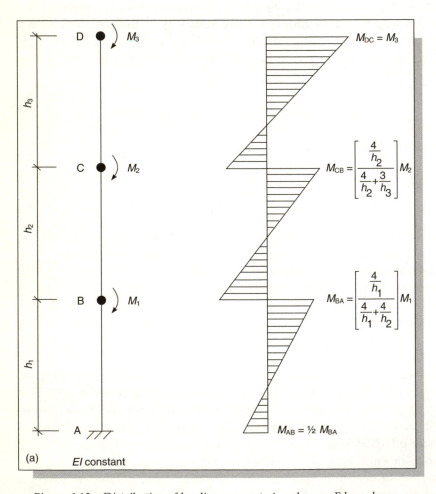

Figure 6.13a: Distribution of bending moments in columns: Edge column.

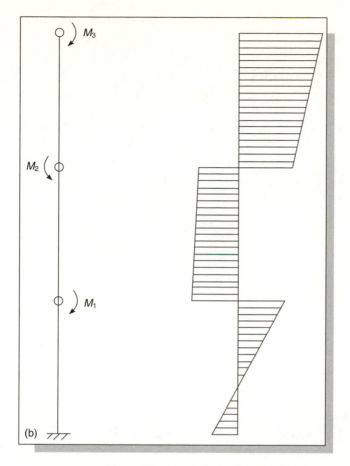

Figure 6.13b: Distribution of bending moments in columns: Internal column.

Unbraced columns: the effective height for columns in framed structures may be taken as the lesser of:

$$l_e = l_o[1.0 + 0.15(\alpha_{c1} + \alpha_{c2})] \qquad 6.12$$

$$l_e = l_o(2.0 + 0.3\alpha_{c,\,min}) \qquad 6.13$$

where α_{c1} and α_{c2} ratio of the sum of the column stiffness to the sum of the beam stiffness at the lower and upper end of a column, respectively. $\alpha_{c,\,min}$ is the lesser of α_{c1} and $\alpha_{c2} \cdot l_o$ is the clear height between end restraints. The stiffness of each member equals I/l_o, where I is the uncracked second moment of area for the section considered. Where a precast beam is simply supported at a column $\alpha_c = 10$. Where a column is designed to resist only nominal moment at a foundation $\alpha_c = 10$. Where the connection between column and foundation is designed to resist bending moments $\alpha_c = 1$.

Account has to be taken of the additional moment M_{add} induced in slender columns by its deflection, and is given by:

$$a_u = \left(\frac{1_e}{b}\right)^2 \frac{Kh}{2000} \qquad\qquad 6.14$$

where h = overall depth of a column in the plane considered, and b = smaller dimension of a column, except that in the case of bending about a major axis, for example where the frame is braced about the smaller dimension of the column, b may be taken as h. K = a reduction factor that corrects the deflection to allow for the influence of axial load and is given by:

$$K = \frac{N_{uz} - N}{N_{uz} - 0.25f_{cu}bd} \qquad\qquad 6.15$$

where

$$N_{uz} = 0.45f_{cu}\, bh + 0.95f_y A_{sc} \qquad\qquad 6.16$$

Then

$$M_{add} = Na_u \qquad\qquad 6.17$$

If the value of a_u is different for columns at any given level or storey, the deflection may be averaged for all columns of number n at that level, i.e.

$$a_{u,ave} = \frac{\Sigma a_u}{n} \qquad\qquad 6.18$$

The code allows any value of $a_u > 2a_{u,ave}$ to be ignored in Eq. 6.18.

The design procedure is therefore iterative as some prior knowledge of A_{sc} and K is required in the determination of M_{add}.

Some additional conditions apply. For bending about a major axis if *either* of the following are satisfied, $h/b > 3$ and $l_e/h > 20$, the column should be designed as bi-axially bent with the initial frame moment zero. This means that the additional moments M_y about the minor axis have to be considered. The design moment M_x is increased to:

$$M'_x = M_x + \beta\frac{h'}{b'}M_y \qquad\qquad 6.19$$

where the coefficient β is given in Table 6.1 (from BS8110, Table 3.24).

Table 6.1: Bi-axial bending coefficients

$\dfrac{N}{f_{cu}bh}$	0	0.1	0.2	0.3	0.4	0.5	≥ 0.6
β		1.00	0.88	0.77	0.65	0.53	0.42	0.30

In addition if $l_e/h > 20$ then M_{add} must be transferred to the members to which the column is attached, e.g. foundation bases, ground beams etc.

6.2.5   Columns in braced structures

Braced columns in so-called 'no-sway' frames must be checked for slenderness; an effective length factor of 1.0 being used in all upper floors and 0.9 between a fixed foundation and the first floor. Base plates and pocket foundations may be considered as fully moment resisting, but the stiffness coefficient $\alpha_c = 1.0$ means that they are not fully rigid in the interpretation of the code.

Columns are classified as slender if $l_e/h > 15$. Braced slender columns are analysed in the usual manner taking into consideration the initial moments $M_e = Ve$ and the additional moments as given above. If M_1 and M_2 are the smaller and larger initial end moments at the ends of the column, the maximum design moment M_t is given as the greater of the following:

(a) M_2

(b) $0.4M_1 + 0.6M_2 + M_{add}$ or $0.4M_2 + M_{add}$

(c) $M_1 + 0.5M_{add}$

(d) $0.05\,Nh$

6.20

Example 6.4.   *Columns in a braced frame*
The four-storey precast column shown in Figure 6.14a supports beams on two opposite sides. The distance from the face of the column to the centre of beam bearing is 80 mm. The construction tolerance is 15 mm. The characteristic beam end reactions (in kN) are:

	Dead load	Live load
Roof beams	96	36
Floor beams	144	120

Determine suitable sizes for the column and main reinforcement. The column cross-section should be square. Assume that the nodal point is at the mid-height of the beams. Use $f_{cu} = 50\,\text{N/mm}^2$ and $f_y = 460\,\text{N/mm}^2$. Cover to centre of bars = 50 mm.

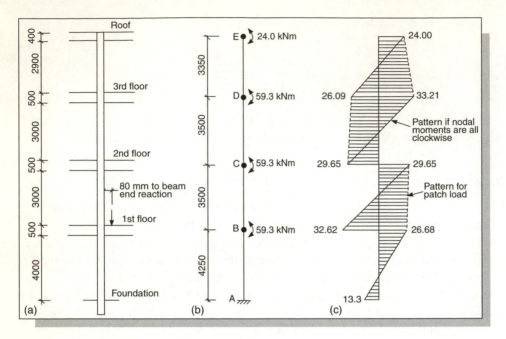

Figure 6.14: Detail to Example 6.4.

Solution

The dimensions to the nodal points are shown in Figure 6.14b. It is necessary to select an initial column size. Try $b = 250\,$mm, $h = 250\,$mm (self weight $= 1.5\,$kN/m height). $d/h = 200/250 = 0.8$.

Load Case 1. Patch loading on adjacent spans

Loads and moments

$e_{max} = 125 + 80 + 15 = 220\,$mm; $e_{min} = 125 + 80 - 15 = 190\,$mm   (*using Eq. 6.6*)

V_{max} at roof $= (1.4 \times 96) + (1.6 \times 36) = 192\,$kN;

V_{min} at roof $= 1.0 \times 96 = 96\,$kN (rounded up to integers)   (*using Eq. 6.7*)

V_{max} at floors $= (1.4 \times 144) + (1.6 \times 120) = 394\,$kN;

V_{min} at roof $= 1.0 \times 144 = 144\,$kN   (*using Eq. 6.7*)

M_{net} at roof $= (192 \times 0.22) - (96 \times 0.19) = 24.0\,$kNm;

M_{net} at floors $= 394 \times 0.22 - 144 \times 0.19 = 59.3\,$kNm   (*using Eq. 6.8*)

Axial loads

At roof $N = 192 + 96 = 288\,$kN

At 3rd floor $N = 288 + 538 + (1.4 \times 1.5 \times 3.3) = 833\,$kN

At 2nd floor $N = 833 + 538 + 7 = 1378\,$kN

At 1st floor $n = 1378 + 538 + 7 = 1923\,$kN

At foundation $N = 1923 + 9 = 1932\,\text{kN}$

Moment distribution factors

At 3rd floor

$$k_{DE} = \frac{\dfrac{3}{3.350}}{\dfrac{3}{3.350} + \dfrac{4}{3.500}} = 0.44 \quad \text{and} \quad k_{DC} = 1 - 0.44 = 0.56$$

At 2nd floor

$$k_{CD} = k_{CB} = 0.5 \text{ by inspection}$$

At 1st floor

$$k_{BC} = \frac{\dfrac{4}{3.500}}{\dfrac{4}{3.500} + \dfrac{4}{4.250}} = 0.55 \quad \text{and} \quad k_{BA} = 0.45$$

At foundation $k_{AB} = 0.5\,k_{BA} = 0.225$ ($=50\%$ carry over)

The resulting bending moment diagram is shown in Figure 6.14c.

Slenderness

Foundation – 1st floor, $l_e = 0.9\,l_o = 0.9 \times 4000 = 3600\,\text{mm}$

$l_e/h = 3600/250 = 14.4 < 15$, therefore, column 'short'

1st to 2nd floor and 2nd to 3rd floor, $l_e = 1.0\,l_o = 1.0 \times 3000 = 3000\,\text{mm} <$ above, therefore, column 'short'

3rd floor to roof, not critical by inspection.

Column design moment

Reader to verify that underside at 1st floor is critical

$$M_t = 26.68\,\text{kNm and } N = 1923\,\text{kN} \qquad (\textit{using Eq. 6.20})$$
$$N/bh = 30.7 \text{ and } M/bh^2 = 1.7$$

Use BS8110, Part 3, Table 47 because $d/h = 200/250 = 0.8$

$$\text{Then } 100A_{sc}/bh = 3.3$$

Load Case 2. Maximum loading on all spans

Loads and moments

M_{net} at roof $= (192 \times 0.22) - (192 \times 0.19) = 5.8\,\text{kNm}$ (or $= 192 \times 0.03 = 5.8$)

(*using Eq. 6.8*)

M_{net} at floors $= 394 \times 0.03 = 11.8\,\text{kNm}$ \qquad (*using Eq. 6.8*)

Axial loads

At foundation $N = 384 + (3 \times 788) + 30 = 2778\,\text{kN}$ (other values are not important)

Column design moment
Reader to verify that foundation is critical

$$M_t = 0.05 \, Nh = 34.7 \, \text{kNm and } N = 2778 \, \text{kN} \qquad (\text{using Eq. 6.20})$$
$$N/bh = 44.45 \text{ and } M/bh^2 = 2.22$$

Then $100A_{sc}/bh = 7.1$
Maximum value $A_{sc} = 7.1 \, bh/100 = \underline{4438 \, \text{mm}^2}$
Use 250 × 250 mm column with 4 no. T32 plus 4 no. T 20 bars (4472).

(It is not practical to use 6 no. bars as the bars placed at mid-face will inevitably clash with the beam-column connectors.)

Alternative
Using 300×300 mm column may (possibly) prove to be more economical and would certainly cause less problems during lifting and pitching. The reader should verify that the design loads are $N = 2794 \, \text{kN}$ (the extra is due to self weight) and $M_t = 0.05 \, Nh = 41.9 \, \text{kNm}$. Then $A_{sc} = 3.2 \, bh/100 = \underline{2790 \, \text{mm}^2}$. *Use 300 × 300 mm column with 4 no. T32 bars (3216).*

6.2.6   Columns in unbraced structures

The stability of unbraced pin-jointed structures is provided entirely by columns designed as cantilevers for the full height of the structure. The line of load application is at the centroid of the flooring system. The distribution of horizontal loading between columns is directly proportional to the second moment of area of the columns in the uncracked condition.

The maximum overturning moment in each column is $\Sigma H_i \, h_i$, where H_i is the floor diaphragm reaction at each column, and h_i is the effective height from a point 50 mm below the top of the foundation to the centroid of the floor plate at the floor level called *i*. (see Figure 6.15). The overturning moment is additive to the frame moments derived under column design. There is no moment distribution into the beams if the connections are pinned, and therefore the columns are designed using an effective length factor of 2.3, according to Eq. 6.13.

Columns are classified as slender if $l_e/h > 10$. Unbraced slender columns are analysed in the

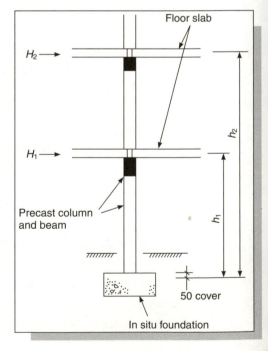

Figure 6.15: Overturning moments in unbraced structures.

usual manner taking into consideration the initial moments $M_e = Ve$ and the additional moments as given above. If M_1 and M_2 the larger and smaller initial end moments at the ends of the column, and M_{add1} and M_{add2} are the corresponding additional moments, the maximum design moment M_t is given as the greater of the following:

$$
\begin{array}{ll}
\text{(a)} & M_1 + M_{add1} \\
\text{(b)} & M_2 + M_{add2} \\
\text{(c)} & 0.05\,Nh
\end{array}
\qquad 6.21
$$

Example 6.5. *Column in unbraced frame*
The three-storey frame shown in Figure 6.16a has the same dimensions and supports the same beam loads as in Example 6.4. The characteristic wind load on the frame is 6.0 kN/m height. Determine suitable sizes and reinforcement for the internal column only. It may be assumed that the frame is braced in the out-of-plane direction and there are *no* additional second-order moments in this plane. It may also be assumed that the horizontal force of 1.5 per cent G_k is not critical.

Solution
The dimensions to the nodal points are shown in Figure 6.16b. It is necessary to select an initial column size.
Try $b = 300$ mm, $h = 500$ mm (self weight $= 3.6$ kN/m). $d/h = 0.9$.
Use BS8110, Part 3, Chart 49.

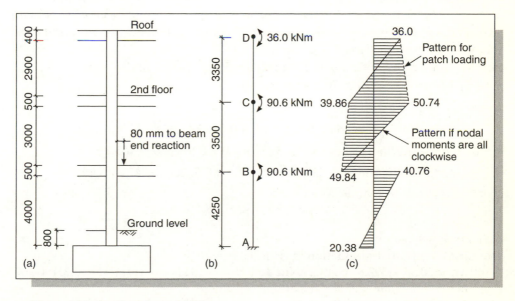

Figure 6.16: Detail to Example 6.5.

Load Case 1. Patch loading on adjacent spans, no wind load

Loads and moments

$e_{max} = 250 + 80 + 15 = 345\,mm$, $e_{min} = 250 + 80 - 15 = 315\,mm$ (*using Eq. 6.6*)

M_{net} at roof $= 192 \times 0.345 - 96 \times 0.315 = 36.0\,kNm$ (*using Eq. 6.8*)

M_{net} at floors $= 394 \times 0.345 - 144 \times 0.315 = 90.6\,kNm$ (*using Eq. 6.8*)

Axial loads

N due to roof load $= 288\,kN$

N due to floor loads $= 538 + (1.4 \times 3.6 \times 3.35) = 555\,kN$

At foundation $N = 288 + (2 \times 555) + 20 = 1418\,kN$

Then $N/bh = 9.45$, which means that $K = 1.0$ irrespective of M/bh^2.

Moment distribution factors

At 2nd floor $k_{CD} = 0.44$; $k_{CB} = 0.56$

At 1st floor $k_{BC} = 0.55$; $k_{BA} = 0.45$

At foundation $k_{AB} = 0.5\,k_{BA} = 0.225 (= 50\%$ carry over$)$

The resulting bending moment diagram is shown in Figure 6.16c.

Slenderness

Foundation – 1st floor, $l_e = 2.3\,l_o = 2.3 \times 4000 = 9200\,mm$

$l_e/h = 9200/500 = 18.4 > 10$, therefore, column 'slender'.

Foundation – 2nd floor, $l_e = 2.3\,l_o = 2.3 \times (4000 + 500 + 3000) = 17\,250\,mm$

$l_e/h = 17\,250/500 = 34.5 > 20$, therefore, bi-axial effects should be considered, but as stated in the question and proven in Example 6.4 there are no additional moments in the minor axis.

Foundation – roof, $l_e = 2.3\,l_o = 2.3 \times (7500 + 500 + 2900) = 25\,070\,mm$

$l_e/h = 25\,070/500 = 50.1 > 20$, as before.

Additional moments

M_{add} due to roof load $= 288 \times 50.1^2 \times 1.0 \times 0.5/2000 = 180.7\,kNm$

M_{add} due to 2nd floor load $= 555 \times 34.5^2 \times 1.0 \times 0.5/2000 = 165.1\,kNm$

M_{add} due to 1st floor load $= 555 \times 18.4^2 \times 1.0 \times 0.5/2000 = 47.0\,kNm$

Total $M_{add} = 392.8\,kNm$ at foundation

Column design moment

Reader to verify that foundation is critical

$M_t = 20.38 + 392.8 = 413.2\,kNm$ and $N = 1418\,kN$ (*using Eq. 6.21*)

$N/bh = 9.45$ and $M/bh^2 = 5.51$

Then $100A_{sc}/bh = 1.7$ and $K = 1.0$

Load Case 2. Maximum loading on all spans, no wind load
Loads and moments

$$M_{net} \text{ at roof} = 5.8 \text{ kNm (or } = 192 \times 0.03 = 5.8) \qquad (using\ Eq.\ 6.8)$$
$$M_{net} \text{ at floors} = 11.8 \text{ kNm} \qquad (using\ Eq.\ 6.8)$$

M at foundation $= 0.225 \times 11.8 = 2.66$ kNm

Axial loads
N due to roof load $= 384$ kN
N due to floor loads $= 788 + 17 = 805$ kN
At foundation $N = 384 + (2 \times 805) + 20 = 2014$ kN (other values are not important).
Then $N/bh = 13.43$, try $K = 0.9$ and verify later

Additional moments
$$M_{add} \text{ due to roof load} = 384 \times 50.1^2 \times 0.9 \times 0.5/2000 = 216.9 \text{ kNm}$$

M_{add} due to 2nd floor load $= 805 \times 34.5^2 \times 0.9 \times 0.5/2000 = 215.6$ kNm

M_{add} due to 1st floor load $= 805 \times 18.4^2 \times 0.9 \times 0.5/2000 = 61.3$ kNm
Total $M_{add} = 493.8$ kNm at foundation

Column design moment
Reader to verify that foundation is critical

$$M_t = 493.8 + 2.66 = 496.5 \text{ kNm and } N = 2014 \text{ kN} \qquad (using\ Eq.\ 6.21)$$

$$N/bh = 13.43 \text{ and } M/bh^2 = 6.62$$

Then $100A_{sc}/bh = 2.8$ and $K = 0.9$

Load Case 3. Maximum loading on all spans with wind load
Loads and moments

$$V_{max} \text{ at roof} = 1.2 \times (96 + 36) = 159 \text{ kN} \qquad (using\ Eq.\ 6.7 \text{ with } \gamma_f = 1.2)$$

$$V_{max} \text{ at floors} = 1.2 \times (144 + 120) = 317 \text{ kN} \qquad (using\ Eq.\ 6.7 \text{ with } \gamma_f = 1.2)$$

$$M_{net} \text{ at roof} = 159 \times 0.03 = 4.8 \text{ kNm} \qquad (using\ Eq.\ 6.8)$$

$$M_{net} \text{ at floors} = 317 \times 0.03 = 9.5 \text{ kNm} \qquad (using\ Eq.\ 6.8)$$

M at foundation $= 0.225 \times 9.5 = 2.14$ kNm

Axial loads
N due to roof load $= 318$ kN
N due to floor loads $= 634 + 15 = 649$ kN
At foundation $N = 318 + (2 \times 649) + 18 = 1634$ kN (other values are not important).
Then $N/bh = 10.89$, which means that $K = 1.0$ irrespective of M/bh^2.

Additional moments

M_{add} due to roof load $= 318 \times 50.1^2 \times 1.0 \times 0.5/2000 = 198.9\,kNm$

M_{add} due to 2nd floor load $= 649 \times 34.5^2 \times 1.0 \times 0.5/2000 = 193.1\,kNm$

M_{add} due to 1st floor load $= 649 \times 18.4^2 \times 1.0 \times 0.5/2000 = 54.9\,kNm$

Total $M_{add} = 446.9\,kNm$ at foundation

Wind loading

Ultimate horizontal wind pressure $= 1.2 \times 6.0 = 7.2\,kN/m$ height
Wind load at each floor taken to act between mid-nodal heights
Wind load at roof $= 7.2 \times 3.35/2 = 12.1\,kN$
Wind load at 2nd floor $= 7.2 \times (3.35 + 3.50)/2 = 24.7\,kN$
Wind load at 1st floor $= 7.2 \times (3.50 + 4.25 - 0.8)/2 = 25.0\,kN$
Moment taken at 50 mm below top of foundation
$M_{wind} = (12.1 \times 11.150) + (24.7 \times 7.800) + (25.0 \times 4.300) = 435.1\,kNm$ per 3 columns
All columns of equal stiffness, then $M_{wind} = 145.0\,kNm$ per column

Column design moment

Reader to verify that foundation is critical

$$M_t = 145.0 + 446.9 + 2.14 = 594.1\,kNm \text{ and } N = 1634\,kN \qquad (using\ Eq.\ 6.21)$$

$$N/bh = 10.89 \text{ and } M/bh^2 = 7.92$$

Then $100A_{sc}/bh = 3.2$ and $K = 1.0$
Maximum value $A_{sc} = 3.2\,bh/100 = \underline{4800\,mm^2}$
Use 500 × 300 mm column with 4 no. T32 plus 4 no. T25 bars (5180).

6.2.7 Columns in partially braced structures

Partially braced structures are used in situations where stability walls are architecturally undesirable, or structurally unnecessary, in a certain part of a structure (see Figure 3.10). The structure is designed as fully braced up to a specified level, and unbraced thereafter. This may not always be the same level throughout the entire building and may be different in the different directions of stability. The columns are cantilevered above this level as in an unbraced structure, but because they are not founded at a rigid foundation, their behaviour is different to ordinary cantilever columns.

According to Eq. 6.13, columns in the unbraced part of the structure are designed as cantilevers with an effective length ratio of 2.3. This is a conservative value because some of the columns immediately adjacent to the stabilizing walls may have an effective length factor of 2.0. This is because their connection to the braced structure can be considered as infinitely rigid (although BS8110 does not recognize such an end condition). The mean value of a_u according to Eq. 6.18 will be used in the calculation of M_{add}. The maximum design moment M_t is calculated according to the above sets of Eqs. 6.20(a–d) or 6.21(a–c).

Example 6.6. *Column in a partially braced frame*
The four-storey frame shown in Figure 6.17 has the same dimensions and supports the same beam loads as in Example 6.4. However, the frame is braced using shear walls up to the second floor. The self weight of the shear wall between 1st and 2nd floors may be taken as 80 kN. The characteristic wind load on the frame is 6.0 kN/m height. Determine suitable sizes and reinforcement for the internal columns only. It may be assumed that the frame is braced in the out-of-plane direction and there are no additional second-order moments in this plane.

Solution
The nodal distances are as shown in Figure 6.14b. The axial loads are the same as in Example 6.4. The column is designed at two positions: (i) at the second floor

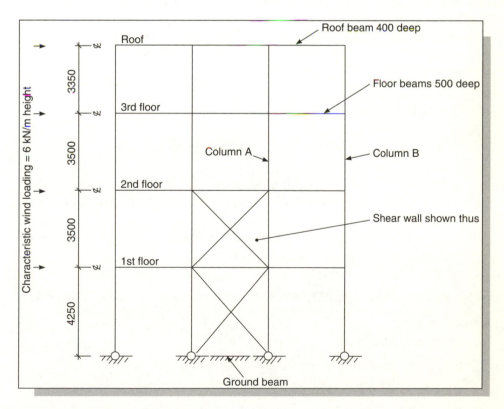

Figure 6.17: Detail to Example 6.6.

where wind and additional moments are greatest; and (ii) at the foundation where the axial load is greatest.

It is necessary to determine the second-order deflections a_u for the internal columns A and external columns B and to calculate the average value $a_{u,ave}$ according to Eq. 6.18. By inspection of Examples 6.4 and 6.5 the critical load condition is not Case 1.

Try column $b = 300$ mm, $h = 300$ mm. $d/h = 0.83$, therefore, use BS8110 Chart 47.

Load Case 2. Maximum load, no wind
(i) At 2nd floor. *Slenderness in Column A.*

2nd to 3rd floor, α_{c1} at bottom $= 1$, α_{c2} at top $= 10$

$$l_e = 2.3\, l_o = 2.3 \times 3000 = 6900 \qquad \text{(using Eq. 6.13)}$$

$l_e/h = 6900/300 = 23.0 > 20$, therefore, column 'slender' and bi-axial effects should be considered, but as stated in the question and proven in Example 6.4 there are no additional moments in the minor axis.

a_u at 3rd floor $= 23.0^2 \times 1.0 \times 0.3/2000 = 0.079$ m (because $N/bh < 10$ and $K = 1.0$ at the 2nd floor level)

2nd floor to roof, $l_e = 2.3\, l_o = 2.3 \times (3000 + 500 + 2900) = 14\,720$ mm

$$l_e/h = 14\,720/300 = 49.1$$

a_u at roof $= 49.1^2 \times 1.0 \times 0.3/2000 = 0.361$ m

Slenderness in Column B
2nd to 3rd floor, α_{c1} at bottom $= \infty$, α_{c2} at top $= 10$

$$l_e = 2.0\, l_o = 2.0 \times 3000 = 6000 \qquad \text{(using Eq. 6.13)}$$

$l_e/h = 6000/300 = 20$ therefore, column 'slender'.

a_u at 3rd floor $= 20.0^2 \times 0.9 \times 0.3/2000 = 0.054$ m (because $N/bh \cong 13$ and $K = 0.9$ at the 2nd floor level)

2nd floor to roof, $l_e = 2.0\, l_o = 2.0 \times (3000 + 500 + 2900) = 12\,800$ mm

$$l_e/h = 12\,800/300 = 42.7$$

a_u at roof $= 42.7^2 \times 0.9 \times 0.3/2000 = 0.246$ m

Average a_u at 3rd floor $= (0.079 + 0.054)/2 = 0.066$ m, and at roof $= 0.304$ m

Additional moments

M_{add} due to roof load $= 384 \times 0.304 = 116.7$ kNm

M_{add} due to 2nd floor load $= 799 \times 0.066 = 52.7$ kNm

Total $M_{add} = 169.4$ kNm at 2nd floor

M due to connector eccentricity $= 0.5 \times 11.8 = 5.9$ kNm

Column design moment

$$M_t = 169.4 + 5.9 = 175.3 \, \text{kNm and } N = 1183 \, \text{kN} \qquad (\textit{using Eq. 6.21})$$

$$N/bh = 13.1 \text{ and } M/bh^2 = 6.5$$

Then $100A_{sc}/bh = 3.1$ and $K = 0.9$

(ii) At foundation. $N = 2794 \, \text{kN}$ (from Example 6.4) plus self weight of wall at 1st floor $= 1.4 \times 80/2 = 2850 \, \text{kN}$. $M_t = 0.05 \, Nh = 42.75 \, \text{kNm}$

$$N/bh = 31.66 \quad \text{and} \quad M/bh^2 = 1.58$$

Then $100A_{sc}/bh = 3.4$

Load Case 3. Maximum load with wind
(i) At 2nd floor. $a_{u,ave}$ as Case 2. Axial forces from Example 6.5 Case 3.

Axial loads
N due to roof load $= 318 \, \text{kN}$
N due to floor loads $= 634 + 9 = 643 \, \text{kN}$
At 2nd floor $N = 318 + 643 + 9 = 970 \, \text{kN}$. Then $N/bh = 10.78$, which means that $K = 1.0$ irrespective of M/bh^2.

Additional moments

M_{add} due to roof load $= 318 \times 0.304 = 96.4 \, \text{kNm}$
M_{add} due to 2nd floor load $= 643 \times 0.066 = 42.4 \, \text{kNm}$
Total $M_{add} = 138.8 \, \text{kNm}$ at 2nd floor
M due to connector eccentricity $= 0.5 \times 9.5 = 4.75 \, \text{kNm}$

Wind loading
Ultimate horizontal wind loading as Example 6.5
At 2nd floor M_{wind} per column $= [(12.1 \times 6.850) + (24.7 \times 3.500)]/4 = 42.3 \, \text{kNm}$ per column

Column design moment

$$M_t = 138.8 + 42.3 + 4.75 = 192.6 \, \text{kNm and } N = 970 \, \text{kN} \qquad (\textit{using Eq. 6.21})$$

$$N/bh = 10.78 \quad \text{and} \quad M/bh^2 = 7.13$$

Then $100A_{sc}/bh = 3.7$ and $K = 1.0$

(ii) At foundation

$N = 318 + (3 \times 643) + 11 = 2258 \, \text{kN}$ plus self weight of wall at 1st floor $= 1.2 \times 80/2 = 2306 \, \text{kN}$.

$$M_t = 0.05 \, Nh = 34.6 \, \text{kNm}$$
$$N/bh = 25.6 \quad \text{and} \quad M/bh^2 = 1.28.$$

Then $100A_{sc}/bh = 1.8$

Maximum value $A_{sc} = 3.7\,bh/100 = \underline{3330\,\text{mm}^2}$
Use $300 \times 300\,mm$ column with 8 no. T25 bars (3928).

6.3   Precast concrete shear walls

When the height of a pin jointed skeletal structure reaches certain limits, typically about three storeys, it is no longer possible to transfer both vertical and horizontal loads to the foundations via columns. This is because of the combined actions of second-order deflections adding to the frame moments due to wind and connector eccentricity. It is clear from Example 6.5 that the bending moments resulting from these actions in an unbraced frame are very large, leading to an uneconomical design – columns are intended to resist compressive loads, not bending moments.

To eliminate overturning moments from the columns, diagonal bracing is used. Horizontal forces are transferred through the structure as shown in Figure 6.18 and, instead of bending the columns, are resisted by an axial diagonal force in the bracing. The bracing can be constructed in many ways, these being, in descending order of popularity:

- precast concrete infill wall, Figure 6.19 (see Section 6.5.1)

- precast concrete cantilever wall, Figure 6.20 (see Section 6.6)

- brickwork (or blockwork) infill wall, Figure 6.21 (see Section 6.5.2)

- precast concrete cantilever cores, Figure 6.22

- steel or precast concrete cross bracing, Figure 6.23.

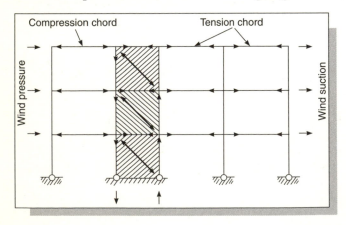

Figure 6.18: Horizontal force transfer in braced structures.

Figure 6.19: Precast concrete infill wall

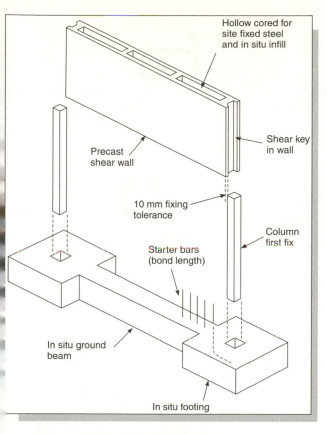

Figure 6.20: Precast hollow core cantilever wall.

Labels within figure 6.20:
- Hollow cored for site fixed steel and in situ infill
- Precast shear wall
- Shear key in wall
- 10 mm fixing tolerance
- Column first fix
- Starter bars (bond length)
- In situ ground beam
- In situ footing

Figure 6.21: Brick infill shear wall.

Figure 6.22: Storey height precast lift shaft units.

The design assumption is that the bracing will always resist diagonal forces in compression rather than tension – the exception to this rule being steel cross bracing where the tension leg is effective. Under the reverse and cyclical action of such as wind loading the diagonal compressive strut changes direction, and so different parts of the bracing are subjected to alternating compression and tension as the direction of the wind changes.

Because the strength of the bracing is very large compared with the bending capacity of columns, it is not necessary to position the bracing everywhere in the structure. The bracing is therefore posi-

Figure 6.23: Concrete strut diagonal bracing.

tioned strategically to maximize its effect, for example near to the ends of the structure or in a central core area as shown in Figure 6.24. A symmetrical pattern reduces torsional sway, and hence reduces shear stress in the floor diaphragm as explained in the following section. The framework in between the bracing may be designed as a pinned-jointed column-beam-slab structure with the columns designed as 'braced'. The floor slab must be capable of transferring horizontal forces – the greater of wind loads or 1.5 per cent G_k – to the bracing elements by so-called 'diaphragm action' (see Chapter 7), otherwise the columns between the bracing will become unbraced. This action must take place throughout 360° although in practice we only design in the two principal orthogonal planes x and y.

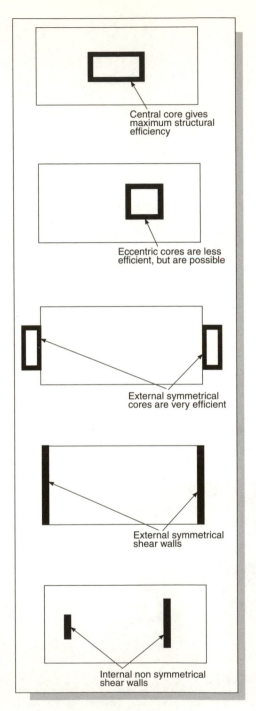

Figure 6.24: Bracing position possibilities for braced structures.

6.4 Distribution of horizontal loading

The amount of horizontal loading carried by the bracing is determined from the position and stiffness of the bracing in the structure. The stiffness of each bracing element is proportional to elastic modulus '*E*' times second moment of area I_u in the uncracked condition (at the serviceability limit state). It is assumed that the floor plate is a rigid diaphragm and the relative deflections of each of the bracing elements is proportional to the distance *a* from the centroid of stiffness to the bracing. In the following analysis, the walls are assumed to be parallel with each other and to the direction of the load.

If there are only two bracing elements as shown in Figure 6.25a the solution is statically determinate irrespective of the shape of the building and the positions x_1 and x_2 and stiffness I_1 and I_2 of the bracing. The structure may be analysed as a beam as shown in Figure 6.25b. Choosing an origin, say O, and taking moments, H_1 and H_2 may be determined using:

$$\frac{qL^2}{2} = H_1 x_1 + H_2 x_2 \qquad\qquad 6.22a$$

$$qL = H_1 + H_2 \qquad\qquad 6.22b$$

where *q* is the line load due to wind loading or 1.5 per cent G_k. Thus, H_1 and H_2 are found independently of the stiffness of the bracing.

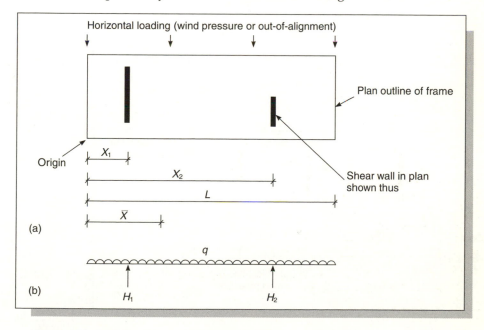

Figure 6.25: Analysis of a two-wall bracing system of bracing.

If there are more than two bracing elements the system is statically indeterminate and the equilibrium of forces and moments and compatibility of deflections must be considered. The first step is to determine the 'shear centre' of the bracing system. This is the position at the centre of stiffness, and is calculated from the origin as:

$$\bar{X} = \frac{\sum E_i I_i x_i}{\sum E_i I_i}$$ 6.23

where $I = tL^3/12$ if the bracing is of thickness t and length L.

If the centre of pressure is coincident with the shear centre, i.e. $\bar{X} = L/2$ in Figure 6.26a, the structure will be subject only to in-plane deflections, called 'translation' in the y direction. The force reaction carried by each bracing element H_n is therefore proportional to the stiffness of that bracing, and is given as:

$$\frac{H_n}{H} = \frac{E_n I_n}{\sum E_i I_i}$$ 6.24

where H is the total applied force qL.

If the centre of pressure is not coincident with the shear centre, as in Figure 6.26b, the structure will be subject to rotations and translation. The centre of rotation lies at the shear centre, where the deflection due to rotation is zero. The eccentricity of the applied load is given as:

$$e = L/2 - \bar{X}$$ 6.25

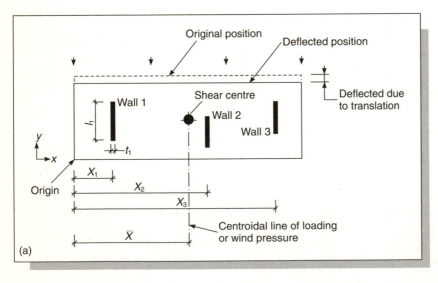

Figure 6.26: Analysis of a multi-wall (more than two walls) system of bracing: (a) Definitions.

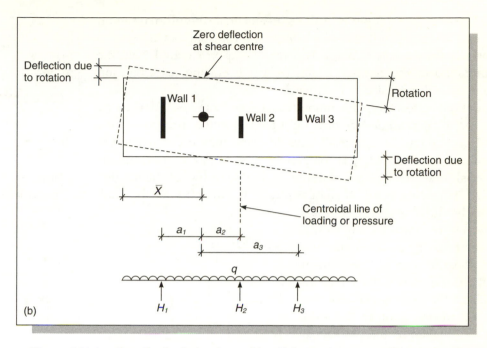

Figure 6.26 (continued): Analysis of a multi-wall (more than two walls) system of bracing: (b) Floor deflections due to rotation.

The force reaction carried by each wall H_i is the sum of two parts – translation and rotation, as:

$$\frac{H_n}{H} = \frac{E_n I_n}{\sum E_i I_i} \pm \frac{e E_n I_n a_n}{\Sigma E_i I_i a_i^2}$$

6.26

where H_n = reaction in bracing n,

H = total applied load,

$E_n I_n$ and $E_i I_i$ = stiffness of bracing n and of all bracing,

a_n and a_i = distance from the centroid of stiffness to bracing n and all bracing.

The $\pm$ sign in Eq. 6.26 means that the component of the reaction due to rotations is *additive* if the bracing is positioned on the opposite side of the shear centre as the centre of load, i.e. $x_i < \bar{X}$ (bracing 1 in Figure 6.26b). The reaction due to e is subtracted if the bracing lies on the same side of the shear centre as the load (bracings 2 and 3 in Figure 6.26b). Of course if $e = 0$ then Eq. 6.26 reverts back to Eq. 6.24.

If a wall is composed of cross walls, forming I, T, U or L configurations the I of the composite elements is used in place of the above providing the vertical joint at the intersections of the legs of the shape are capable of resisting the vertical interface shear force. If the walls are discrete components separated by columns no interaction between the legs is considered.

In cases where 1.5 per cent G_k is greater than the ultimate wind force, e is taken as the distance from the centroid of stiffness to centre of the dead load mass. This may be approximated from the summation of the centres of masses of the bracing (external and internal) and floor slab at each level.

Torsional effects in non-symmetrical systems may be balanced by bracing at right angles (or near-right angles) to the direction of the load, as shown in Figure 6.27. This situation may occur where the front or side of a building is completely open or glazed, for example. At least three bracings are required, with at least two of them, often called 'balancing walls' at right angles to the direction of the load. Provided there is shear continuity between the bracings, any statical method may be used to determine the shear centre of the system, and the reactions in the balancing walls. If the bracings are not connected to one another the shear centre is taken at the centroid of the main bracing parallel to the load. If this is a distance e from the centre of pressure, then referring to Figure 6.27:

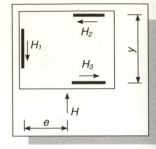

Figure 6.27: Analysis of out-of-plane wall system of bracing.

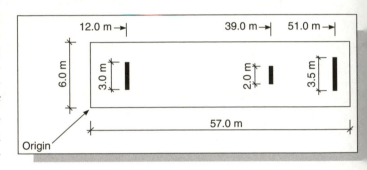

Figure 6.28: Detail to Example 6.7.

$$H_1 = H \quad \text{and} \quad H_2 = -H_3 = He/y \qquad\qquad 6.27$$

Example 6.7. *Distribution of horizontal loading*
Calculate the percentage reactions in each of the shear walls in Figure 6.28. The walls are all of equal thickness and stiffness.

Solution
$E_i I_i = L^3$ because walls are all of equal t and EI.
Choose left-hand side as origin for $x = 0$.

Calculation for determining the wall stiffness

Wall Ref	I	x	Ix	$a = x - \bar{X}$	Ia	Ia^2
A	27.0	12	324	24.24	654	15864
B	8.0	39	312	2.76	22	61
C	42.9	54	2187	14.76	633	9340
Totals	77.9		2823			25265

$$\bar{X} = 2823/77.9 = 36.24\,\text{m} \qquad (\textit{using Eq. 6.23})$$
$$e = 57.0/2 - 36.24 = 7.74\,\text{m} \qquad (\textit{using Eq. 6.25})$$

Now $x_1 < \bar{X}$, then the $\pm$ sign in Eq. 6.26 is positive for wall A
And x_2 and $x_3 > \bar{X}$, then the $\pm$ sign is negative for walls B and C.
The fraction of the total force in each of the walls is Eq. 6.26:
At Wall A

$$H_A = \left[\frac{27}{77.9} + \frac{7.74 \times 654}{25\,265} \right] H = 0.550H$$

At Wall B

$$H_B = \left[\frac{8}{77.9} - \frac{7.74 \times 22}{25\,265} \right] H = 0.096H$$

At Wall C

$$H_C = \left[\frac{42.9}{77.9} - \frac{7.74 \times 633}{25\,265} \right] H = 0.354H$$

6.5   Infill shear walls

Infill shear walls are reinforced flat panels that are positioned between columns, and sometimes but not always between beams, to literally 'infill' the opening in the framework. Infill shear walls rely on composite action with the pin-jointed column–beam structure for strength and stiffness. This is shown in the load response sequence in Figure 6.29. Because the beam–column structure is flexible and the infill panel very stiff (large in-plane *EI*) there is a paradox in the fundamental use of infill structures. Theoretically, the problem is similar to analysing stiff beams on elastic foundations, in that, resistance to horizontal loading is affected by the amount of deformation of the frame, and the interaction between the wall and frame. The origins of the following design method may be found in Ref. 2.

On first application of a horizontal load there may be full composite action between the frame and wall if these are bonded together (Figure 6.29a). At a comparatively early stage however, cracks will develop at the interface around the wall, except in the vicinity of two of the corners where the infill panel will lock into the frame and there will be transmission of compressive forces into the concrete wall (Figure 6.29b). The wall acts as a compression diagonal within the frame, the effective width of which depends on the relative stiffness λ of the two components and on the ratio of the height h' to the length L' of the panel. Failure results from the

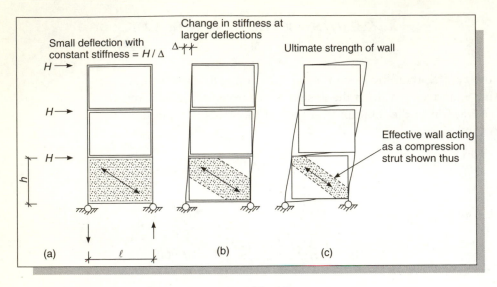

Figure 6.29: Load vs sway mechanism in infill walls.

loss of rigidity of the infill as a result of these diagonal cracks, or from local crushing or spalling in the region of the concentrated loads (Figure 6.29c).

Referring to Figure 6.30, the contact length α between the wall and column depends on their relative stiffness and on the geometry of the wall, and is given as:

$$\frac{\alpha}{h} = \frac{\pi}{2\lambda h} \qquad\qquad 6.28$$

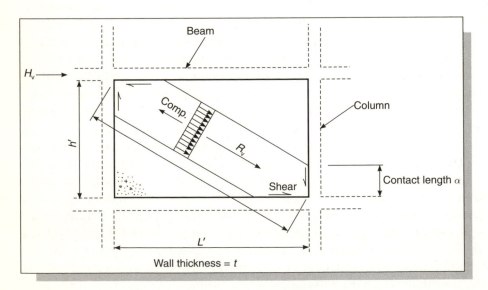

Figure 6.30: Contact zones and stresses in infill walls.

in which λh is a non-dimensional parameter expressing the relative stiffness of the frame and infill, where:

$$\lambda = \sqrt[4]{\frac{E_i t \sin 2\theta}{4E_c Ih'}}$$ 6.29

where E_i = infill modulus

E_c = concrete frame modulus

t = infill thickness

θ = slope of infill = h'/L'

I = minimum second moment of area of beams or columns

h' = height of infill

L' = length of infill

Given α and all other material properties, the resistance of any type of infill wall may be calculated as in the following sections for precast concrete and in situ masonry infill.

6.5.1 Precast concrete infill walls

Precast infill walls are preferred by the industry because they keep this important design and erection work under the control of the precast manufacturer and designer. They are used to brace structures of between two and 12–15 storeys in height. At greater heights the transfer of force to the column becomes uncontrollably large, unless there are numerous positions to locate these walls in a building, which is generally unlikely in commercial office developments.

Concrete walls are considered to be plain walls, according to BS8110, Part 1, Clause 3.12.5 because the minimum area of reinforcement is provided only for lifting purposes. The concrete is grade C40. The wall is built in on all sides and is therefore braced. The strength of the mortar used to pack the gap between the wall and frame, shown for example in Figure 6.31, should be of equal strength. If the corners of the wall are temporarily supported by angles or cleats, the cut out around the angle should likewise be properly filled and compacted.

The ultimate horizontal force is resisted by a diagonal compressive strut across the infill wall. Mainstone[3] found that the width of the strut may be conservatively taken as 0.1 times the diagonal length of the wall = $0.1w'$. The horizontal component of this force must also be resisted by interface shear along the horizontal interfaces. These may or may not be reinforced. The force must also be resisted by the vertical equilibrium reaction between the wall and column. These three criteria will now be determined.

The concrete is unconfined in the third dimension such that the limiting compressive failure stress is $0.3f_{cu}$. In Figure 6.30, the strength of the strut R_v is given by:

$$R_v = 0.3\,f_{cu}0.1w'(t - 2e_x) \qquad 6.30a$$

where $e_x = 0.05t$

The horizontal resistance is given by:

$$H_v = R_v\cos\theta = 0.03f_{cu}L'(t - 2e_x) \qquad 6.31$$

Where infill slenderness ratio $w'/t > 12$, equation 6.30a is modified in accordance with BS8110, Part 1, Clause 3.9.4.16. As the wall is built in on two sides at the corner, the effective length of the wall is taken as $L_e = 0.75w'$. The diagonal reistance R_v is modified to:

$$R_v = 0.3\,f_{cu}0.1w'(t - 1.2e_x - 2e_{add}) \qquad 6.30b$$

where $e_{add} = L_e^2/2500t$, with a limit on the slenderness ratio $0.75w'/t < 30$.

The length of contact at the corners is $\alpha = \pi/2\lambda$ along the column. Thus, if the limiting horizontal shear stress between unreinforced concrete surfaces in compression is $0.45\,\text{N/mm}^2$, according to BS8110, Part 1, Clause 5.3.7, the diagonal resistance R_v is limited by the following:

$$R_v\sin\theta = 0.45\alpha t \qquad 6.32a$$

Figure 6.31: Dry packing between precas concrete walls.

If the applied load $H > 0.45\alpha t\cot\theta$ the residual vertical shear force $(H - 0.45\alpha t\cot\theta)\tan\theta$ is carried by one of two mechanisms:

1   If the wall is bearing onto a beam the residual force may be transferred to the beam-to-column connector. If the connector has a shear capacity V_{beam} then Eq. 6.32a is modified to:

$$R_v\sin\theta = 0.45\alpha t + V_{beam} \qquad 6.32b$$

2   If the wall is bearing onto another wall, as shown in Figure 6.32, the residual force must be transmitted to the column in a manner that exceeds the capacity derived from Eq. 6.32a. In this case short welded connections were made at regular intervals along the wall-to-column interface. It is quite likely that the resistance derived from Eq. 6.32a is ignored such that the entire vertical force $R_v\sin\theta$ is carried by the weld.

Figure 6.32: Horizontal interface shear reinforcement and vertical welded joints in an infill wall.

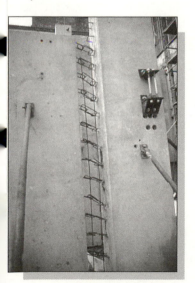

Figure 6.33: Vertical interface shear resistance by projecting loops from wall and column.

Figure 6.33 shows an alternative method of resisting the vertical shear by projecting loops. The gap between the wall and column is restricted to about 100 mm to avoid the large shear lag across the gap. The reinforcing loops of area A_v projecting from both the wall and column are essentially in shear such that:

$$A_v = \frac{R_v \sin \theta}{0.6 f_y} \qquad 6.33$$

If the precast manufacturer is not confident in providing interlocking loops an intermediate loop is provided as shown in Figure 6.33. This has a slight weakening effect on the joint due to the possibility of shear-tension developing across the joint, but a longitudinal (=vertical) 'lacer' bar is provided (based on a 45° stress component) of area A_l such that:-

$$A_l = R_v \sin \theta \qquad 6.34$$

The horizontal force $R_v \cos \theta$ is resisted by a stress of $0.45 \, \text{N/mm}^2$ along the contact length L' along the beam, then

$$R_v \cos \theta = 0.45 L' t \; (\text{N, mm units only}) \qquad 6.35$$

If $H > 0.45 L' t$ the residual horizontal shear may be taken through added interface reinforcement passing through beams and grouted into holes in the wall or otherwise fixed to wall panels such that the area of interface reinforcement A_v is given by:

$$A_v = \frac{H - 0.45 L' t}{0.6 \times 0.95 f_{yv}} > 0.15\% \times \text{contact area} \qquad 6.36$$

The length of embedment of the dowels in both the beam and wall should be 8 bar diameters, although a minimum practical length of 300 mm (inclusive of bends) is used.

The bearing stress under the corners of the wall is checked against Clause 5.2.3.4 where f_{cu} = weakest concrete:

$$R_v \sin \theta = \frac{0.6 f_{cu} L' t}{2} \qquad 6.37$$

Finally, the horizontal resistance $H_v = R_v \cos \theta$.

Example 6.8. *Precast concrete infill wall capacity*

Calculate the ultimate horizontal capacity of the 200-mm thick precast concrete infill wall shown in elevation in Figure 6.34. The beam end ultimate shear capacity may be taken as 200 kN.

Use $f_{cu} = 40\,\text{N/mm}^2$ and $E_c = 28\,\text{kN/mm}^2$.

Solution

$$I_{min} = 300 \times 300^3/12 = 675 \times 10^6\,\text{mm}^4$$
$$\theta = \tan^{-1} 3000/5000 = 31°$$

$$\lambda = \sqrt[4]{\frac{28000 \times 200 \times 0.882}{4 \times 28000 \times 675 \times 10^6 \times 3000}} = 0.00216\,\text{mm}^{-1} \qquad (\text{using Eq. 6.29})$$

$$\alpha = \pi/2\lambda = 727\,\text{mm}$$

Diagonal length $w' = \sqrt{5000^2 + 3000^2} = 5831\,\text{mm}$

$w'/t = 29.15 > 12$, therefore wall is slender and effective length is
$L_e = 0.75 \times 5831 = 4373\,\text{mm}$

$L_e/t = 21.9 < 30$

$e_x = 0.05 \times 200 = 10\,\text{mm}$

$e_{add} = 4373^2/2500 \times 200 = 38.2\,\text{mm}$

$R_v = 0.3 \times 40 \times 0.1 \times 5831 \times (200 - 12 - 76.4) \times 10^{-3}$

$\quad = 780.9\,\text{kN} \qquad (\text{using Eq. 6.30b})$

$R_v = [(0.45 \times 727 \times 200 \times 10^{-3}) + 200]/0.514 = 516.4\,\text{kN} \qquad (\text{using Eq. 6.32b})$

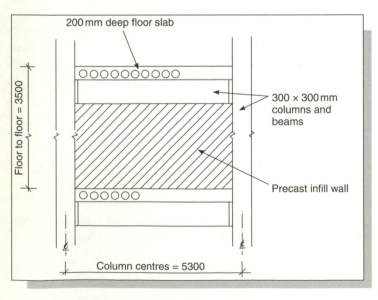

Figure 6.34: Detail to Example 6.8.

$$R_v = \frac{0.45 \times 5000 \times 200 \times 10^{-3}}{0.857} = 525.1\,\text{kN} \qquad (using\ Eq.\ 6.35)$$

$$R_v = \frac{0.6 \times 40 \times 5000 \times 200 \times 10^{-3}}{2} = 12\,000\,\text{kN} \qquad (using\ Eq.\ 6.34)$$

Lowest value $R_{v,\text{min}} = 516.4\,\text{kN}$
$H_v = 516.4 \times 0.857 = \underline{442.5\,\text{kN}}$

Example 6.9. *Precast concrete infill wall thickness*
Calculate the minimum thickness of the wall in Example 6.8 if the ultimate horizontal shear force is 400 kN.

Solution
$R_v = 400/0.857 = 466.7\,\text{kN}$

$$0.3 \times 40 \times 0.1 \times 5831 \times [t - 0.06t - (4373^2/1250t)] \geq 466.7 \times 10^3$$
$$(using\ Eq.\ 6.30b)$$

Hence, $t \geq 167.9\,\text{mm}$

$$t \geq \frac{[(466.7 \times 0.514) - 200] \times 10^3}{0.45 \times 727} \geq 121.9\,\text{mm} \qquad (using\ Eq.\ 6.32b)$$

$$t \geq \frac{466.7 \times 10^3 \times 0.857}{0.45 \times 5000} \geq 177.7\,\text{mm} \qquad (using\ Eq.\ 6.35)$$

$$t \geq \frac{466.7 \times 10^3 \times 0.514 \times 2}{0.6 \times 40 \times 5000} \geq 4\,\text{mm} \qquad (using\ Eq.\ 6.37)$$

Largest value $t = 177.7\,\text{mm}$, rounded to 180 mm

6.5.2 *Brickwork (or blockwork) infill walls*

Brick (or block) infill walls are an excellent alternative to precast infill walls and are used in buildings of up to about five storeys in height. Limitations on strength tend to come from a rather low horizontal shear capacity, whilst the diagonal compressive strut capacity is surprisingly large. The main practical drawback is the speed at which the wall can be built to keep pace with the progress of the precast frame, and the shared design and construction responsibility between the precaster and brick builder.

As with concrete walls, the ultimate horizontal force is resisted by a diagonal compressive strut across the infill. The criterion is based on the stiffness factor λ as before. The method is the same as for concrete infill walls except a factor k replaces the constant 0.1 for the width of the diagonal strut, and the crushing strength of the concrete is replaced with f_k the strength of brickwork in compression obtained from Table 6.2 (see BS5628, Part 1). A further criterion is the local

Table 6.2: Characteristic compressive strength of brick masonry f_k in N/mm^2

Mortar designation	Mortar proportion by volume*	Brick strength 20 N/mm²	Brick strength 27.5 N/mm²	Brick strength 35 N/mm²	Brick strength 50 N/mm²	Brick strength 70 N/mm²
i⁺	1:1/4:3	7.4	9.2	11.4	15.0	19.2
ii⁺	1:1/2:4	6.4	7.9	9.4	12.2	15.1
iii	1:1:5	5.8	7.1	8.5	10.6	13.1
iv	1:2:8	5.2	6.2	7.3	9.0	10.8

Note:
* cement:lime:sand
⁺ preferred mortar.

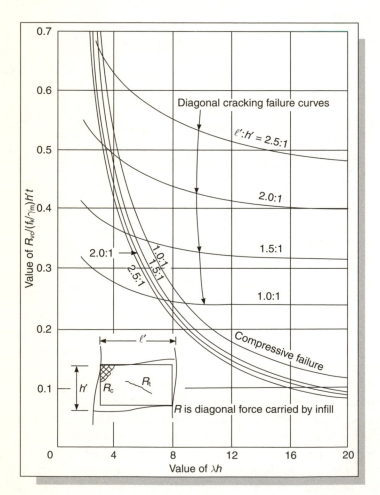

Figure 6.35: Infill wall design graph for compressive resistance.[2]

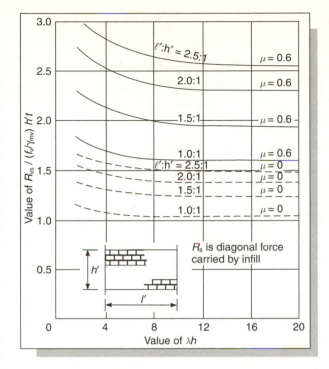

Figure 6.36: Infill wall design graph for shear resistance.[2]

crushing resistance of the brickwork at the corners. The k factors for these two modes of failure are given in the design graph in Figure 6.35.[2]

Horizontal shear failure takes place in the mortar courses just above the end of the contact zone, of height α. The horizontal shear limit is obtained from Figure 6.36 with $\mu = 0.6$ for solid (or unperforated) bricks, because vertical compressive stress acts at the same time as the shear. In these calculations Young's modulus for brick $= 450f_k$ (N/mm^2 units).

The compression limit is given as:

$$R_{vc} = \left(\frac{f_k}{\gamma_m}\right) h't \times \text{value from the}$$

design graph in Figure 6.35   6.38

The shear strength is given as:

$R_{vs} = (f_v/\gamma_{mv})h't \times$ value from the design graph in Figure 6.36   6.39

where f_v from BS5628 $= 0.35\,N/mm^2$ for mortar designations (i) and (ii), and $0.15\,N/mm^2$ for mortar designations (iii) and (iv). The bricks should have a minimum crushing strength of $20\,N/mm^2$, $\gamma_m = 3.5$ and $\gamma_{mv} = 2.5$.

Example 6.10. *Brick infill wall capacity*
Calculate the ultimate horizontal capacity of infill wall in Example 6.8 if the wall is constructed in 215 mm standard format brick of unit strength $20\,N/mm^2$ in grade (ii) mortar. Use $E_i = 450f_k$ (N/mm^2 units).

Solution

$$f_k = 6.4\,N/mm^2 \text{ and } E_i = 450 \times 6.4 = 2880\,N/mm^2 \text{ (from Table 6.2)}$$

$$\lambda = \sqrt[4]{\frac{2880 \times 215 \times 0.882}{4 \times 28\,000 \times 675 \times 10^6 \times 3000}} = 0.001\,246\,mm^{-1} \quad (using\ Eq.\ 6.29)$$

$$\alpha = \pi/2\lambda = 1261\,mm$$

$$\lambda h = 0.001\,246 \times 3500 = 4.36$$

$$l'/h' = 5000/3000 = 1.67$$

Figure 6.35, $k = R_{vc}/(f_k/\gamma_m)h't = 0.44$ for the diagonal cracking failure, and 0.42 for the compressive crushing failure

$$R_{vc} = 0.42 \times 6.4 \times 3000 \times 215 \times 10^{-3}/3.5 = 495.4 \, \text{kN} \qquad \text{(using Eq. 6.38)}$$

Figure 6.36, using $\mu = 0.6$, $k = (R_v/f_v/\gamma_{mv})\,h't = 2.25$

$$R_{vs} = 2.25 \times 0.35 \times 3000 \times 215 \times 10^{-3}/2.5 = 203.2 \, \text{kN} \qquad \text{(using Eq. 6.39)}$$

Lowest value $= 203.2 \, \text{kN}$

$$H_v = 203.2 \times 0.857 = \underline{174.1 \, \text{kN}}$$

6.6 Cantilever walls

Cantilever walls are used in precast skeletal frames up to 15–20 storeys in height. They are tied together vertically using site placed continuity reinforcement (Figure 6.20) or other mechanical connections such as welded plates or bars. The walls act alone and do not rely on composite action with columns. There are no beams between them, and this leads to complications in narrow walls where floor slabs require a bearing as shown in Figure 6.37. Cantilever walls are usually manufactured in storey height units, but if this is prohibitive due to handling and transportation restrictions the wall may be joined, or 'spliced' at mid-storey height.

The walls contain hollow voids, orientated vertically and rectangular in section which receive the site fixed reinforcement and cast in situ concrete. Both the precast wall and cast in situ infill should be grade C40 minimum. Hollow core cantilever walls have lost favour in recent years to the precast infill wall because of the large amount of wet concrete required and the (relatively) labour intensive steel fixing. Site placed reinforcement includes starter bars cast in the foundation (the length of bar projecting is $1.0 \times$ tension anchorage length) to which additional bars are lapped to provide the holding down (tension) force.

The wall panels are designed as reinforced (not plain) concrete walls in the conventional r.c. manner according to Clause 3.9.3 in BS8110, Part 1. However, this code only offers equations for the design of walls subjected to axial load, suggesting that where bending moments exist the wall is designed using 'statics alone' (=by elastic analysis). Where horizontal loads are carried by several walls the proportion allocated to each wall should be in proportion to its relative stiffness in the uncracked condition, i.e. using Eq. 6.26. In the following paragraph, column analogy will be used to design walls for combined axial force and in-plane moments.

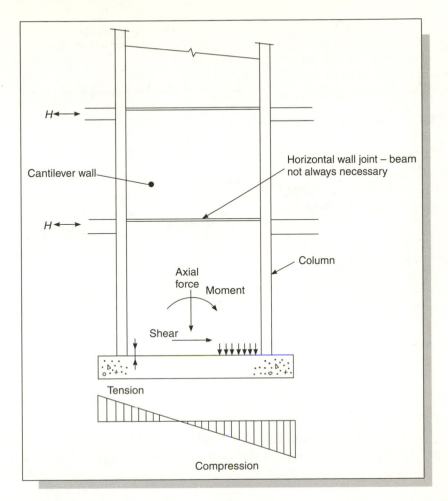

Figure 6.37: Analysis of precast cantilever walls.

Each floor-to-floor storey height should be considered for slenderness. The effective height is assessed as if the wall is 'plain' according to Clause 3.9.4. This is because the slab-wall connections transmitting axial forces to the wall are simply supported. Cantilever walls are nearly always braced in the other direction – it is hard to imagine a case where cantilever walls would be used to brace a structure in one plane only. Therefore, the effective height of the wall l_e is equal to the distance between centres of lateral supports, i.e. the actual storey height, not the clear height. If the wall happens to be unbraced in the other direction and supports slabs perpendicular to the wall, then $l_e = 1.5l_o$, where l_o is the clear height between lateral supports. Otherwise $l_e = 2.0l_o$. BS8110 offers no value for the limiting slenderness of stocky walls, but a value of 12 is taken. The limiting slenderness ratio $l_e/t < 40$, but this is rarely critical.

Providing the wall is loaded uniformly with slabs on either side of the wall whose spans do not differ by more than 15 per cent, the axial capacity of short walls is given as:

$$n_w = 0.35f_{cu}A_c + 0.7A_{sc}f_y \qquad 6.40$$

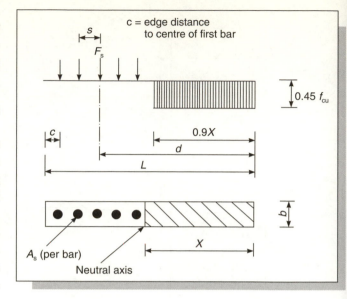

If the conditions are not satisfied, the section should be designed for axial forces and transverse bending moments using a concrete stress of $0.45f_{cu}$.

The design for axial forces and in-plane moments is as follows. In Figure 6.38 (ignore the compressive forces in the bars in the compression zone), the tensile force in the reinforcing bars F_s and the compressive force in the concrete F_c are given by:

Figure 6.38: Design principles for cantilever walls.

$$F_s = 0.95nA_sf_y \qquad 6.41$$

$$F_c = 0.45f_{cu}b\,0.9X \qquad 6.42$$

where n = number of bars in tension zone, i.e. in the zone $2(d - X)$ at a spacing s, where d is the distance to the centroid of the steel, and X is the depth to the NA. The ultimate axial load capacity is given by:

$$N = F_c - F_s = 0.405f_{cu}b\,X - \frac{1.9(d - X)A_sf_y}{s} \qquad 6.43$$

The ultimate in-plane moment of resistance is given by the least of:

$$M_R = 0.405f_{cu}b\,X(d - 0.45X) - N(d - 0.5L) \qquad 6.44$$

$$M_R = \frac{1.9(d - X)A_sf_y(d - 0.45X)}{s} + N(0.5L - 0.45X) \qquad 6.45$$

depending on whether the steel or concrete is critical.

From these equations X and d may be determined for given geometry, materials and loads. The minimum length of wall is given by:

$$L = d + (d - X) + \text{cover to centre of first bar (usually } 75–150\,\text{mm)} \qquad 6.46$$

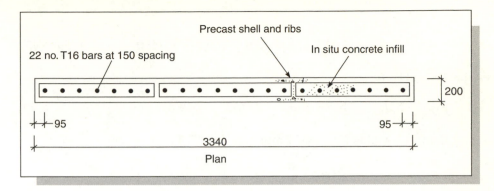

Figure 6.39: Detail to Example 6.11.

If this length is greater than the available length either the value of b, f_{cu} or A_s/s must be increased until it is less than the available length. The usual option is to decrease the bar spacing s so that the same size of bar may be used in all walls.

Example 6.11.   *Cantilever wall capacity*
Calculate the ultimate moment of resistance of the cantilever hollow core wall shown in plan cross-section in Figure 6.39 for the following ultimate axial forces:
(a) $N = 0$; (b) $N = 5000\,\text{kN}$; and (c) $N = 0.45\,f_{cu}\,b\,L$, i.e. the 'squash' load.
    The reinforcing bars are T16 at 150 mm spacing, and the cover to the centre of the bar nearest to the end of the wall is 95 mm. Use $f_{cu} = 40\,\text{N/mm}^2$ and $f_y = 460\,\text{N/mm}^2$.

Solution
(a) $N = 0$

$$0 = 0.405 \times 40 \times 200\,X - 1.9 \times 201 \times 460(d - X)/150 \qquad (\textit{using Eq. 6.43})$$

Then $3240\,X = 1171\,(d - X)$

$$\text{therefore,}\quad d = 3.767\,X \tag{1}$$

$$d + (d - X) + 95 = 3340 \qquad (\textit{using Eq. 6.46})$$

$$2d - X = 3245 \tag{2}$$

Solving (1) and (2) yields $X = 496.7\,\text{mm}$ and $d = 1871\,\text{mm}$

$$M_R = 0.405 \times 40 \times 200 \times 496.7 \times (1871 - 0.45 \times 496.7) \times 10^{-6} = \underline{2651\,\text{kNm}}$$
$$\textit{(using Eq. 6.44)}$$
$$M_R = 1.9 \times 1374.3 \times 201 \times 460 \times (1871 - 0.45 \times 496.7) \times 10^{-6}/150 = \underline{2651\,\text{kNm}}$$
$$\text{(of course)} \qquad (\textit{using Eq. 6.45})$$

(b) $N = 5000\,\text{kN}$

$$5000 \times 10^3 = 0.405 \times 40 \times 200\,X - 1.9 \times 201 \times 460(d - X)/150 \qquad \text{(using Eq. 6.43)}$$

Then $3240\,X = 1171(d - X) + 5000 \times 10^3$

$$\text{Therefore,}\quad d = 3.767X - 4269.8 \tag{3}$$

Solving (3) and (2) yields $X = 1803.7\,\text{mm}$ and $d = 2524\,\text{mm}$

$$M_R = [[0.405 \times 40 \times 200 \times 1803.7 \times (2524 - 0.45 \times 1803.7)]$$
$$- [5000 \times 10^3 \times (2524 - 3340/2)]] \times 10^{-6} = \underline{5737\,\text{kNm}} \qquad \text{(using Eq. 6.44)}$$

(c) $N = 0.45 \times 40 \times 200 \times 3340 \times 10^{-3} = 12024\,\text{kN}$ (*contd. on* p. 203)

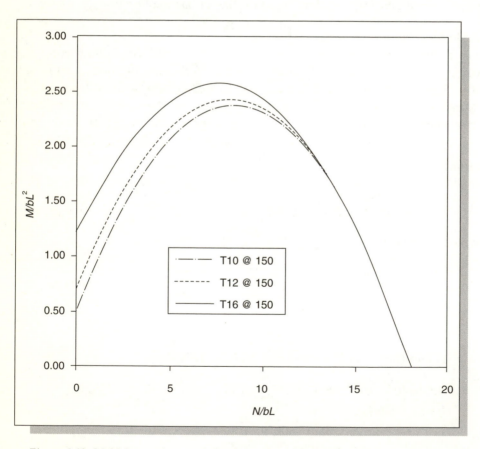

Figure 6.40: M–N interaction graph for the wall used in Example 6.11.

The steel cannot contribute in tension and the line of action of the axial force is coincident with the centre of concrete pressure. Therefore, the moment of resistance is (theoretically) zero.

For completeness, the following table gives the *M–N* data for the wall used in Example 6.11 for three different values of bar diameter. Figure 6.40 presents the same information graphically. Note the diminishing contribution of using additional reinforcement when $N/bL > 12$.

N (kN)	*T10 @ 150 M* (kNm)	*T12 @ 150 M* (kNm)	*T16 @ 150 M* (kNm)
0	1132	1588	2651
1000	2477	2858	3747
2000	3561	3874	4603
3000	4384	4635	5221
4000	4945	5141	5598
5000	5245	5393	5737
6000	5284	5390	5637
7000	5061	5132	5297
8000	4577	4619	4718
9000	3831	3852	3899
10000	2824	2830	2842
11000	1556	1553	1545
12000	26	21	9
12024	0	0	0

Example 6.12. *Cantilever wall reinforcement*
Calculate the reinforcement requirements if the wall used in Example 6.11 is subjected to $N = 3000\,\text{kN}$ and $M = 4600\,\text{kNm}$.

Solution
Because both A_s and s are unknown variables, compute a value of A_s/s and choose appropriate reinforcement.
Let $\rho = A_\text{s}/s$

$$3000 \times 10^3 = 3240X - 874\rho\,(d - X) \qquad (\textit{using Eq. 6.43}) \tag{1}$$

$$4600 \times 10^6 = 874\rho\,(d - X)(d - 0.45X) + 3000 \times 10^3(1670 - 0.45X)$$
$$(\textit{using Eq. 6.45}) \tag{2}$$

$$X = 2d - 3245 \qquad (\textit{using Eq. 6.46}) \tag{3}$$

Substituting (3) into (1) yields,

$$\rho = \frac{13\,514 \times 10^3 - 6480d}{874d - 2836 \times 10^3} \tag{4}$$

Substituting (3) into (2) yields,

$$\rho = \frac{2.7 \times 10^6 d - 4790 \times 10^6}{-87.6d^2 - 0.99 \times 10^6 d + 4140 \times 10^6} \tag{5}$$

Setting (4) = (5) and solving the resulting cubic equation in d yields:
$d = 2188 \, \text{mm}$

$$X = 4376 - 3245 = 1131 \, \text{mm} \qquad (using \ Eq. \ 6.44)$$

$$\rho = 0.72 \, \text{mm}^2/\text{mm} \qquad (using \ Eq. \ (4))$$

Use T12 at 157 mm spacing, rounded to 150 mm, or T16 at 280 mm spacing.

References

1   Cook, N., *The Designer's Guide to Wind Loading of Building Structures*, Butterworth, 1985.
2   Stafford-Smith, B. and Carter, C., A Method of Analysis for Infill Frames, Paper 7218, *Proc. Instn. Civ. Engrs.*, September 1969.
3   Mainstone, R. J., On The Stiffnesses and Strengths of Infill Frames, Building Research Establishment Paper CP2/72, February 1972.

7 Horizontal floor diaphragms

7.1 Introduction to floor diaphragms

The stability of precast concrete buildings is provided in two ways. First the horizontal loads due to wind are transmitted to shear walls or moment resisting frames by the floor (and/or the roof) acting as a horizontal deep beam. Any type of floor construction may be designed and constructed to function in this way, but particular problems arise in precast concrete floors which comprise individual units, such as hollow cored or double-tee floors, because of the localized manner in which they are connected together, as shown in Figure 7.1. If the floor is a solid construction, such as composite plank, these localized areas do not exist and the horizontal forces are spread right across the floor area.

Second, the horizontal reaction forces resulting from the floor at each floor level are transmitted to the foundation via columns or bracing elements (see Chapter 6). Where the distance between the bracing elements is large, say more than 6–10 m, the floor has to be designed as a plate, or so called 'diaphragm', which must sustain shear forces and bending moments. A 'ring beam', or series of 'ring beams', as shown in Figure 7.2, is formed around the precast floor units using small quantities of cast in situ concrete to

Figure 7.1: Precast concrete hollow core floor diaphragm in multi-storey construction (courtesy Reinforced Concrete Council, UK).

effectively clamp the slabs together to ensure the diaphragm action.

The way in which the diaphragm behaves depends on the plan geometry of the floor, shown in Figure 7.3a. The diaphragm may behave either as a Virendeel girder, Figure 7.3b, or more usually as a deep horizontal beam having a compression arch and tensile chord as shown in Figure 7.3c. It is assumed that no out-of-plane deformations are present; in this case these would be vertical deflections, and therefore seismic behaviour is excluded from the analysis presented here. As well as wind loading, the floor diaphragm may also be subjected to additional horizontal forces, such as:

1   horizontal forces due to lack of verticality of the structure which may be manifest as small restoring forces between the stabilizing walls. This is equal to 1.5 per cent G_k at each floor level;

2   temperature and shrinkage effects; and

3   in-plane, or catenary, forces as a consequence of abnormal loading, accidental damage, etc. (see Chapter 10).

Examples of bracing positions are possible, as shown in Figure 7.4a. The diaphragm bending moments and shear forces distributions calculated for the cases cited in Figure 7.4a are as shown in Figure 7.4b

Figure 7.2: Precast floor diaphragm without structural topping (courtesy Trent Concrete Ltd).

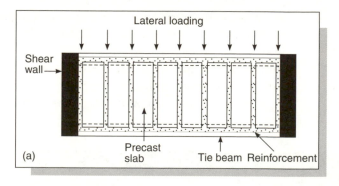

(a)

Figure 7.3: Structural models used to analyse floor diaphragms: (a) Diagonal truss; (b) Virendeel girder.

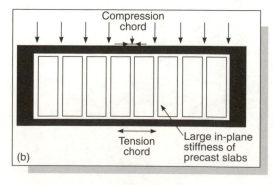

(b)

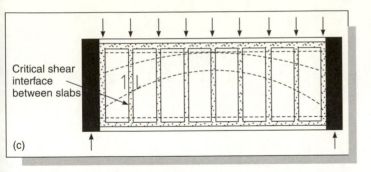

Critical shear interface between slabs

for uniform horizontal pressure. The diaphragm must be checked in both orthogonal directions because the complementary shear stresses perpendicular to the direction of loading may be equally as important as those in the direction of the load.

The reactions in the bracing elements, hereafter referred to as shear walls, are calculated

Figure 7.3 (continued): Structural models used to analyse floor diaphragms: (c) Deep horizontal beam.

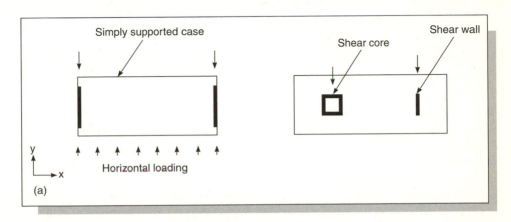

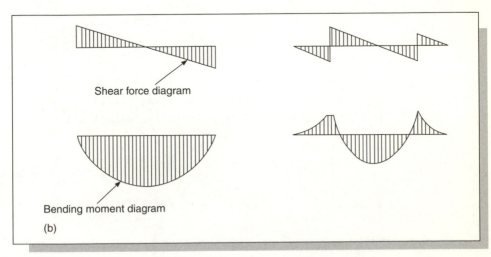

Figure 7.4: (a) Plan geometry of floor diaphragms; (b) Floor diaphragm shear force and bending moments diagrams.

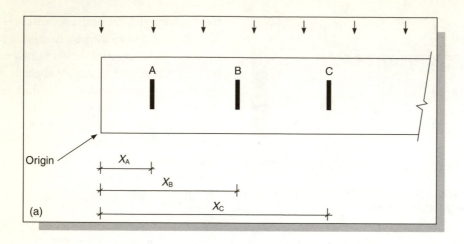

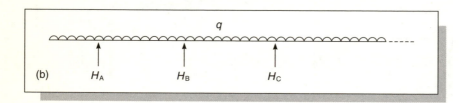

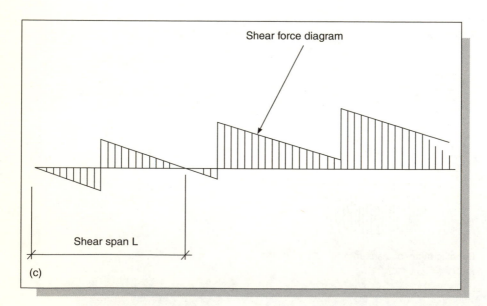

Figure 7.5: Distribution of reactions, bending moments and shear forces in a rigid floor diaphragm.

according to Eq. 6.26 and are given as $H_1 \ldots H_n$ in Figure 7.5. If the applied horizontal load acting at a floor level is q (kN/m run) the diaphragm shear forces are:
At wall A:

$$V_{AO} = qx_A; \quad V_{AB} = H_A - qx_A$$

At wall B:

$$V_{BA} = qx_B - H_A; \quad V_{BC} = H_A + H_B - qx_B \qquad 7.1$$

and so on.

The maximum diaphragm bending moment occurs where the shear force is zero (viz. $\partial M / \partial x = -V$). Then at wall A, the moment towards the free end of the diaphragm is:

$$M = qx_A^2 / 2 \qquad 7.2a$$

Providing that the shear force is zero at some point between walls A and B the distance L to the point of zero shear is:

$$x_A < L = \frac{H_A}{q} < x_B \qquad 7.3a$$

The bending moment at this point is:

$$M = qL^2 / 2 - H_A(L - x_A) \qquad 7.2b$$

If Eq. 7.3a is not satisfied, the point of zero shear is searched for between walls B and C such that:

$$x_B < L = \frac{H_A + H_B}{q} < x_c \qquad 7.3b$$

and the corresponding moment is:

$$M = qL^2 / 2 - H_A(L - x_A) - H_B(L - x_B) \qquad 7.2c$$

...and so on to the end of the diaphragm. There may be several points of zero shear and hence all cases must be examined. However experience will soon show that maximum moments exist between those walls that are furthest apart.

Shear forces not only exist in the longitudinal y direction between the floor units, but also in the transverse x direction in the end joints at beams. Except in the case of a single floor bay where the shear V_x is zero (and the maximum shear V_x occurs in the body of the precast units), the transverse shear may be greater than the longitudinal shear. In multi-bay floors, where the slabs are spanning parallel to the direction of the applied load, as in Figure 7.6a, the shear V_x at interior support between span L_1 and L_2 is given as:

$$V_x = \frac{V_y S}{I} = \frac{6V_x(B-L_1)L_1}{B^3} \qquad 7.4$$

for $L_1 > L_3$, where V_x is the applied design shear force, and S and I are the first and second moments of area at the diaphragm at the joint concerned.

Where the slabs span perpendicular to the direction of the applied load (Figure 7.6b), the maximum transverse shear between the slabs V_y occurs at the NA of the entire floor diaphragm is given as:

$$V_x = \frac{V_y S}{I} = \frac{1.5V_y}{L} \qquad 7.5$$

If the floor diaphragm is subjected to horizontal bending, as shown in Figure 7.7, the internal equilibrium is maintained by tension and compression chords. The forces in the chords may be mobilized either by adding in situ tie steel in the joints between the precast units, or may be provided as part of the supporting floor beams, e.g. steel UB, precast beam, RC beam, but ONLY if a mechanical connection is made between the slabs and beams. The tie force resisting bending moments is given as:

$$T_b = \frac{M_h}{z} \qquad 7.6$$

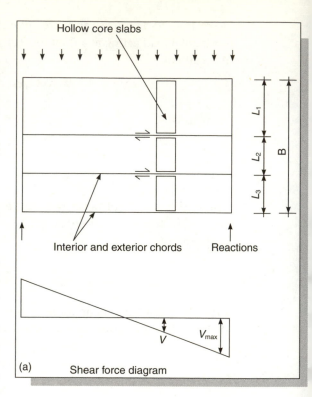

(a)   Shear force diagram

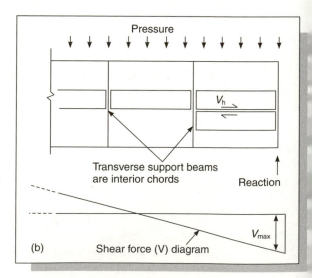

(b)   Shear force (V) diagram

Figure 7.6: Shear stresses in floor diaphragms: (a) Floor span parallel to load; and (b) Floor span perpendicular to load.

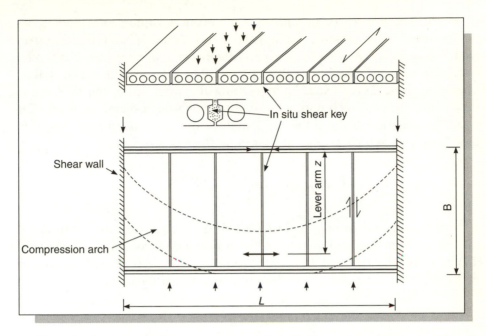

Figure 7.7: Forces acting in a precast concrete floor diaphragm (according to deep beam theory).

where M_h is the applied diaphragm moment obtained from Eq. 7.2, and z is the lever arm. The value of z depends on the aspect ratio for the floor, and on the magnitude of the bending moment. Maximum values for z/B at the points of maximum bending are as follows:[1]

B/L	z/B
<0.5	0.9
0.5 <1.0	0.8

Walraven[2] proposes a constant value for $z/B = 0.8$, which is used in this chapter. Where the aspect ratio $B/L > 1$ the behaviour will be closer to the strut and tied arch model than either the deep beam or truss models. The tie force is given by:

$$T_b = \frac{0.5V}{B/L} \qquad\qquad 7.7$$

where the diaphragm is subjected to shear forces the tie force is generated by the action of shear wedging and shear friction trying to pry the floor units apart. The relationship between shear force V and tie force due to shear T_q is given as:

$$T_q = \frac{V}{\mu'} \qquad\qquad 7.8$$

where μ' is an effective coefficient of shear friction and wedging combined. μ' has been determined experimentally[3,4] and by calculation models[5] and found to vary between 5 and 26 for various hollow core diaphragm configurations. It would seem sensible to take a minimum value of $\mu' = 5.0$ in any subsequent calculation.

If there are a number of floor bays resisting the shear force, the tie force is shared equally between each number of end joints between them, i.e. $n + 1$. The lever arm is taken as 0.8 times the full length of the floor diaphragm. Then, in combined bending and shear, the maximum tie force is given as:

$$T = T_b + T_q = \frac{V}{(n+1)\mu'} + \frac{M_h}{z} \qquad 7.9$$

where n = number of floor bays resisting the shear force. Where two walls are at the ends of the floor the maximum force T occurs at a distance X from the nearest stabilizing wall. The moment $M = qX(L - X)/2$ and $V = q(L/2 - X)$. It can be shown by differentiation of Eq. 7.9 with respect to X that the distance X is:

$$X = \frac{L}{2} - \frac{z}{(n+1)\mu'} \qquad 7.10$$

where L is the distance between the stabilizing walls.

For multi-wall systems where the reaction force in the end wall will not be $qL/2$, the distance to $X < x_B$ from wall A is:

$$X = \frac{H_A}{q} - \frac{z}{(n+1)\mu'} \qquad 7.11a$$

and in between walls A and B where $x_B < X < x_c$

$$X = \frac{H_A + H_B}{q} - \frac{z}{(n+1)\mu'} \qquad 7.11b$$

Example 7.1
Calculate the maximum diaphragm reinforcement required in the floor layout shown in Figure 7.8a. The floor diaphragm is subject to an ultimate horizontal uniformly distributed load of 4 kN/m run. The thickness and material specification for all walls are identical. The floor slab is of hollow cored units of 200 mm depth. Use $f_y = 460\,\text{N/mm}^2$.

Solution
Calculate the horizontal reactions in the walls. If t and E_c are same, then $I = L^3$. Take origin $x = 0$ at left-hand end. Referring to Eq. 6.26 then:

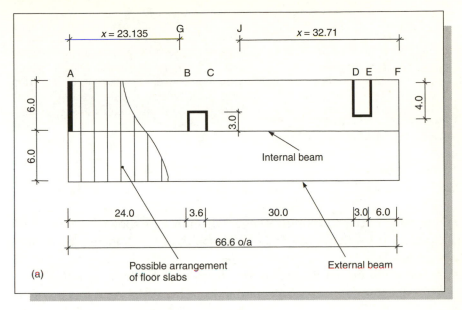

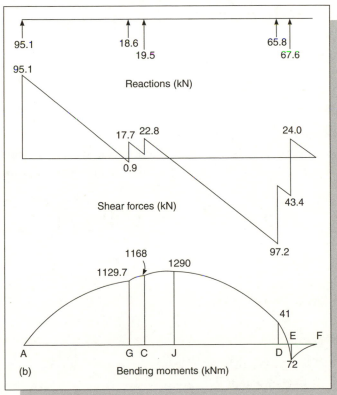

Figure 7.8: Detail to Example 7.1, (Figure 7.8a, dimensions in m).

Wall	I	x	Ix	a	eIa	Ia²	H_i/H	H_i/H
A	216	0	0	22.5	52488	109350	0.543−0.186	0.357
B	27	24.0	648	1.5	437	61	0.068 + 0.002	0.070
C	27	27.6	745	5.1	487	702	0.068 + 0.005	0.073
D	64	57.6	3686	35.1	24261	78848	0.161 + 0.086	0.247
E	64	60.6	3878	38.1	26334	92903	0.161 + 0.093	0.254
Totals	398	–	8958	–	–	281865	–	1.001

$H = 4.0 \times 66.6 = 266.4\,\text{kN}$

$H_A = 0.357 \times 266.4 = 95.1\,\text{kN},\ H_B = 18.6\,\text{kN},\ H_C = 19.5\,\text{kN},\ H_D = 65.8\,\text{kN},$
$H_E = 67.6\,\text{kN}$

The resulting shear force and bending moment diagrams are shown in Figure 7.8b. The lever arm $z = 0.8 \times 12.0\,\text{m} = 9.6\,\text{m}$. No. of slab bays $n = 2$, therefore, number of tie bars groups $(n + 1) = 3$.

Between A and B, maximum tie force is at X from Wall A where:

$X = 95.1/4.0 - 9.6/3 \times 5.0 = 23.135\,\text{m}$ from A (from Eq. 7.11a). Call this point G.

$M_G = 4 \times 23.14^2/2 - 95.1 \times 23.14 = 1129.7\,\text{kNm}$ (note this is not M_{max}) and

$V_G = 2.6\,\text{kN}$ (*using Eq. 7.2b*)

$$T_G = \frac{2.6}{3 \times 5.0} + \frac{1129.7}{9.6} = 117.8\,\text{kN}$$ (*using Eq. 7.9*)

At point C, $M_C = 1168\,\text{kNm}$ and $V_C = 22.8\,\text{kN}$, then $T_C = 123.2\,\text{kN}$

Between C and D, the maximum tie force is at X from F where Equation 7.11b (but with the notation from F) $X = 67.6 + 65.8/4.0 - 9.6/3 \times 5.0 = 32.71\,\text{m}$ from F. Call this point J.

$M_J = 4 \times 32.71^2/2 - 68.5 \times 23.71 - 67.6 \times 26.71 = 1290\,\text{kNm}$ and

$V_J = 2.56\,\text{kN}$ (*using Eq. 7.2c*)

$T_J = 134.5\,\text{kN}$ (*using Eq. 7.9*)

Then $T_{\text{max}} = 134.5\,\text{kN}$

$$A_{\text{hd}} = \frac{134.5 \times 10^3}{0.95 \times 460} = 308\,\text{mm}^2$$

Use 2 no. T16 bars (402).

7.2 Shear transfer mechanism

Various structural models can be applied to model the shear transfer mechanism between the diaphragm and the bracing elements. The behaviour of a precast hollow core floor diaphragm is different to that of a solid slab, because the precast unit has large in-plane stiffness relative to that of the joints. Small shrinkage cracks appear at the interface between the precast and in situ concretes, and the width of this crack influences the effective shear area in the joint. An initial crack width of 0.1 to 0.2 mm may be adopted for use in the calculations if the width of the hollow core unit is no more than 1.2 m (see Table 7.1).

Shear resistance R (see Figure 7.9) is a combination of:

1 Aggregate interlock in cracked concrete, by so-called 'wedging action' and 'shear friction'. (Figure 7.9a). The elasticity of the tie steel enables a normal stress σ_n to 'clamp' the units together.

2 Dowel action through kinking and shear capacity of tie bars placed in the chords. (Figure 7.9b).

The precast units should be prevented from moving further apart by the placement of two types of steel bars shown in Figure 7.10a:

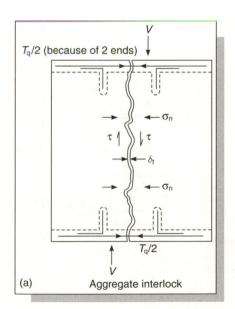
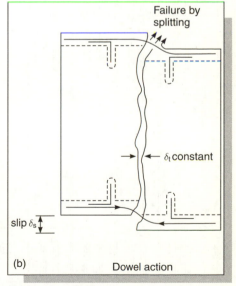

Figure 7.9: Definitions of shear transfer mechanism: (a) Aggregate interlock mechanism; and (b) Dowel action mechanism.

1 Tie bars in the tension chord, which should cross the longitudinal joint between the slabs and be fully anchored, either by bond or by mechanical means, on either side of the joint.

2 Shear friction bars, often called 'coupling bars', which connect the precast units to transverse tie beams and resist the force V_y in the Eqs. 7.4 and 7.5.

The ties generate the clamping forces which ensure shear friction in the joints. Although HT deformed bar ($f_y = 460\,\mathrm{N/mm^2}$) is used mainly for the tie bars in the tension chord, it is becoming increasingly popular to reinforce the in situ perimeter strip using 7-wire helical prestressing strand ($f_{pu} = 1750\,\mathrm{N/mm^2}$).

Shear strength and stiffness is provided by aggregate interlock, and the structural integrity by dowel action of the reinforcing bars crossing the cracked interface. Aggregate interlock may be separated into two distinct phases: (1) 'shear wedging' where the inclined surfaces on either side of the crack are in contact; and (2) 'shear friction' where the contact surfaces are being held in contact by the normal stress generated by the transverse tie force T_q.

Shear wedging relies on the adhesion and bond at the precast-in situ concrete interface and is exhausted when the width of the interface cracks is sufficient to

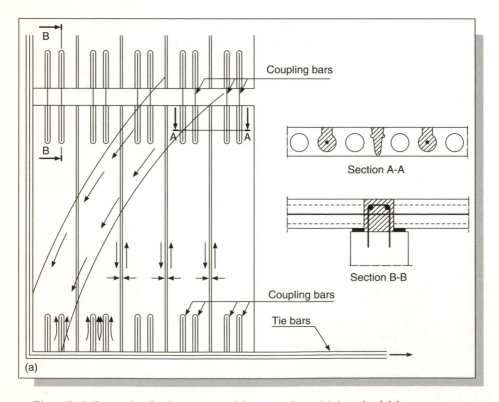

Figure 7.10: Connection details to structural frame members: (a) At ends of slab.

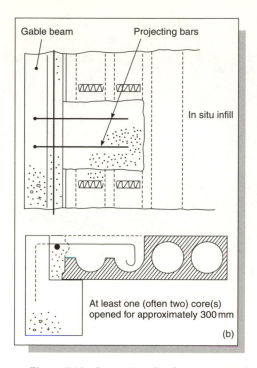

Figure 7.10: Connection details to structural frame members: (b) At edges of slab.

cause an increase in the tie force T_q. It is influenced by the surface roughness of the slabs and shrinkage of the in situ infill concrete in the joints. Shear friction also provides a high shear resistance. It is present when the crack width and the tie force T_q are increasing. It is also influenced by the surface roughness of the slabs, but more by the amplitude of the crevices than by the profile. Shear friction is exhausted when the crack width exceeds a certain value; experimental tests show that this limit corresponds to about 2–3 mm, and is roughly equal to the amplitude of the surface crevices.[3] Figure 7.9b now applies.

Dowel action provides lower strength than the above but greater deformation capacity and ductility. It is influenced by the ability of the tie steel in the chords to resist shear forces by bending and kinking, and is dependent on the manner in which the tie steel is anchored to the precast slab and tied into the floor diaphragm. The edge profile of the precast slab has no influence on dowel action.

7.3 Edge profile and tie steel details

The most important feature of the floor design for ensuring horizontal diaphragm action is the edge profile, shown in Figure 7.11. The edge is not made deliberately rough, but the drag of the casting machine on the semi-dry mix creates a surface roughness vital to diaphragm action. These units have edge profiles which permit the placement of grade C25–C30 in situ concrete (or mortar) in the longitudinal joints between adjacent units. The joints may be considered 'plain' (i.e. uncastellated) and 'unreinforced'.

Tie steel is usually concentrated as one or two bars, placed at the mid-height of the slab (see Figure 7.10). L-shape coupling bars should be tied to the continuous reinforcement, otherwise U-shape or straight bars should pass over the top of the tie bars. One leg of the coupling bar should be concreted into milled slots in the slabs, at positions which coincide with the second (or third) hollow core from the edges of the unit. A full bond length of at least 30 × diameter of bar should be specified, and hooked ends used if the length of slot is excessively large, say more than 600 mm long. Details at longitudinal edges are shown in Figure 7.10b where cut outs need to be made in the edges of the precast units at centres of 1.0–2.0 m, depending on the overall frame stability tie force requirements.

Figure 7.11: Edge profile of slip-formed hollow core slab.

7.4   Design of floor diaphragm

7.4.1   *Limiting stresses*

The calculation of the shear resistance involves the separate actions of:

1   Shear wedging and shear friction; and

2   Dowel action.

The combined mechanism of aggregate interlock and dowel action may be used to predict the response of cracked concrete subjected to a shearing force. However, when calculating the areas of tie steel A_s and A_{sv}, the two effects are not additive if the design shear stress, measured over the full length of the longitudinal joint, is *greater* than the limiting value τ_u given in National Design Codes, e.g. in BS8110 the limiting ultimate shear stress is 0.23 N/mm². Only then is the shear capacity calculated according to the dowel action resistance alone. This is given in Section 7.4.2. If it is lower than the limiting value, the parameters are determined according to Section 7.4.1.

In computing the shear force resistance in the longitidinal joints, the effective depth of the diaphragm can only be taken to the depth of the in situ-precast interface, i.e. $D - 30$ mm, in most types of hollow core slabs. This is because it is found that grout loss occurs in the bottom of the joint and that the lower 10–15 mm remains not grouted. Secondly, the lip at the bottom of the units, typically 10–15 mm deep, prevents full penetration (see Figure 3.25). The value also recognises that differential camber will be present, further reducing the net contact depth. Then, in the longitudinal joint (i.e. y direction):

$$\tau_y = \frac{V_y}{B(D - 30)} \text{(in N, mm units only)} < \tau_u \qquad 7.12$$

If Eq. 7.12 is violated, transverse bars must be provided across the ends of the units according to Eq. 7.18. The bars must be continuous and properly anchored at their ends. In the transverse x direction:

$$\tau_x = \frac{V_x}{1.0D} < 0.23 \, \text{N/mm}^2 \qquad 7.13$$

If Eq. 7.13 is violated then coupling bars are placed across the ends of the hollow cored units. The area of steel A_{hd} is as given by Eq. 7.18. They are placed into the opened cores or into the longitudinal joints as shown in Figure 7.10a. Coupling bars are L-shape at edge beams and often straight (if the opened cores align) at internal beams.

In many cases the point of maximum shear will coincide with minimum bending and therefore the full length of the slab B may be used in computing the average value for τ. However, where the maximum moment and shears coincide the breadth of the diaphragm is reduced to z to allow for the decay in shear stress in the compression zone. Then Eq. 7.12 is modified to:

$$\tau_y = \frac{V_y}{z(D - 30)} \text{(in N, mm units only)} < \tau_u = 0.23 \, \text{N/mm}^2 \qquad 7.14$$

Equation 7.13 is not affected by the effects of bending.

The tie force T in the bars is given by Eqs. 7.7–7.9, depending on geometry and whether bending moments exist. To ensure that the force is fully effective, the elastic elongation of the tie bar l_s is calculated as follows:

$$l_s = \frac{TL_s}{A_s E_s} + \delta_{ti} < \delta_{t,\max} \qquad 7.15$$

The maximum transverse crack width $\delta_{t,\max}$ just *before* the commencement of interface shear failure has been found by full scale experimentation to be in the region of 1–2 mm. However, the test results show that a certain amount of non-linear behaviour takes place, and so $\delta_{t,\max}$ is limited to 0.5 mm. The area of tie bars

A_s should be increased if this does not comply, which may often be the case where large bending moments exist.

L_s is the least of:

$$L_s = 30d \frac{A_s \text{ calculated}}{A_s \text{ provided}} \text{ for high tensile ribbed bar, or} \qquad 7.16a$$

$$L_s = 0.8W, \qquad 7.16b$$

where W = the width of the hollow core unit (typically 1.2 m) and d is the bar diameter. Equation 7.16a is an anchorage requirement found by experiments to be approximately $8d$ to $15d$ on either side of the crack, making a total of $30d$ at worst. Equation 7.16b is a conservative value that allows for interaction of tie forces between adjacent longitudinal joints.

The initial crack width δ_{ti} has been found by experimental measurements taken across the longitudinal joint between two hollow cored slabs positioned side by side. δ_{ti} depends on the time after casting of the precast unit to when the longitudinal joint is filled, and is given in Table 7.1.

Example 7.2
Calculate the horizontal shear stresses in the longitudinal and transverse joints in Example 7.1. Check the limiting values according to BS8110.

Solution
Maximum shear force (at point D) $V_{Dy} = 97.2$ kN (Figure 7.8b). It is in the presence of bending $M_D = 41$ kNm, therefore Eq. 7.12 does not apply. Longitudinal:

Table 7.1: Initial crack widths between precast units

Age of precast unit when joint is filled (days)	Width of precast unit (mm)	Width of longitudinal joint (mm)	Initial crack width δ_{ti} (mm)
<7	1200	25	0.215
	1200	50	0.230
	600	25	0.115
	600	50	0.130
28	1200	25	0.135
	1200	50	0.150
	600	25	0.075
	600	50	0.090
>90	1200	25	0.095
	1200	50	0.110
	600	25	0.055
	600	50	0.070

$$\tau_y = \frac{97.2 \times 10^3}{9600 \times (200 - 30)} = 0.06 \, \text{N/mm}^2 \qquad (\textit{using Eq. 7.14})$$

Transverse:

$$V_x = \frac{6 \times 97.2 \times (12.0 - 6.0) \times 6.0}{12.0^3} = 12.15 \, \text{kN/m run} \qquad (\textit{using Eq. 7.4})$$

$$\tau_x = \frac{12.15 \times 10^3}{1000 \times 200} = 0.06 \, \text{N/mm}^2 \qquad (\textit{using Eq. 7.13})$$

Maximum stress $<0.23 \, \text{N/mm}^2$ from BS8110, therefore no dowel action reinforcement required.

Example 7.3
Check that the maximum crack width is not violated anywhere in the floor diaphragm in Example 7.1. The age of the units when the 50 mm wide longitudinal joints were filled is 28 days. Use $E_s = 200 \, \text{kN/mm}^2$.

Solution
Maximum tie force (at point J) $= 134.5 \, \text{kN}$ is resisted using 2 no. T16 bars $= 402 \, \text{mm}^2$

$\quad L_s = $ lesser of $30 \times 16 \times 308/402 = 368 \, \text{mm}$ or $0.8 \times 1200 = 960 \, \text{mm}$

$\quad \delta_{ti}$ from Table 7.1 $= 0.15 \, \text{mm}$

$$l_s = \frac{134.5 \times 10^3 \times 368}{402 \times 200 \times 10^3} + 0.15 = 0.77 \, \text{mm} > 0.5 \, \text{mm} \qquad (\textit{using Eq. 7.15})$$

Increase bars to 3 T16 ($603 \, \text{mm}^2$), $L_s = 245 \, \text{mm}$ and $\delta_t = 0.27 + 0.15 = 0.42 \, \text{mm} < 0.5 \, \text{mm}$.

7.4.2 Reinforcement design

The maximum horizontal bending moment M_h is calculated from equilibrium of external wind pressures and the reactions from the shear walls obtained from Eq. 7.2. Referring to Figure 7.7 and given that the breadth of the diaphragm is B, the diaphragm reinforcement A_{hd1} to be positioned in the chord elements over the top of the beams at the ends of the floor slab is determined as:

$$A_{hd1} = \frac{M_h}{0.8B0.95f_y} \qquad\qquad 7.17$$

where the term 0.8B is the lever
arm between the compressive
zone and the diaphragm steel.
A_{hd1} should not exceed:

$$A_{hd1} = \frac{0.45f'_{cu}0.4dB}{0.95f_y} \qquad 7.18$$

where d is the depth of the precast floor slab $= D - 30$ mm,
and f'_{cu} is the strength of the
in situ grout (or concrete) in
the longitudinal gaps between
the floor units.

It is necessary to collect the
floor diaphragm tie steel A_{hd1}
in, or above, the chord elements as shown in Figure

Figure 7.12: Ties in floor diaphragm.

7.12. (Note the coupling bars into the milled slots in the hollow core slabs.) The
force must be continuous, but the means of achieving may utilize different items,
for example:

1 By utilizing the reinforcement already provided in the chord elements, such as
   edge beams, and by providing a positive non-slip tie between the beams. This tie
   may be continuous through the column connector; or

2 By placing tie steel additional to the steel provided in the chord members. (This
   steel may also be used as part of the stability tie steel determined in Section 10.4.)
   Some designers prefer to pass the reinforcement through small holes in the
   column. Alternatively if the beam is wider than the column, the bars are placed
   symmetrically on either side of the column.

Diaphragm reinforcement may be curtailed according to the usual rules governing
lap lengths and anchorage. The force due to dowel kinking action (note this is not
the same as the tie force T_q in Eq. 7.8, which is the force due to prising the slabs
apart) in each of the groups of bars is:

$$T = \frac{V}{\mu(n+1)} \qquad 7.19$$

where μ is a coefficient of friction, active in the longitudinal joints between the
slabs. Values for μ are given in codes of practice, e.g. BS8110, Part 1, Table 5.3 is

Table 7.2: Values for μ for concrete connections

Type of surface	μ
Smooth surface, as in untreated concrete. This is the case of the edges of hollow core units unreinforced except for continuous ties across their ends.	0.7
Roughened or castellated joint without continuous in situ strips across the ends of joints. 'Roughened' implies exposure of coarse aggregate without damaging the matrix. 'Castellated' joints must have a root depth of 10 mm × 40 mm long.	1.4
Roughened or castellated joint with continuous in situ strips across the ends of joints.	1.7

reproduced and annotated with respect to precast floor units in Table 7.2. (Note that the value of μ used in this calculation, to take account of the dowel action of the reinforcement A_{hd2} crossing the cracked plane, is considerably lower than $\mu' = 5$ used in Eqs 7.8 and 7.9 where shear friction and wedging are active.)
The steel area A_{hd2} due to dowel action is given as:

$$A_{hd2} = \frac{T}{f_s} = \frac{V}{\mu(n+1)0.6 \times 0.95 f_y} \qquad 7.20$$

(note that the 0.6 factor is the shear stress reduction factor $1/\sqrt{3}$).
The total steel $A_{hd} = A_{hd1} + A_{hd2}$. The minimum tie steel $A_{hd,min}$ (mm^2) is provided as a perimeter tie and given as:

$$A_{hd,min} = \frac{40 \times 10^3}{1.0 f_y} \qquad 7.21$$

Example 7.4
Repeat Example 7.2 using $V = 300$ kN and $D = 150$ mm

Solution
Longitudinal:

$$\tau_y = \frac{300 \times 10^3}{9600 \times (150 - 30)} = 0.26 \, \text{N/mm}^2 > 0.23 \, \text{N/mm}^2 \qquad (using \ Eq. \ 7.14)$$

$$A_{hd} = \frac{300 \times 10^3}{0.7 \times 3 \times 0.6 \times 0.95 \times 460} = 545 \, \text{mm}^2 \qquad (using \ Eq. \ 7.20)$$

$$A_{hd,min} = 40 \times 10^3 / 460 = 87 \, \text{mm}^2 \qquad (using \ Eq. \ 7.21)$$

Use 3 no. T16 bars (603).

Transverse:

$$V_x = \frac{6 \times 300 \times (12.0 - 6.0) \times 6.0}{12.0^3} = 37.5\,\text{kN/m run} \qquad (using\ Eq.\ 7.4)$$

$$\text{and } \tau_x = 0.25\,\text{N/mm}^2 > 0.23\,\text{N/mm}^2 \qquad (using\ Eq.\ 7.13)$$

Coupling bars are required. For 1.2 m wide unit $A_{\text{hdc}} = (1.2 \times 37.5 \times 10^3)/$ $(0.7 \times 0.6 \times 0.95 \times 460) = 245\,\text{mm}^2$

If the coupling bars are spaced at, say, 400 mm, there will be 3 no. per hollow core unit $= 82\,\text{mm}^2$ per bar.

Use 1 no. T12 coupling bar (113) at 400 mm spacing.

7.5 Shear stiffness

The shear stiffness of a single longitudinal joint is given as:

$$K_s = \frac{V}{\delta_s} \qquad\qquad 7.22$$

where δ_s is the longitudinal slip between two adjacent units, see Figure 7.9 (='slip'). As explained earlier, longitudinal slip is accompanied by a transverse displacement, or crack width δ_t. The relative displacements of δ_s and δ_t depend on many factors including the surface roughness of the edges of the precast units and the initial crack widths. Experimental results, shown in Figure 7.13, have found that

$$\delta_t = e^{(\beta\delta_s)} + \delta_{ti}$$

where β is the gradient of a log δ_t *vs* log δ_s graph and may be conservatively taken as 3.0.[3] Therefore:

$$\delta_s = \frac{\ln\frac{\delta_{t,\text{max}}}{\delta_{ti}}}{3.0} \qquad\qquad 7.23$$

Thus, if $\delta_{t,\text{max}}$ is limited to 0.5 mm prior to shear friction failure, and δ_{ti} is known from Table 7.1, then δ_s and K_{si} may be determined.

The shear stiffness of the entire diaphragm is determined by the addition of the shear displacements in Eq. 7.23 in each longitudinal joint as shown in Figure 7.14. The shear displacement in joint i is:

$$\delta_{si} = \frac{V_i}{K_{si}} \qquad\qquad 7.24$$

where V_i is the shear force for the joint in question, and K_{si} is the shear stiffness for the same joint. The shear displacement in the precast unit and in situ infill is negligible.

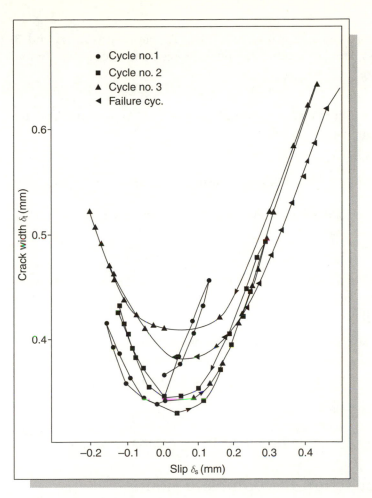

Figure 7.13: Actual relationship between longitudinal slip δ_s and transverse displacement δ_t.[3,4]

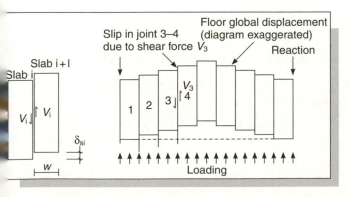

Figure 7.14: Shear displacements in the total floor diaphragm.

The total horizontal deflection of the floor diaphragm is therefore the sum of the individual displacements δ_{si}. The effect of this displacement can be taken into account by calculating an 'effective shear modulus' G' for the diaphragm as:

$$G' = \frac{\tau}{\gamma} = \frac{K_{si}W}{B(D-30)} \quad \text{in N, mm}$$

units only   7.25

Example 7.5

Calculate the shear displacements in Example 7.1 to 7.3 between walls A and B. Assume that 3 no. T16 bars are used as the tie reinforcement throughout. Ignore the bending deflections.

Solution

The maximum crack width is 0.42 mm (from Example 7.3). As $\delta_{ti} = 0.15$ mm the corresponding shear slip is:

$$\delta_s = \frac{\ln\dfrac{0.42}{0.15}}{3} = 0.343 \text{ mm} \quad (using \ Eq. \ 7.23)$$

The shear stiffness of the joint is based on a maximum shear force of 97.2 kN. Then

$$K_s = \frac{97.2}{0.343} = 283.4 \text{ kN/mm} \quad (using \ Eq. \ 7.24)$$

Between walls A and B there are 20 no. $\times$ 1.2 m wide units. The shear force in each joint i (counting the joint in contact with the shear wall as $i = 0$) is $V_i = 95.1 - 4.0 \times 1.2i$ (kN)

Therefore, $V_0 = 95.1$ kN, and $\delta_{s,0} = 95.1/283.4 = 0.335$ mm

Therefore, $V_1 = 95.1 - 4.8 = 90.3$ kN, and $\delta_{s,1} = 90.3/283.4 = 0.319$ mm

Therefore, $V_2 = 95.1 - 9.6 = 85.5$ kN, and $\delta_{s,2} = 85.5/283.4 = 0.302$ mm

and so on to $V_{20} = 95.1 - 96.0 = -0.9$ kN,

and $\delta_{s,0} = -0.9/283.4 = -0.003$ mm

Total shear deflection $\delta_s = 0.335 + 0.319 + 0.302 + \cdots + (-0.003) = \underline{3.5 \text{ mm}}$

7.6   Diaphragm action in composite floors with structural toppings

Certain types of precast floor units are not capable of carrying horizontal forces in the floor diaphragm, because:

1   They are too thin, in the case of double-tee units where the flange may be 50–75 mm;

2   The ends of the units cannot be tied to the supporting structure to transmit diaphragm forces to the vertical bracing elements; and

3   There is no horizontal shear transfer mechanism between individual floor units, as with beam-and-block floors.

If either of these conditions apply, the diaphragm forces must be transmitted by other means. The obvious choice is a structural topping (see Section 4.4 for full details of the topping). The resulting floor is designed on the basis that the precast

Figure 7.15: Laying a structural topping onto a precast double-tee floor.

flooring units provide restraint against lateral (in this case vertical) buckling in the relatively thin topping. In other words, the precast floor is acting as permanent shuttering. The shear is carried entirely by the reinforced in situ concrete topping (Figure 7.15). The minimum thickness of a structural topping is 40 mm, although the more common minimum thickness is 50–75 mm. Where the horizontal bending moment is zero the design ultimate shear stress is given as:

$$v = \frac{V}{Bh_s} \leq 0.45 \, \text{N/mm}^2 \qquad\qquad 7.26$$

for grade C25 concrete (BS8110, Part 1, Table 3.9) and the effective depth of the topping is measured at the crown (thinnest part) of the prestressed flooring unit. Where there is combined shear and bending, B is replaced by $0.8B$.

Although there is no published design method or guidance as to limiting stresses, etc. the reinforcement is designed according to bending theory using a rectangular compressive stress block of depth 0.4 times the breadth of the diaphragm. Referring to Figure 7.16, the ultimate compressive force in the topping of thickness h_s is given as:

$$F_c = 0.45 f_{cu} \, 0.4 B h_s \qquad\qquad 7.27$$

The reinforcement is uniformly distributed over the full area of the diaphragm (usually as welded fabric or mesh). All of the reinforcing bars which do not lie near to the compression zone, which may be taken as $0.6B$ from the compression

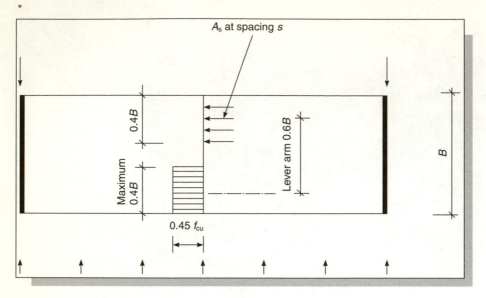

Figure 7.16: Design method for composite floor diaphragm.

surface, are fully stressed. If the spacing between the bars is s the ultimate force in the steel bars is:

$$F_s = \frac{0.95f_yA_s0.4B}{s}$$ 7.28

The lever arm $= 0.6B$.
Then

$$M_R = 0.108f_{cu}h_sB^2 \text{ (based on concrete)}$$ 7.29a

and

$$M_R = \frac{0.228f_yA_sB^2}{s} \text{ (based on reinforcement)}$$ 7.29b

Continuity of reinforcement in a structural topping is always extended to the shear walls or cores and it is safe to assume that the shear capacity of in situ diaphragms will not be the governing factor in the framing layout. Designers are careful not to allow large voids near to external shear walls, and to ensure that if an external wall adjacent to a prominent staircase is used then a sufficient length of floor plate is in physical contact with the wall. Although there is no interface shear between the precast units and the topping – the precast unit is ignored in design, small loops as shown in Figure 4.29 are often (not obligatory) provided. The minimum area of reinforcement is 0.15 per cent concrete area.

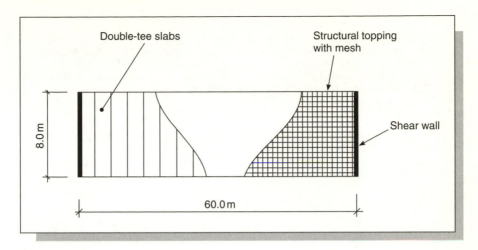

Figure 7.17: Detail to Example 7.6.

Example 7.6

Calculate the required strength of concrete and the reinforcement requirements in the composite floor diaphragm shown in Figure 7.17. The minimum depth of topping is 50 mm. The ultimate horizontal uniformly distributed load is 4.0 kN/m. Use $f_y = 460\,\text{N/mm}^2$.

Solution

$$M_u = 4.0 \times 60.0^2/8 = 1800\,\text{kNm}$$
$$V_u = 4.0 \times 60.0/2 = 120\,\text{kN}$$

Flexure

$$f_{cu} > \frac{1800 \times 10^6}{0.108 \times 50 \times 8000^2} = 5.2\,\text{N/mm}^2 < \text{C25 minimum allowed} \quad (\textit{using Eq. 7.29a})$$

$$\frac{A_s}{s} = \frac{1800 \times 10^6}{0.228 \times 460 \times 8000^2} = 0.27\,\text{mm}^2/\text{mm} \quad (\textit{using Eq. 7.29b})$$

Use B283 mesh (283 mm²/m).

Shear

$$V = 120\,\text{kN} \quad \text{and} \quad M = 0,$$

$$v = \frac{120 \times 10^3}{8000 \times 50} = 0.03\,\text{N/mm}^2 < 0.45\,\text{N/mm}^2 \text{ allowed} \quad (\textit{using Eq. 7.26})$$

References

1 Bruggeling, A. S. G. and Huyghe, G. F., *Prefabrication With Concrete*, Balkema, Rotterdam, 1991, 380p.
2 Walraven, J. C., Diaphragm Action in Floors, Prefabrication of Concrete Structures, International Seminar, Delft University of Technology, Delft University Press, October 1990, pp. 143–54.
3 Davies, G., Elliott, K. S. and Wahid Omar, Horizontal Diaphragm Action in Precast Hollow Cored Floors, *The Structural Engineer*, Vol. 68, No. 2, January 1990, pp. 25–33.
4 Elliott, K. S., Davies, G. and Wahid Omar, Experimental and Theoretical Investigation of Precast Hollow Cored Slabs Used as Horizontal Diaphragms, *The Structural Engineer*, Vol. 70, No. 10, May 1992, pp. 175–87.
5 Cholewicki, A., Shear Transfer in Longitudinal Joints of Hollow Core Slabs, *Concrete Precasting Plant and Technology*, B + FT, Wisenbaden, Germany, Vol. 57, No. 4, April 1991, pp. 58–67.

8   *Joints and connections*

8.1   Definitions

The design and construction of joints and connections is the most important consideration in precast concrete structures. Their purpose is to transmit forces between structural members and/or to provide stability and robustness. There may be several different ways of achieving a satisfactory connection, e.g. bolting, welding, or grouting, but whichever is used the method should be simple and must convey unambiguous messages to the site operatives. The joints should not only be designed to resist applied serviceability and ultimate loads, which are relatively straightforward to predict and calculate, but they should be adequate in cases of abnormal loads due to fire, impact, explosions, subsidence, etc. Failure of the joint should not, under any circumstances, lead to structural instability. It is therefore unfortunate to have to report that information on the design of joints for abnormal loading conditions in precast concrete structures is scarce – provisions to guard against this are only provided in the form of continuous column and floor ties (see Chapter 10), which in many cases bypass the joints.

Within a single connection there may be several load transmitting joints, and so it is first necessary to distinguish between a 'joint' and a 'connection'. A 'joint' is the action of forces (e.g. tension, shear, compression) that takes place at the interface between two (or more) structural elements. In many instances there may be an intermediate medium, such as rubber, steel, felt, cementitious mortar, epoxy mortar, etc. The design of the joint will be greatly influenced by how much these materials differ from concrete. This is explained in Figure 8.1.

The definition of a 'connection' is the action of forces (e.g. tension, shear, compression) and/or moments (bending, torsion) through an assembly comprising one (or more) interfaces. The design of the connection is therefore a function of both the structural elements and of the joints between them. This is explained in Figure 8.2 where the zone of the connection may extend quite far from the mating

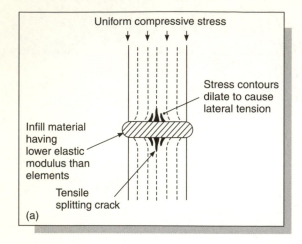

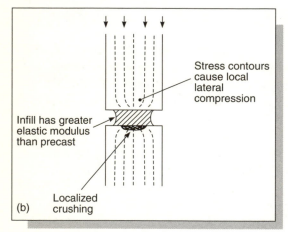

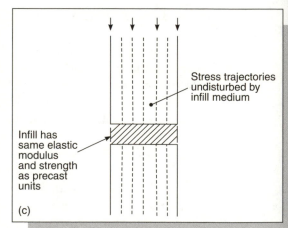

Figure 8.1: Stress contours through mediums of different stiffness and size.

surfaces. In addition to the actions of forces, connection design must consider the hazards of fire, accidental damage, effects of temporary construction and inaccurate workmanship, and durability.

In this chapter, the methods of jointing will be studied separately, then in Chapter 9 it will be shown how some of these joints are used to form the major structural connections, such as at beam-to-column, column-to-foundations, etc.

8.2   Basic mechanisms

Here the term *mechanism* means the action of forces between structural elements (not in the kinetic sense). It is used to illustrate the differences between a

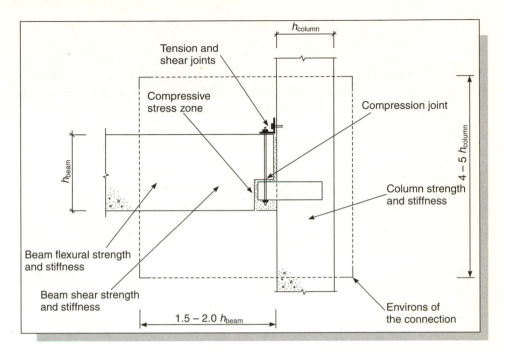

Figure 8.2: Definition of 'joint' and 'connection'.

monolithic cast in situ connection and a site jointed precast concrete one. Additional forces unique to a precast structure are generated owing to relative displacements and rotations between elements. These movements must be properly assessed and designed for – even though the relevant information may not be available in codes of practice.

There is an important division between precast elements which are considered to be *non-isolated* and those which are *isolated*. Non-isolated elements are connected to other elements with a secondary means of load transfer, which would sustain loads in the event of failure in the primary support. For example, hollow core flooring units, which are grouted together, would distribute shear forces to adjacent members in the event of a failure at the beam support and would be classed as non-isolated. On the other hand, a stairflight unit seated on to a dry corbel is an isolated element.

The most commonly used methods of connection analysis are:

1 Strut and tie (Figure 8.3a), for the transfer of bearing forces;

2 Coupled joint (Figure 8.3b), for the transfer of bearing forces and/or bending and/or torsional moments; and

3   Shear friction or shear wedging (Figure 8.3c), for the transfer of shear with or without compression.

Although the figures essentially show 2D behaviour, the design should include the effects of the 3D, particularly in narrow sections where lateral bursting forces or eccentric loads may result in reduced bearing capacity. Simple rules are observed as follows:

1   'Cover' concrete outside the reinforcement is ignored.

2   Allowances for construction tolerances and permitted manufacturing dimensional inaccuracies, given by Δ in Figure 8.3, are always made. The usual allowance for units upto 6 m long is 15 mm, plus 1 mm per 1 m additional length.

3   Rotations, given by ϕ in Figure 8.3a, of approximately 0.01 radian should be allowed for in the realignment of loads and the design of bearing pads, etc.

4   Where $H > V \tan 20°$ (Figure 8.4a) lateral reinforcement in the top of the supporting member or continuity reinforcement to prevent splitting in the supported member is provided.

5   For effective force transfer the angle (θ in Figure 8.3a) between a compressive strut and tensile tie should ideally be between 40 and 50°, and not less than 30°.

6   Full anchorage of ties by mechanical means, e.g. using an anchoring device, must not interfere with compressive stress regions, as shown in Figure 8.4b.

7   The pressure zones X in a coupled joint must not exceed 0.9 times of the net depth h of the section (see Figure 8.3b).

8   Shear friction joints are not used in isolated elements or in situations where direct tension may develop without the provision of a tensile tie.

8.3   Compression joints

The main types of compression bearings, summarized in Figure 8.5, are:

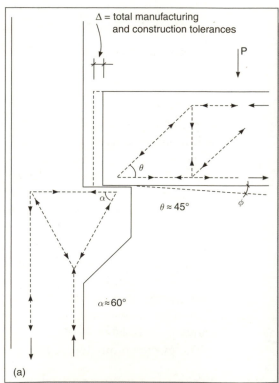

Figure 8.3: Force paths and deviations in connections (a) Beam on corbel.

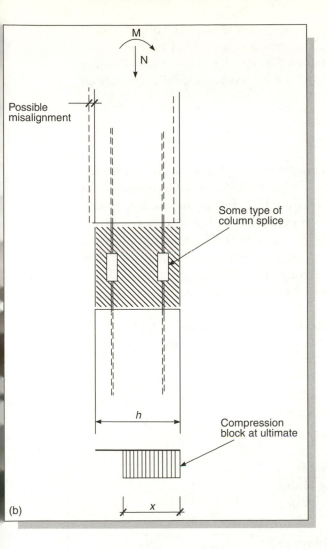

(b)

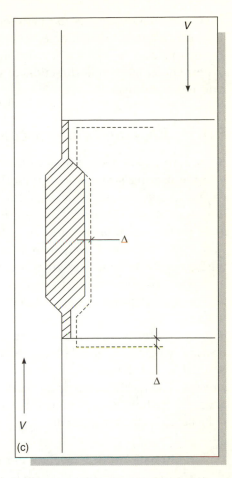

(c)

Figure 8.3 (continued): Force paths and deviations in connections: (b) Column splice; and (c) Shear key.

1 Dry bearing of precast-to-precast or precast-to-in situ concrete;

2 Dry packed bearing where elements are located on thin (3–10 mm thick) shims and the resulting small gap is filled using semi-dry sand/cement grout;

3 Bedded bearing where elements are positioned onto a prepared semi-wet sand/cement grout;

4 Elastomeric or soft bearing using neoprene rubber or similar bearing pads;

5 Extended bearings where the temporary bearing is small and reinforced in situ concrete is used to complete the connection;

6 Steel bearing using steel plates or structural steel sections.

There is a major sub-division between compression joints in plain (unreinforced) concrete and those in reinforced concrete. The reason is because of the free lateral expansion due to Poisson's ratio and the ability for the joint to resist lateral (=perpendicular to the direction of applied forces) loads internally (= without external restraint).

8.3.1   Bearing in plain concrete

Bearing in plain concrete may be used where the bearing is uniformly distributed and the bearing stress is low, typically $f_b < 0.2f_{cu}$ to $0.3f_{cu}$. Reinforcement is not required, even where a horizontal force H is applied, unless the bearing area is less than about $12\,000\,\text{mm}^2$, where it is recommended in the *PCI Design Handbook*[1] that nominal lateral steel $A_s = H/0.95f_y$ is provided. The minimum bearing dimension is 50 mm. Plain concrete also applies to such units as hollow core floors where reinforcement is provided in one direction only. The importance of checking these distances is illustrated in Figure 8.6 where less than 10 mm bearing length was recorded in some places.

The design method, which is given in Section 4.3.6, considers the bearing stresses in both of the abutting precast elements and the sandwiched jointing material (if any). Bearing lengths l_b and bearing widths l_w are defined in Section 4.3.6, together with bearing stresses. The ineffective bearing width, i.e. allowances for spalling, are given in Table 4.4. The ultimate bearing stress is given as:

$$f_b = \frac{\text{ultimate support reaction per member}}{\text{effective bearing length} \times \text{net bearing width}}$$

$$8.1$$

The only additional bearing stress condition to *add* to the list on page 88 in Section 4.3.6, appropriate to isolated beam and column bearings is:

4   steel bearing of size b_p cast into member or support and not exceeding 40 per cent of the concrete dimension $b : 0.8f_{cu}$. At the edges the steel bearing should not extend to a distance equal to the spalling allowances. Higher bearing stresses may be used only if proved by adequate testing. For larger bearing plates the allowable bearing stress f_b is given as follows (see Ref. 7.1):

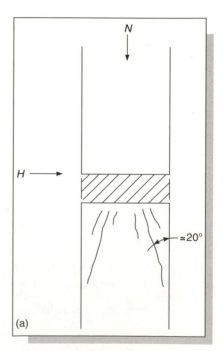

Figure 8.4: Force and stress limitations in compression joints.

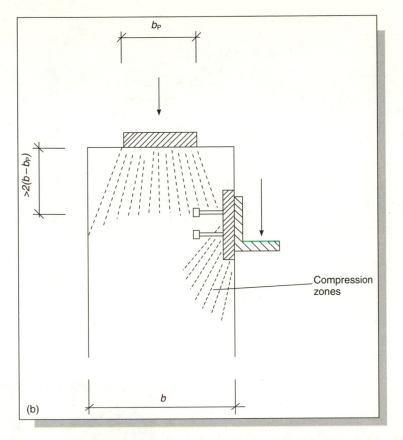

Figure 8.4 (continued): Force and stress limitations in compression joints.

$$f_b = \frac{1.5f_{cu}}{1 + \dfrac{2b_P}{b}}$$

8.2

This reduced stress is to cater for diagonal tension directly beneath the insert and close to the outside face of the column.*

In calculating compressive strengths the area of concrete or mortar filler in horizontal joints between load bearing walls and/or columns, and for solid floors onto beams, should be the greater of (Clause 5.3.6.):

1   The area of the in situ concrete ignoring the area of any intruding element, but not greater than 90 per cent of the contact area; and

2   75 per cent of the contact area.

---

*Equation 8.2 is not given in BS8110, but is found, in various versions, in most European literature.

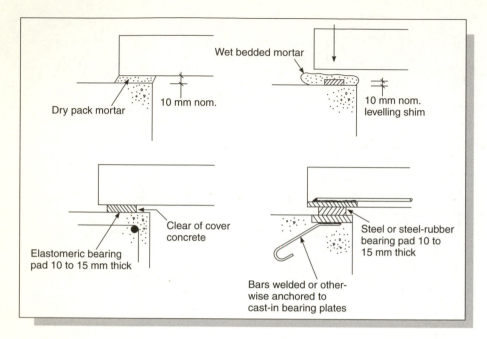

Figure 8.5: Types of bearings.

Example 8.1

Calculate the load bearing capacity of a 6.0 m long × 1.2 m wide prestressed hollow core floor unit supported by a precast beam. The nominal bearing length is 75 mm. The hollow core unit is grade C50 concrete and the tendons reach the ends of the units. The beam is grade C40 concrete and more than 300 mm deep. Ignore any horizontal frictional forces and assume that the hollow core unit is otherwise tied to the beam.

Solution

Ineffective bearing length = 15 mm in beam and zero in floor unit (from Table 4.4)

Clear distance between support faces = 6.0 − 0.075 − 0.075 = 5.85 m

Constructional inaccuracy = 3 × 5.85 = 17.6 mm > 15 mm

Net bearing width = 75 − 15 − 0 − 17.6 = 42.4 mm > 40 mm

Effective bearing length = least of

Figure 8.6: Inadequate bearing lengths.

- actual length, i.e. width of hollow core unit $= 1200\,mm$

- $1200/2 + 100 = 700\,mm$

- $600\,mm$

$F_b = 0.4 \times 40 \times 600 \times 42.4 \times 10^{-3} = 407\,kN$ per 1.2 m wide unit   (*using Eq. 8.1*)

Example 8.2
Calculate the bearing capacity of a precast beam supported by steel plates on a $300 \times 300\,mm$ column. The end of the beam is reinforced using T20 hooked-end bars. The top of the column is reinforced horizontally using T10 links.
   Use $f_{cu} = 40\,N/mm^2$ and 30 mm cover.
Note: internal radius to HT rebars is 3 diameters.

Solution
Ineffective bearing width in column $\doteq$ cover to T10 bars $= 30\,mm$ (from Table 4.4)
Ineffective bearing in beam $= 30 + 20 + 3 \times 20 = 110\,mm$
Net bearing width $= 300 - 30 - 110 = 160\,mm$
Bearing length $\leq 0.4 \times 300 = 120\,mm$

$$F_b = 0.8 \times 40 \times 160 \times 120 \times 10^{-3} = \underline{614\,kN} \qquad (\textit{using Eq. 8.1})$$

8.3.2 *Concentrated loads in bearing*

The effect of a concentrated load is shown in Figure 8.7. The dimensions of the applied loading platen (subscript p) $b_p \times h_p$ are smaller than the dimensions of the supporting member $b \times h$. The distribution of compressive stress across the section is uniform at a distance from the load known as the characteristic length l_e. This distance depends on Poisson's ratio (taken as $\nu = 0.2$ for concrete) and on the ratio b_p/b and h_p/h. However as a general rule $l_e \approx b$ or h. The change in stress within this region (=the gradient of the stress curves in Figure 8.7) gives rise to a shear stress, which, in conjunction with lateral expansion due to ν, produces lateral tension in the concrete commencing at about $y/b = 0.15–0.20$ and extends for a distance of about $y = 1.2b$ (although in theory it extends to infinity). Lateral tension stress is greatly influenced by the ratio b_p/b, and distance from the bearing pad as shown in Figure 8.8. (In some literature an effective platen dimension of $b_p + 3t$ or $h_p + 3t$ is used, where t is platen thickness.)
   In the case of a bearing plate $b_p \times h_p$ embedded in a section $b \times h$ the allowable bearing stress f_b is given as:[1]

$$f_b = 0.6 f_{cu} \sqrt{\frac{bh}{b_p h_p}} < 2.0 f_{cu} \qquad\qquad 8.3$$

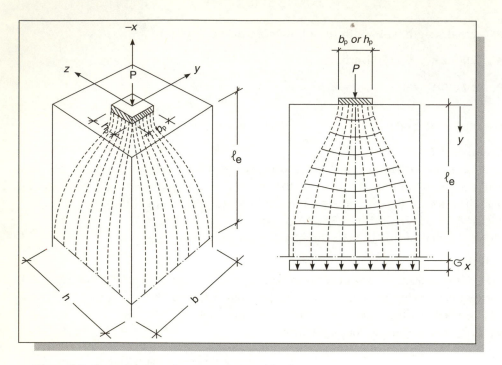

Figure 8.7: Stress contours due to concentrated loads.

In design, the lateral stress in Figure 8.8[3] is integrated to give a lateral tensile force, which must either be resisted by the tensile strength of the concrete or by reinforcement in the form of closed links in the region $y/b = 0.1-1.0$. The ratio of the lateral bursting force, known as F_{bst} to the applied force F is called the lateral bursting coefficient ζ. BS8110, Part 1, Table 4.7 gives values for ζ_{bst} in terms of the 2D ratio y_{po}/y_o, where y_{po} is half the side of the loaded dimension $= b_p/2$ or $h_p/2$ in Figure 8.7, and y_{po} is half the side of the block $= b/2$ or $h/2$ in Figure 8.7, whichever is the least. Although the notation is different, and the data in Table 4.7 are derived from end block theory in post-tensioned concrete, this is the best information available to determine F_{bst}. In rectangular shapes ζ is worked out for the lesser of b_p/b or h_p/h. Circular shapes are treated as square of equivalent area. It is reproduced here in Table 8.1.

Table 8.1: Lateral bursting force coefficients (adapted from BS8110, Table 4.7)

b_p/b or h_p/h	≤ 0.3	0.4	0.5	0.6	≥ 0.7
$\zeta = F_{bst}/F$	0.23	0.20	0.17	0.14	0.11

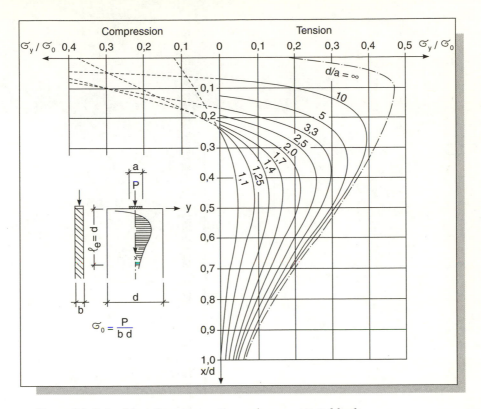

Figure 8.8: Lateral bursting stress ratios under concentrated loads.

The area of reinforcement to resist the lateral tension is:

$$A_{bst} = \frac{F_{bst}}{0.95 f_y} \qquad\qquad 8.4$$

Eccentric loads are dealt with by considering an effective breadth of the supporting member as shown in Figure 8.9. The stress is uniform at a distance of twice the edge distance u. The terms b and h in Eq. 8.3 and Table 8.1 are respectively replaced with:

$$b = b - 2e_b \quad \text{and} \quad h = h - 2e_h \qquad\qquad 8.5$$

However, $e_b \leq 0.5b$ and $e_h \leq 0.5h$ should be checked to prevent the bursting of the side walls.

Example 8.3
Calculate the maximum concentrated load that may be applied to a 200×300 mm concrete block through a steel plate. Calculate the minimum lateral bursting force and choose suitable bursting reinforcement.

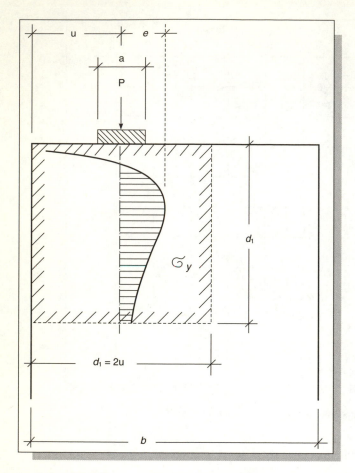

Figure 8.9: Effect of eccentric concentrated load.

$$\text{Use } f_{cu} = 40 \, \text{N/mm}^2 \text{ and } f_y = 460 \, \text{N/mm}^2.$$

Solution

Let $b = 200 \, \text{mm}$ and $h = 300 \, \text{mm}$

$$f_b = 0.6 \times 40 \sqrt{\frac{200 \times 300}{b_p h_p}} \leq 2 \times 40 \, \text{N/mm}^2 \qquad (\textit{using Eq. 8.3})$$

Then $h_p b_p = 5400 \, \text{mm}^2$.

To ensure that the minimum bursting ratio is found, $h_p/b_p = 300/200 = 1.5$. Plate dimensions $h_p = 90 \, \text{mm}$ and $b_p = 60 \, \text{mm}$. Then $F = 80 \times 60 \times 90 \times 10^{-3} = 432 \, \text{kN}$.

Ratio b_p/b and $h_p/h = 0.3$

$\zeta = 0.23$ (from Table 8.1)

$$F_{bst} = 0.23 \times 432 = 99.4\,\text{kN}$$

$$A_{bst} = 227\,\text{mm}^2 \qquad (using\ Eq.\ 8.4)$$

Use 3 no. T10 (235) distributed from 20 mm to 200 mm below the bearing plate.

Example 8.4

Calculate the maximum eccentric concentrated load that may be applied to a 300 mm wide × 400 mm deep column through a 100 × 100 mm steel plate as shown in Figure 8.10. Design the bursting reinforcement in the column.
Use $f_{cu} = 40\,\text{N/mm}^2$ and $f_y = 460\,\text{N/mm}^2$.

Solution

For eccentricity $e_h = 50 + 100/2 = 100\,\text{mm}$, $e_b = 0$

Effective $h = 400 - 2 \times 100 = 200\,\text{mm}$ \qquad (*using Eq. 8.5*)

Effective $b = 300\,\text{mm}$ \qquad (*using Eq. 8.5*)

$$f_b = 0.6 \times 40\sqrt{\frac{300 \times 200}{100 \times 100}} = 58.8\,\text{N/mm}^2 \le 80\,\text{N/mm}^2 \qquad (using\ Eq.\ 8.3)$$

$$F = 58.8 \times 100 \times 100 \times 10^{-3} = \mathbf{588\,kN}$$

Ratio $b_p/b = 100/300 = 0.33$ and $h_p/h = 100/200 = 0.5$

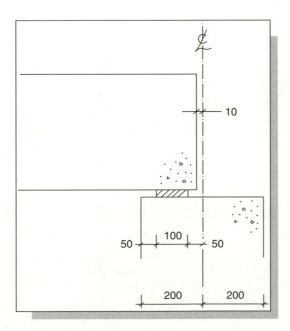

Figure 8.10: Detail to Example 8.4.

$\zeta_{max} = 0.22$ (from Table 8.1)

$$F_{bst} = 0.22 \times 588 = 129.3 \, \text{kN}$$
$$A_{bst} = 296 \, \text{mm}^2 \qquad (using \ Eq. \ 8.4)$$

Use 4 no. T10 (314) distributed from 30 mm to 300 mm below the bearing plate.

8.3.3 Reinforced and plate reinforced concrete bearings

If the applied bearing stress exceeds the values of Eqs. 8.2 and 8.3, a bearing plate and designed reinforcement are required in the bearing area. Referring to Figure 8.11 in the presence of a vertical force V, a horizontal force H exists due to the frictional restraint against thermal movement and shrinkage, where:

$$H = \mu V \qquad\qquad\qquad 8.6$$

where μ is static coefficient of friction. Values for μ are given in Table 8.2.

The reinforcement is designed by considering shear-friction across the cracked plane, as shown in Figure 8.11, that is assumed to extend across the entire end of the supported member of cross-sectional area A_c. In the case of cracked concrete bound by reinforcement an 'effective' shear friction factor μ' is used (similar to the factor in Eq. 7.8) where:

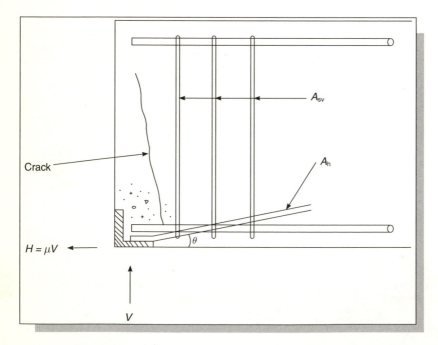

Figure 8.11: Structural mechanism at reinforced end bearing.

Table 8.2: Values for coefficient of friction μ

Interface materials	μ
Steel to concrete	0.4
Concrete to concrete (both hardened)	0.7
Concrete to hardened concrete	1.0
Monolithic concrete	1.4

$$\mu' = \frac{7A_c}{H} \text{(N, mm units only)} \qquad 8.7$$

(Note Eq. 8.7 is dimensional.) If the yield strength of the bars A_h is f_{yh} (taken as $250\,\text{N/mm}^2$ if they are welded to the bearing plate), then:

$$A_h = \frac{V}{0.95 f_{yh} \cos \theta \mu'} \qquad 8.8$$

and

$$A_{sv} = \frac{H}{0.95 f_{yv} \mu'} = \frac{A_h f_{yh} \cos \theta \mu}{f_{yv}} \qquad 8.9$$

The thickness of the bearing plate (or angle section) should be sufficient to resist the force H, but this will be found to be rather small and therefore a minimum thickness of 10 mm is used. (Corrosion protection requires a minimum thickness of 4 mm.)

Example 8.5
A precast beam 6 m long × 400 mm deep × 300 mm wide is supported over a length of 150 mm on a dry bearing onto a precast wall. If the ultimate end beam reaction is 400 kN, determine whether or not a bearing plate is required, and if so calculate its thickness. Calculate also the area of longitudinal reinforcement as inclined at 20° to horizontal, and the area of vertical reinforcement.

Use $f_{cu} = 40\,\text{N/mm}^2$, $f_{yh} = f_{yv} = 250\,\text{N/mm}^2$, $p_y = 275\,\text{N/mm}^2$ and cover to reinforcement = 30 mm.

Solution
Net bearing width (=perpendicular to span of beam) is least of: (a) 300 mm; (b) $300/2 + 100 = 250$ mm; (c) 600 mm
Net bearing length (=parallel with span of beam) $= 150 -$ (tolerances) $15 -$ (ineffective bearing) $15 -$ (ditto) $30 = 90$ mm

$$f_b = \frac{400 \times 10^3}{250 \times 90} = 17.8\,\text{N/mm}^2 > 0.4 f_{cu} \qquad (using\ Eq.\ 8.1)$$

Use bearing plate with end cover = 30 mm.

Maximum width of plate = 300 − (2 × cover) 60 = 240 mm

Try 200 mm × 90 plate

$$f_b = \frac{400 \times 10^3}{200 \times 90} = 22.2 \, \text{N/mm}^2 < \frac{1.5 \times 40}{1 + \frac{2 \times 200}{300}} = 25.7 \, \text{N/mm}^2 \ (\textit{using Eqs 8.1 and 8.2})$$

To determine horizontal resistances

$$H = \mu V = 0.7 \times 400 = 280 \, \text{kN} \qquad (\textit{using Eq. 8.6})$$

$$\mu' = \frac{7 \times 400 \times 300}{280 \times 10^3} = 3.0 \qquad (\textit{using Eq. 8.7})$$

Plate thickness $t_w = \dfrac{400 \times 10^3}{275 \times 200 \times 3.0} = 2.42 \, \text{mm}$

Use 200 × 90 × 10 mm mild steel plate.

$$A_h = \frac{400 \times 10^3}{0.95 \times 250 \times \cos 20° \times 3.0} = 597 \, \text{mm}^2 \qquad (\textit{using Eq. 8.8})$$

Use 3 no. R16 bars (603).

$$A_{sv} = \frac{280 \times 10^3}{0.95 \times 250 \times 3.0} = 393 \, \text{mm}^2 \qquad (\textit{using Eq. 8.9})$$

Use 2 pairs of R12 bars (452).

The final details are shown in Figure 8.12.

8.3.4 *Bearing pads*

Bearing pads are used to distribute concentrated loads and to allow limited horizontal and rotational movement. They also prevent direct concrete–concrete contact, which may lead to unsightly spalling and/or reduction of the effective cover to reinforcement – both of which require some remedial repair work. Their use is not as widespread as the technical literature leads us to believe, in that they are used only where a dry or wet bedding (on mortar) is not practical. Their main use is supporting double-tee floor units and long-span beams where the end rotations may be quite large, typically 0.02–0.03 radians. A typical size is 150 × 150 mm. The length (or breadth) of the pad should be 5 times its thickness, which should not be less than 6 mm for floor units and 10 mm for beams and rafters.

Bearing pads may be manufactured using hard natural rubbers, synthetic rubbers, such as neoprene (sometimes called chloroprene), lead, steel or felt.

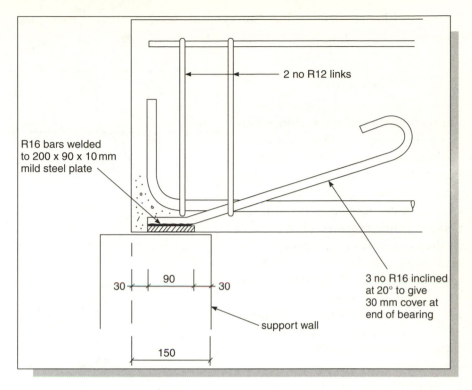

Figure 8.12: Detail to Example 8.5.

Composite sandwich pads reinforced with thin steel plate or randomly orientated fibres have been successfully used in some of the more highly loaded situations. The hardness of the rubber, and hence its ability to deform laterally under normal stress, is given by *Shore A* measurement of between 40 and 70.[1] Limiting compressive strength for rubbers is taken as 7–10 N/mm^2, although some materials will sustain greater stresses – manufacturer's test data are available. Stresses due to permanent dead loads should not exceed 3.5 N/mm^2. The compressive strain limit is taken as 0.15. Young's modulus of elasticity is not quoted because these materials are non-linear at stresses in excess of 1 N/mm^2. Design calculations are carried out at service loads. No data exist for ultimate conditions. The important behavioural characteristics are shown in Figure 8.13 (see also Table 2.4).

The ability of bearing pads to behave satisfactorily depends on their ability to deform evenly and laterally. Lateral expansion is resisted in two ways: (1) either by interface friction (Figure 8.13a); or (2) by internal tensile reinforcement, e.g. vulcanized or sandwich as shown in Figure 8.13b. Frictional restraint depends on the loaded area, as given by the 'shape factor' S. If the bearing pad is infinitely long and the contact dimension is b_l, the shape factor is given as $S = b_l/t$, where t

is the original thickness. Figure 8.13c. However, in the case of an isolated pad where the contact dimensions are b_l and b_p, this is:

$$S = \frac{b_l b_p}{2t(b_l + b_p)} \qquad 8.10$$

This information may be used with Figure 8.14[1] where the strain vs stress response is given for bearing pads of various hardness and shape factor. S should be >2 for floor units and >3 for beams and rafters.

Combined compression and bending are dealt with by superposition of axial and bending stress, as shown in Figure 8.13d. The eccentricity of the load, $e = M/N$, should not exceed $1/6$ of the bearing pad breadth (=middle third rule). The cross-sectional area and section modulus are calculated on the effective contact dimensions b_l and b_p. It is, however, quite rare for compressible bearing pads to be used where bending moments exist, as joint rigidity is usually the whole purpose of having a moment resisting connection. The more likely situation is where rotations are present. Where the rotation θ is known the deformation $\Delta t = \pm 0.5 b_l \theta$ (the positive value at the leading edge and the negative value at the rear edge of the pad). The *PCI Handbook*[1] limits the rotation of unreinforced and random fibre elastomerics to $0.3t/b$. BS8110 offers no guidance.

Example 8.6

A double-tee floor unit is simply supported over an effective span of 11.0 m, and carries UDLs of 10 kN/m live load and 5.0 kN/m dead load. The self weight of the unit is 7.8 tonnes. The floor unit is supported on a $150 \times 150 \times 6$ mm unreinforced rubber bearing pad of hardness 60 Shore. Determine the maximum and minimum deformation of the

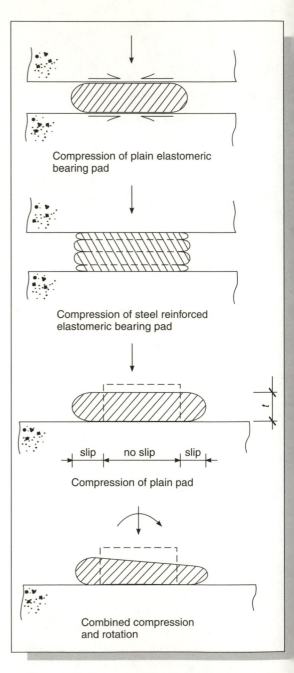

Compression of plain elastomeric bearing pad

Compression of steel reinforced elastomeric bearing pad

Compression of plain pad

Combined compression and rotation

Figure 8.13: Behaviour of elastomeric bearing pads.

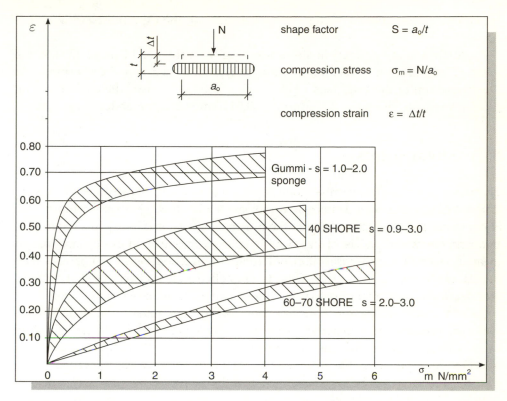

Figure 8.14: Strain vs stress relationship for bearing pads of different hardness.[1]

bearing pad. Check the shape factor, the PCI recommended limiting rotation, and the permanent dead stress limit. The flexural stiffness of the floor unit may be obtained using $I = 8800 \times 10^6\,\text{mm}^4$ and $E_c = 15\,\text{kN/mm}^2$ (long-term value).

Solution

Shape factor $= (150 \times 150)/(2 \times (150 + 150) \times 6) = 6.25 > 2$, OK

Total ultimate load $W = [(5.0 + 10.0) \times 11.0] + 78 = 243\,\text{kN}$

End reaction per support $= 243/4 = 60.75\,\text{kN}$

Compressive stress $\sigma = (60.75 \times 10^3)/(150 \times 150) = 2.7\,\text{N/mm}^2$

From Figure 8.14, for 60–70 Shore and $\sigma = 2.7\,\text{N/mm}^2$, $\epsilon = 0.2$, then $\Delta t = 0.2 \times 6.0 = 1.2\,\text{mm}$ (due to compression). End rotation of a simply supported member $\theta = WL^2/24E_c I$

$$\theta = \frac{243 \times 11000^2}{24 \times 15 \times 8800 \times 10^6} = 0.0093\,\text{rads}.$$

PCI limit $= 0.3 \times 6/150 = 0.012\,\text{rads}. > 0.0093\,\text{rads}$. Complies.

$\Delta t = \pm 0.5 \times 150 \times 0.0093 = \pm 0.7\,\text{mm}$ (due to rotation)

Maximum deformation $\Delta t = 1.2 + 0.7 = 1.9\,\text{mm} < 6\,\text{mm}$.

Minimum deformation $\Delta t = 1.2 - 0.7 = 0.5\,\text{mm} > 0$, then the full bearing pad is active.

To determine the permanent stress resulting from the dead load of 133 kN, the stress resulting from a maximum strain of $1.9/6.0 = 0.32$ is $\sigma = 4.6\,\text{N/mm}^2$. As the proportion of the dead load to the total $= 133/243 = 0.55$, the corresponding dead stress $= 0.55 \times 4.6 = 2.5\,\text{N/mm}^2 < 3.5\,\text{N/mm}^2$ recommended maximum.

8.4   Shear joints

The action of a shear force across a joint is seldom alone. In most cases shear forces are transferred across concrete surfaces in combination with direct or flexural compression. Shear transfer is never considered in the presence of tension. Shear joints occur most frequently between 'panels' of significantly large surface area. In this context, panels means floor units, as in the case of horizontal diaphragm action (Figure 7.3), or walls, as in the case of shear walls (Figure 6.29). Shear transfer often occurs between precast elements and cast in situ infill or topping, as in the case of composite floors or beams (Figure 5.19).

Shear transfer is a complex phenomenon owing to its reliance on small surface textures, physical and material properties, stress patterns and workmanship. Designers are rightly cautious for two reasons. Firstly, although the shear force interface may be extremely large, several sq.m. in some cases, the critical force transfer interface may be very small, say less than 5 per cent of this area, and less than 1 or 2 mm in thickness. Hydration of the cement paste, the suction of free water and the cleanliness of mating surfaces play important roles in shear force transfer. Secondly, the failure mode for shear is brittle and is not recoverable elastically. For these reasons partial safety factors are quite large as the margins between experimental test results and code values testify.

Shear forces can be transferred between concrete elements by one, or more, of the following methods:

1   Adhesion and bonding;

2   Shear friction;

3   Shear keys;

4   Dowel action; and

5   Mechanical devices.

8.4.1   Shear adhesion and bonding

When cast in situ concrete is placed against a precast concrete surface, adhesive bond develops in the fresh cement paste, in the tiny crevices and pores in the mature concrete. The bond stress depends largely on workmanship and the cleanness of the mature surface, particularly where oil and dust have gathered. Although the bond is quite strong in shear alone, the presence of small tensile stress in the absence of any transverse retraint will cause a rapid shear failure. For these reasons shear bond is not relied on, and in general is not allowed to act alone.

8.4.2 Shear friction

As with shear bonding, shear friction relies on the nature of the interface between contact surfaces. When a joint has certain roughness, shear will be transmitted by friction even if the interface is cracked to a value less than a critical width, typically 0.5–2.0 mm depending on how the surface has been prepared. There is nearly always an intermediate cast in situ infill (grout or mortar), although dry contact surfaces will also produce large frictional resistance. In either case a normal, or transverse, force N must be mobilized in order to develop shear friction force V, and is given as:

$$V = \mu N \qquad 8.11$$

where μ = coefficient of friction across the joint, as given in Table 7.2 for specific surfaces, and in Table 8.2 generally.

After dividing by the contact area A_c the average shear stress is given as:

$$\tau = \mu \sigma \qquad 8.12$$

where σ is the compressive stress across the joint $= N/A_c$.

The shear friction mechanism is idealized in Figure 8.15 as a 'saw tooth' model in which the tooth inclination is given as $\mu = \tan \phi$. The height, length and inclination of the teeth will not be equal, and so the larger teeth will be loaded first and may reach their yield capacity long before the attainment of maximum shear force. However, each tooth contributes to the total shear force such that an average shear stress is found.

Normal forces may be generated internally, e.g. by reinforcement (rebars or fibres), as shown in Figure 8.16. For increasing shear slip s the shear force increases until a limiting separation w_{max} is found. Although the code of practice does not evaluate the magnitudes of s and w_{max}, the full yield stress of the transverse bars will be mobilized if a minimum area of reinforcement of 0.15 per cent A_C is provided and the bars are adequately anchored. If $\alpha > 45°$ Eq. 8.12 is modified to:

$$V = \mu N + A_s f_y (\mu \sin \alpha + \cos \alpha) \qquad 8.13$$

The shear capacity increases with increasing quantity of transverse reinforcement up to a limit where concrete crushing takes place. Therefore if there is no external

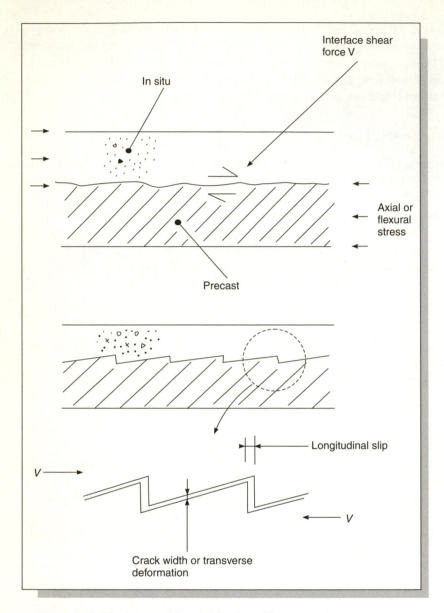

Figure 8.15: The 'saw tooth' model for shear friction.

normal force ($N = 0$) but the contact surfaces are otherwise restrained from moving apart, the ultimate shear stress $\tau = V/A_c \leq 0.23\,\text{N/mm}^2$ according to BS8110, Part 1, Clause 5.3.7a. If N exists then according to Clause 5.3.7b $\tau \leq 0.45\,\text{N/mm}^2$. However, this clause does not specify the magnitude of the compressive stress, stating only that joints should be under compression in all design conditions. If the above values are exceeded, shear friction is then ignored in design and the shear resist-

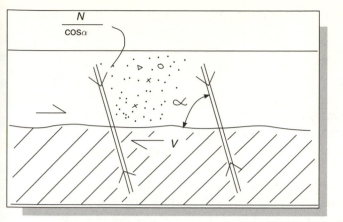

Figure 8.16: Normal force N increases shear friction resistance.

ance must be provided entirely by other means e.g. keyed joints, dowel action or mechanical means.

8.4.3 Shear keys

Shear keys are also known as 'castellated joints' owing to the shape of the cut out as shown in Figure 8.17. Shear keys rely on mechanical interlock and the development of a confined diagonal compressive strut across the shear plane. A taper is usually provided to aid removal from formwork. This also assists in confining the concrete in the second direction. The minimum length of the key should be 40 mm, and the root depth should be at least 10 mm. The length/depth ratio should not exceed 8. The angle α of the compressive strut depends on the dimensions of the keyed surface and may easily be determined whether or not the shear keys align as shown in Figure 8.17b. A resulting normal opening force N has a magnitude $N = V \cot \alpha$, which must be carried either by reinforcement, of area A_s, crossing the interface, a pre-compres-

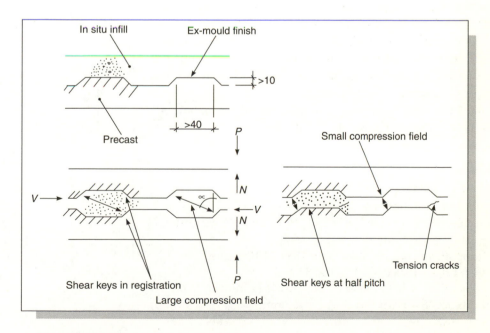

Figure 8.17: Shear keys for shear resistance.

sion P from external sources (post-tensioning for example), or a combination of both such that:

$$A_s = \frac{V \cot \alpha - P}{0.95 f_y}$$ 8.14

Failure is generally ductile due to the warnings given by concrete cracking in this way. When the shear key effect decreases due to dislocation, the behaviour changes to a frictional mode with considerable shear slip along the cracked interface. Also, as with shear friction methods, the interfaces should be prevented from moving apart, either explicitly by external forces, or implicitly by placing reinforcement according to Eq. 8.13 across the shear plane. Providing that this is done BS8110, Part 1, Clause 5.3.7c states that no shear reinforcement is required if the ultimate shear stress is less than $1.3 \, \text{N/mm}^2$, when calculated on the minimum root area. If this value is exceeded transverse steel must be provided and the interface shear capacity is based on the shear strength of the reinforcement alone.

Example 8.7

Figure 8.18 shows the detail of the castellations along a 3000 mm long × 200 deep joint between two precast units subjected to a shear force and possible normal compression. At the ends of the units there is provision to place transverse tie steel. Calculate the ultimate shear force capacity, the average ultimate shear stress, and the maximum ultimate shear stress across the root of the castellations if: (a) There is no tie steel, but the units are otherwise prevented from moving apart; (b) Tie steel comprises 2 no. T16 bars at each end of the joint; and (c) Case (b) plus an external compressive stress of $0.25 \, \text{N/mm}^2$.

Check the local compressive stress across the castellations. Use $f_{cu} = 25 \, \text{N/mm}^2$ for the infill and $f_y = 460 \, \text{N/mm}^2$. Ignore dowel action.

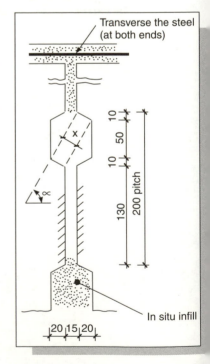

Solution

Pitch of castellations $= 200 \, \text{mm}$
Number of castellations $= 3000/200 = 15$
From geometry $\alpha = \tan^{-1} 60/35 = 59.7°$ and $x = 22 \, \text{mm}$
(a) τ_{max} by code limitation $= 1.3 \, \text{N/mm}^2$
    Root shear area $= 70 \times 200 \times 15 = 210\,000 \, \text{mm}^2$
    $V = 1.3 \times 210\,000 \times 10^{-3} = 273 \, \text{kN}$
    $\tau_{ave} = (273 \times 10^3)/(3000 \times 200) = 0.45 \, \text{N/mm}^2$
(b) Equation 8.4 with $P = 0$, $A_s = 452 \, \text{mm}^2$
    $V = 0.95 \times 460 \times 452 \times 10^{-3} / \cot 59.7° = 338 \, \text{kN}$
    $\tau_{ave} = 0.56 \, \text{N/mm}^2$ and $\tau_{max} = 1.61 \, \text{N/mm}^2$

Figure 8.18: Detail to Example 8.7

(c) $P = 0.25 \times 3000 \times 200 \times 10^{-3} = 150\,\text{kN}$

   $V = 338 + 150/\cot 59.7° = 585\,\text{kN}$

   $\tau_{ave} = 0.98\,\text{N/mm}^2$ and $\tau_{max} = 2.83\,\text{N/mm}^2$

Compression strut force per castellation $= \dfrac{595}{15 \times \sin 59.7°} = 45.9\,\text{kN}$

$$f_c = \frac{45.9 \times 10^3}{200 \times 22} = 10.4\,\text{N/mm}^2 < 0.6f_{cu} = 15\,\text{N/mm}^2$$

8.4.4   Dowel action

Where reinforcing bars, bolts, studs, etc., are placed across joints, shear forces may be transmitted by so-called 'dowel action' of the bars. In this context the bar is called a dowel. (This subject was introduced in Section 7.4.) Where it is used to determine the shear capacity of a joint, dowel action acts alone, i.e. shear friction and shear key effects are ignored. The 'dowel' is loaded by a shear force acting in the concrete in which the dowel is embedded, as shown in Figure 8.19a. Failure can occur by local crushing of the concrete in front of the dowel, which may lead to an increase in the bending arm of the embedded dowel, as shown in Figure 8.19b. This may lead to a plastic (=ductile) bending failure in the dowel – a brittle shear failure is extremely unlikely unless the separation gap, w in Figure 8.19b, is kept small by external compression. The length of embedment should be the lesser of 30 × dowel diameter ϕ or 300 mm, including hooks and bends. Splitting reinforcement, typically R8 loops, may be placed around the dowel to increase dowel resistance, although the code of practice does not recognize its presence in the following equation. The shear capacity of a dowel which is loaded without eccentricity e ($w \rightarrow 0$ in Figure 8.19b) is given as:

$$V_d = 0.6f_y A_s \cos \alpha \qquad\qquad 8.15$$

If a dowel is loaded in shear and bending such that $e > \phi/8$, bending action will cause yielding of the dowel somewhere along the embedded length. The resultant bearing stress of the concrete beneath the dowel has a maximum value of around $2f_{cu}$. An empirical equation, which is not included in BS8110 but has been well proven in tests,[4] gives the dowel capacity V_d as:

$$V_d = 1.15\,\phi\,0.67f_{cu}\sqrt{12e^2 + \frac{0.95f_y\phi^2}{0.67f_{cu}}} - 4e\,\phi\,0.67f_{cu} \qquad\qquad 8.16$$

Example 8.8

Calculate the shear capacity of a 16 mm dowel embedded into a precast concrete element. The dowel is connecting a steel section, 8 mm thick. The gap between the steel section and the face of the concrete is 10 mm. Check the bearing capacity of

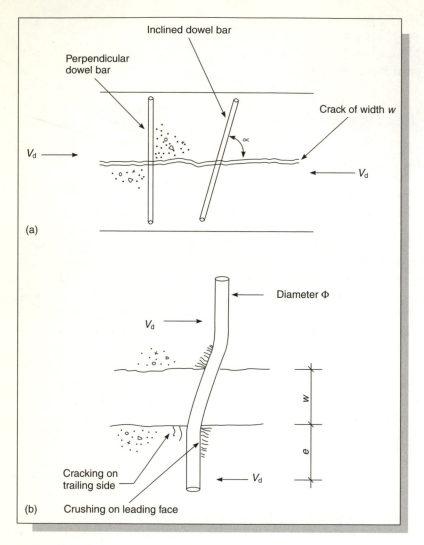

Figure 8.19: Dowel action for shear resistance.

the dowel in the steel section if the edge distance to the hole is 50 mm. Use $f_{cu} = 40 \, \text{N/mm}^2$, $f_y = 460 \, \text{N/mm}^2$, $f_{bs} = 460 \, \text{N/mm}^2$.

Solution

$$e = 10 + 8/2 = 14 \, \text{mm}$$

$$V_d = (1.15 \times 16 \times 0.67 \times 40 \sqrt{12 \times 14^2 + \frac{0.95 \times 460 \times 16^2}{0.67 \times 40}}$$
$$- (4 \times 14 \times 16 \times 0.67 \times 40) \times 10^{-3} = 15.8 \, \text{kN} \qquad (\textit{using Eq. 8.16})$$
$$V_d = 0.6 \times 460 \times 201 \times 10^{-3} = 55.5 \, \text{kN} \qquad (\textit{using Eq. 8.15})$$

Bearing capacity of dowel.

BS5950, Part 1, Clause 6.3.3.3.

$P_{bs} = 16 \times 8 \times 460 \times 10^{-3} = 58.9\,\text{kN}$ or $P_{bs} = 0.5 \times 50 \times 8 \times 460 \times 10^{-3} = 92.0\,\text{kN}$

Limiting capacity $= 15.8\,\text{kN}$

8.4.5 Mechanical shear devices

Shear transfer may be achieved locally using mechanical shear joints. The design must be very carefully considered because to ensure high shear stiffness the joint is made either by site-welding embedded plates, or by tightly clamping using friction-grip bolts. Thus, there is no inherent flexibility in a joint which cannot tolerate out-of-plane forces. The most common form of mechanical connection is the welded plate or bar shown in Figure 8.20. The effects of thermal expansion of the embedded plate must be considered to prevent cracking in the surrounding concrete. A small slit (e.g. made by diamond tip wheel) at either end of the plate will suffice. Steel angle sections anchored with headed studs are often used. The top leg of the angle should contain air bleed hole(s). Bolted connections are rarely used, except for friction-grip (or similar) bolts, because of the potential for sliding in the oversized hole reducing the initial stiffness. There is some difficulty in achieving the correct torque in every bolt in a bolt group owing to the flexibility of the embedded plate.

Typical dimensions for the welded plate detail are $100 \times 100 \times 6\,\text{mm}$ mild steel site plate, and $150 \times 75 \times 10\,\text{mm}$ mild steel embedded plates. Plates larger than this should contain air bleed holes to prevent air pockets forming. The holding bars are typically T10 or T12, and are welded to the underside of the embedded plate for a distance of 60–70 mm. Cast-in angles are typically $75 \times 50 \times 6$ rolled section $\times$ 100–150 mm long. The studs are typically 100 mm long $\times$ 10 or 12 mm diameter headed studs, attached using the semi-automatic welding process.

The ultimate shear capacity of the welded plate joint is the least of: (a) the pull out resistance of the embedded plate; (b) the weld capacity of the holding bars to the embedded plate; or (c) the shear capacity of the intermediate plate or bar.

The strength of the bar must be down-rated by a factor of 2 to allow for possible eccentric bending due to the inclined position of the bar relative to the plate. This factor of 2 assumes that the bar is welded as close to the start of the bend as possible and that β and $\gamma \approx 20°$. Referring to Figure 8.20 the pull-out capacity (a) is given as:

$$V = n\,0.95A_s\,0.5f_y\cos\beta\cos\gamma \qquad\qquad 8.17$$

where n is the number (typically 1) and β and γ are the inclinations (typically 20–30°) of the holding bars to the horizontal and vertical. The embedment resistance of the plate itself is ignored.

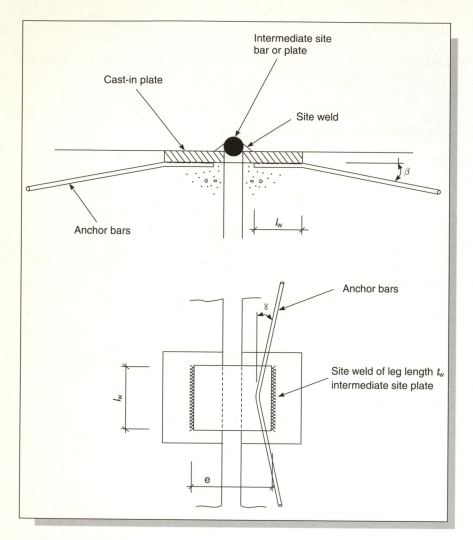

Figure 8.20: Welded plate for shear resistance.

The weld capacity (b) of the holding bars is given as:

$$V = n p_w l_w t_w \qquad\qquad 8.18$$

where p_w is the strength of the weld, taken as $215\,\text{N}/\text{mm}^2$ for grade E43 electrodes and mild steel bars, l_w is the actual weld length $-2t_w$, and t_w is the throat thickness ($=\text{weldsize}/\sqrt{2}$). The weld size is usually between 3 and 6 mm for bars upto 25 mm in diameter.

The plate capacity (c) is determined as follows. The site weld is subject to a shear force V plus a horizontal moment $M = Ve$, where e is the distance between welds and is equal to the width of the intermediate plate (see Figure 8.20). Shear deformations and bending of the plate are negligible. If the net length of the weld is l_w and leg length t_w, the maximum ultimate stress in the weld σ_w is given as:

$$\sigma_w = \frac{V}{l_w t_w} + \frac{4Ve}{l_w^2 t_w} < p_w \qquad\qquad 8.19$$

If $\sigma_w = p_w$ in the limit, then

$$V = \frac{p_w t_w l_w}{1 + \dfrac{4e}{l_w}} \qquad\qquad 8.20$$

The ultimate shear capacity of the studded angle joint is, in addition to the above, given by the shear capacity of the headed stud embedded in concrete. BS8110 (and most other 'concrete' codes) does not give data on this. It is found in BS5950, Part 3.1, Table 5.

Example 8.9
Determine the ultimate shear capacity of the embedded plate joint shown in Figure 8.21.

Solution

$V = 1{\times}0.95{\times}201{\times}0.5{\times}250{\times}\cos 20°{\times}\cos 20°{\times}10^{-3} = 21.1\,\text{kN}$ *(using Eq. 8.17)*

$V = 1{\times}215{\times}(4/\sqrt 2){\times}2{\times}(60 - \text{run-outs } 12){\times}10^{-3} = 58.4\,\text{kN}$ *(using Eq. 8.18)*

$V = \dfrac{215 \times 4.24 \times 88 \times 10^{-3}}{1 + \dfrac{4 \times 60}{88}} = 21.5\,\text{kN}$ *(using Eq. 8.20)*

Minimum shear resistance $V = 21.1\,\text{kN}$

8.5   Tension joints

Lapping of reinforcement bars or loops is often used to connect precast members as shown in Figure 8.22. The precast units have projecting bars, which are embedded in situ after erection. A full anchorage length is provided for the embedded bar, and this is calculated according to the same rules as in situ concrete. The projecting bars are usually hooked a full 180°, as shown in Figure 8.22a, otherwise the lap becomes unacceptably large. The length of the overlap is $2r + 3\phi + \phi$. If $r = 8\phi$, according to Eq. 9.32, then the overlap length is 20ϕ. Allow

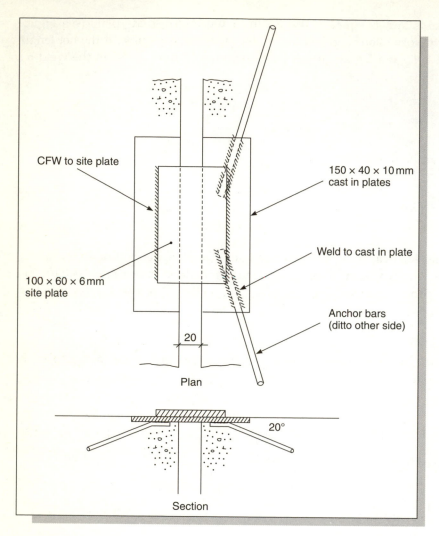

Figure 8.21: Detail to Example 8.9.

20 mm clearance from the face of the precast unit to the tip of the loop. If the loops cannot nestle together (=touching) the maximum distance, above one another, between the loop should be 4ϕ to enable a compressive strut to form between neighbours. The transverse component of the diagonal strut must be resisted by transverse bars which have a force of $0.2N_y$, where N_y is the axial force in the loops. The transverse bars must themselves be anchored – this often causes problems in shallow joints where the loops are situated near to the bottom (or top) of the section. In this case loops may be orientated in a direction perpendicular to the smaller dimension of the joint.

Where a loop of bend radius r is embedded over a length $l_p = 20\phi + 20\,\text{mm}$, the pull-out resistance of the loop N_y is given by the following empirical equation (this does not appear in BS8110 but is validated in tests[4]):

$$N_y = (1.2rl_p + 0.7l_p^2)(0.6\sigma_n + 1.1f_t)$$

or

$$N_y = 2 \times 0.95f_yA_s \qquad\qquad 8.21$$

where σ_n is a normal stress and f_t is the tensile strength of the concrete taken as $f_t = 0.24\sqrt{f_{cu}}$.

Despite the full anchorage provided for the bars embedded in the precast and in situ, concrete bond stresses quickly break down close to the interface and the two halves of the joint may be considered separately. The flexibility of each half of the interface may be determined using the data in Table 8.3[6] – this is not specified in BS8110 as this type of analysis is not a requirement for design. However, it enables designers to calculate the total flexibility of a tension lap and determine crack widths, etc.

A tensile crack resulting from elastic deformation in the bar and slippage is formed in the interface and the joint's tension deformability may be calculated in the same manner as for the compression joint. The main problem with vertical lapping is to ensure that the in situ concrete forms a full and positive bond with the steel bars. Pressurized grout is inserted through a hole beneath the level of the lap, and the appearance of the grout at a vent hole above the top of the lap is used as an indication of complete filling as illustrated in Figure 8.22b. The annulus should be at least 6 mm clear on all sides of the bars. The grout should be non-shrinkable and be sufficiently flowable to allow pressure grouting through a 20 mm diameter nozzle using a manually powered hand pump. A 2:1 sand cement mix containing a proprietary expanding agent is used to give a 24-hour strength of $20\,\text{N}/\text{mm}^2$ and a 28-day strength of around $60\,\text{N}/\text{mm}^2$.

Bolting is used extensively to transfer tensile and shear forces. Anchorages such as bolts, threaded sockets, rails or captive nuts attached to the rear of plates

Table 8.3: Deformability λ_o of different joints in tension $(\text{mm}/\text{N}) \times 10^{-5}$

Type of joint	Diameter (mm) of deformed bar				Diameter (mm) of plain bar			
	6	8	10	16	6	8	10	16
Straight bar	4.0	2.5	2.0	1.3	5.2	3.2	2.6	1.7
180° U-bar or loop	3.2	2.0	1.6	1.0	4.2	2.6	2.1	1.4
90° hooked end bar	6.4	4.0	3.2	2.1	8.4	5.2	4.2	2.8

are anchored in the precast units. Tolerances are provided using over-sized or slotted holes in the connecting member. The tensile capacity of bolted connections should be governed by the yield strength of the bolt, as this gives a ductile failure. In most types of bolted joints tension is accompanied by shear. Shear capacities are governed by the local bearing strength of the concrete in contact with the shank of the threaded socket. Shear bolt failures are brittle and should be avoided.

Welding is used to connect elements through projecting bars, fully anchored steel plates or rolled steel sections, etc. The joint can be made directly between the projecting plates or bars as shown in Figure 8.23a, but is more commonly made

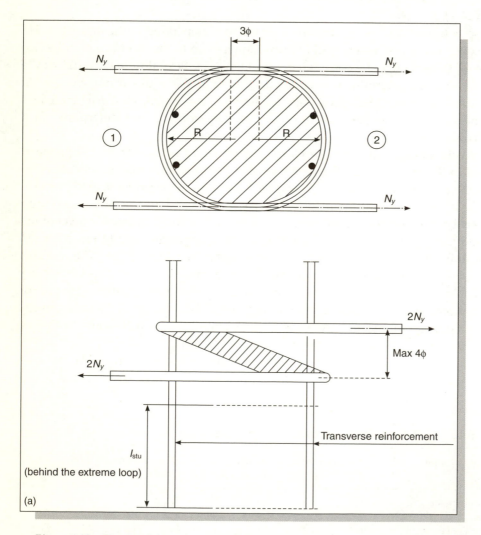

Figure 8.22a: Tension joint using direct lapping loops.

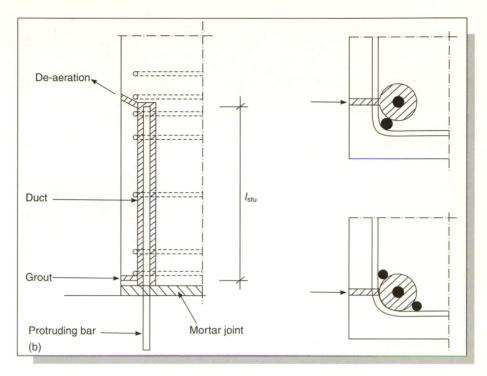

Figure 8.22b: Tension joint using bond resistance.

indirectly using an intermediate bar or plate. Figure 8.23b gives some guidance as to sizes, and Table 8.4 is used to obtain the relationship between bar diameter and weld size.

Figure 8.23a: Welded connection between rebar and plate.

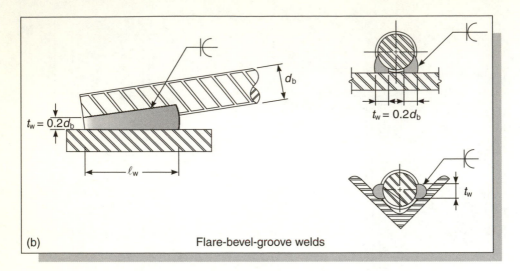

Figure 8.23b: Dimensions for welded rebar to plate or angle.

Table 8.4: Minimum weld lengths to develop full strength in lapped bars

Bar diameter (mm)	Weld depth (mm)	Weld length (mm)	Nominal length (mm)
12	3	25	25
16	4	33	35
20	5	42	45
25	6	52	55

Note:
$f_y = 250\,\text{N/mm}^2$; $\gamma_{ms} = 1.15$; $p_{weld} = 215\,\text{N/mm}^2$ for grade E43 electrode.

Post-tensioning is used to resist tension and shear forces by the application of clamping forces across the joint. Cable ducts are inserted into the precast concrete elements, or in the spaces around the elements, and, after erection, the cables are placed in the ducts and post-tensioned. Tensile capacities are computed from the state of stress in the post-tensioned elements, and shear resistance is calculated using the shear friction hypothesis.

Example 8.10
Calculate the pull-out resistance and displacement at maximum force of a 10 mm diameter 180° loop embedded into a cast in situ joint. All external forces on the joint may be ignored.

$$\text{Use } f_{cu} = 25\,\text{N/mm}^2, f_y = 460\,\text{N/mm}^2.$$

Solution

(a) *Strength*

$$r = 8 \times 10 = 80 \, \text{mm}$$
$$l_p = 10 + 80 + 30 + 80 + 10 + 20 = 230 \, \text{mm}$$
$$f_t = 0.24 \times \sqrt{25} = 1.2 \, \text{N/mm}^2$$

$$N_y = (1.2 \times 80 \times 230 + 0.7 \times 230^2) \times 1.1 \times 1.2 \times 10^{-3} = 78.0 \, \text{kN} \qquad (\textit{using Eq. 8.21})$$
$$N_y = 2 \times 0.95 \times 460 \times 78.5 \times 10^{-3} = 68.6 \, \text{kN}$$

Resistance = 68.6 kN, i.e. the bars will yield first.

(b) *Flexibility*

  Deformability coefficient $= 1.6 \times 10^{-5} \, \text{mm/N}$ (from Table 8.3)

  Ultimate force in bars = 68.6 kN

$$\delta = 1.6 \times 10^{-5} \times 68.6 \times 10^3 = 1.1 \, \text{mm}$$

8.6 Pinned-jointed connections

Pinned connections are used extensively in precast structures as they may be formed in the simplest manner by element-to-element bearing. The very nature of precast construction lends itself to forming simply supported connections in order to avoid flexural continuity across the ends of individual elements. For this reason they are often referred to as 'joints' as they tend to involve one bearing surface only. Although they are much criticized by adventurous engineers attempting to create structural frame continuity, pinned connections are seen by most precast manufacturers as safe and economical. In many cases, for example, in prestressed beams, there is no advantage in providing continuity of hogging moment as the beam is already fully stressed in that mode. In other cases, such as hollow core floors, there is no provision for continuity reinforcement and very special construction details have to be contrived. However, this is not to say that pinned connections should only be considered – each connection should be judged on its merits bearing in mind all manufacturing and erection aspects.

Pinned connections transfer shear and axial forces, both for the (dominant) gravity forces and possible uplifting forces due to overturning. By definition they cannot transfer moment and torsion, although in reality there is no such thing as a pinned joint. Small moments of resistance may develop due to the interlocking effects of infill grouting, shear friction, etc., but these will be rapidly exceeded in service, thus rendering the connection 'pinned'. Recent research has attempted to harness this stiffness and strength and develop design methods for semi-rigid connections (Ref. [3.3]),[7,8,9] but the results have been largely ignored by precast frame designers.

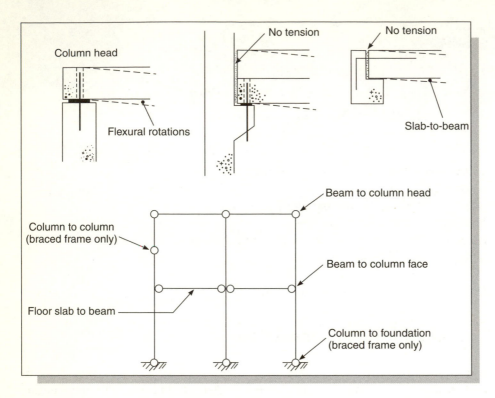

Figure 8.24: Positions of pinned-jointed connections in skeletal structures.

Pinned connections occur mainly where vertical and horizontal elements are joined. Figure 8.24 shows where pinned connections may be used in a skeletal frame and explains the primary function of pinned connections. They should allow for the rotation of the bearing element without spalling or cracking to either of the joining elements. The connection should also enable relative horizontal movement without initiating large interface forces which cannot be resisted by reinforcement (or similar) in the elements. The dimensions of the connection should allow for construction tolerances without jeopardizing the strength of the connection and/or encroaching into the cover concrete at the edges of elements. Therefore, the detailing of the interface and of the contact/end zones in both the supported and supporting elements is important.

8.6.1 *Pinned connections between vertical and horizontal elements*

The most common situation occurs at the top of a column where one or two beams bear directly as shown in Figure 8.25[10]. The bearing may be made either dry or by wet mortar bedding – the former being preferred to speed up erection. If a bearing

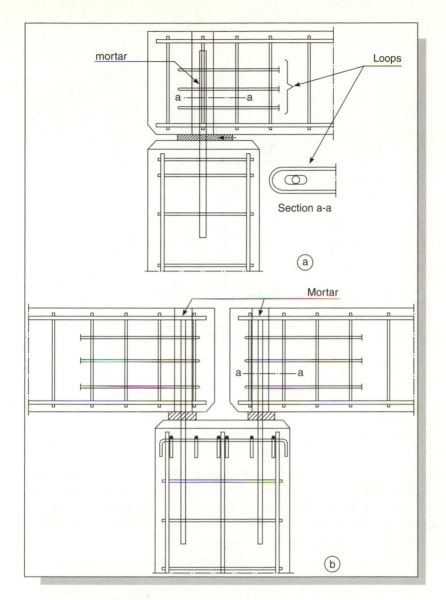

Figure 8.25: Reinforcement details for beam-to-column head connection.

pad is used in accordance with Section 8.3.1 the net concrete bearing stress calculated on the area of the pad should not exceed $0.4f_{cu}$, where f_{cu} is for the weaker concrete element. Bearing pads are typically 100 mm to 150 mm square × 6 mm thickness. Wet bedding area is usually equal to the area of the interface and therefore the limiting stress is $0.6f'_{cu}$, where f'_{cu} is the strength of the mortar. Bedding thickness is harder to control but is usually around 3–5 mm.

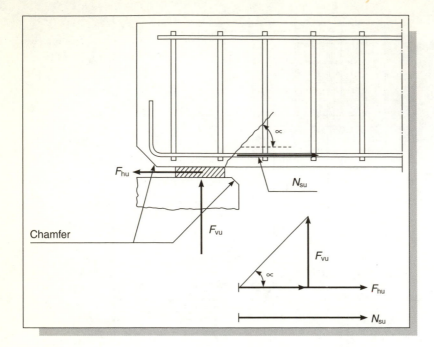

Figure 8.26: Structural mechanism at reinforced beam end connection.

Some form of mechanical joint is provided, e.g. projecting dowel(s), for horizontal restraint. Corresponding holes in the beam are site filled using flowable grout. The shear capacity of the dowels should be capable of resisting horizontal forces due to friction, μV, as defined in Section 8.3.3. Horizontal reinforcement of area A_s is placed around the dowel hole in the form of small diameter loops, such that $A_s = \mu V / 0.95 f_y$. The bending strength of the dowel should resist any overturning moment due to construction eccentricities, e.g. unequal loading on beams.

In beam design, reinforcement is provided to resist tension across an oblique shear crack at 45° to the axis of the beam, commencing at the edge of the bearing, as shown in Figure 8.26[10]. The ultimate force in the bottom reinforcement is given as:

$$F_h = \frac{V}{\tan \theta} + \mu V \qquad\qquad 8.22$$

The bars resisting this force must extend a full bond length beyond the face of the support. The bar may be formed into a horizontal loop providing that the internal radius of the bar is according to Eq. 9.32, and the start of the loop does not commence beyond the face of the support.

This bond length is not possible in prestressed beams where the pretensioning steel is cut off at the end of the beam. However, compressive forces are exerted due to the pretensioning steel such that the bottom of the beam is in a state of compression. Even though the full prestressing force may not develop at this point it is not possible for the oblique crack to develop in this region. Although BS8110 offers no guidance as to this design, it is found by testing that the transmission length is sufficiently developed to act as a bond length, and that no additional anchorage reinforcement is necessary. This design must not be confused with ultimate shear where vertical shear links (or inclined bars) are required.

The vertical force V gives rise to a lateral horizontal bursting force $H_{bst} = \zeta V$, where ζ is obtained from Table 8.1. Horizontal hairpins are provided across the end face of the beam such that

$$A_{bst} = \frac{H_{bst}}{0.95 f_y} \qquad 8.23$$

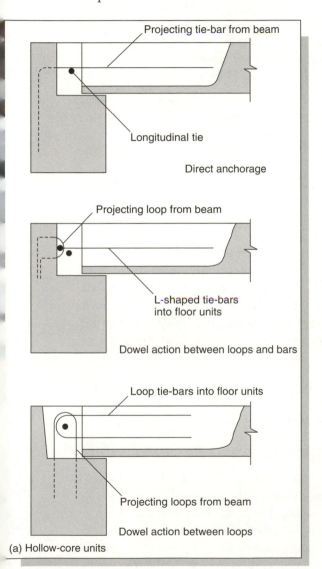

(a) Hollow-core units

Figure 8.27: Floor slab to external beam connections using hollow core slabs.

8.6.2 Simply supported slabs on beams or walls

Slab connections using hollow core and double-tee floors are designed as simple supports despite the presence of reinforced in situ concrete strips cast in the ends of the units. Hollow core floors are usually laid dry directly onto the shelf provided by the boot of the beam, but neoprene bearing pads or (less frequently) wet bedding onto grout is also used in certain circumstances, e.g. double-tee floor slabs. Wet bedded bearings are sometimes used on refurbished beams with uneven surfaces.

Hollow core units are laid directly onto a dry precast beam seating. (Figure 8.27). The design ultimate bearing stress $0.4 f_{cu}$ is rarely critical. A nominal bearing length of 75 mm results in a net length of 60 mm after spalling allowances have been deducted. Rigid neoprene strips or wet mortar bedding, which have been used in special circumstances, e.g. in refurbished buildings or on masonry bearings, ensure that a uniform bearing is made between them. Openings are made in the top flanges of the units (during manufacture) to permit the placement of structural (grade C25 minimum) concrete on site. Continuity of reinforcement is achieved either by direct anchorage between the precast beam and in situ strips, dowel action between loops, or between loops and other bars. Bars may be placed in the longitudinal gaps between the slabs providing that the width of the gap is at least twice the size of the bar or more than 30 mm (Ref. 4.5). The length of embedment is taken as the greater of one anchorage bond length or the equivalent of the transfer length of the prestressing force in the precast unit.

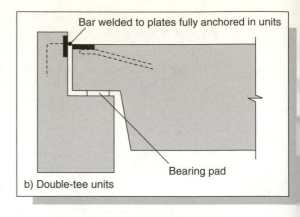

Figure 8.28: Floor slab to external beam connections using double-tee slabs.

A welded connection is made at the end of double-tee floors, as shown in Figure 8.28. (See also Figure 5.2 where a steel plate is waiting to be welded to a plate in the ends of double-tee slabs.) These bearings require greater consideration because it is vital that the loads should be equally shared between the four bearing points. A nominal bearing length of 150 mm minimum is recommended and the units should always be seated on rigid 100×100 mm neoprene (or similar) pads of about 6 mm to 10 mm thickness. No design guidance is given in BS8110, but according to the *PCI Manual*,[1] which considers a potential shear failure crack inclined at 20–25° to the vertical free edge of the bearing, the minimum bearing capacity (based on a design stress of 7N/mm^2 for neoprene) is in the order of 170 kN/m length of bearing. This is rarely critical.

8.7   Moment resisting connections

8.7.1   *Design philosophy for moment resisting connections*

A moment resisiting connection (MRC) is capable of transferring, to some degree, in-plane bending moments. Although torsional moments (=out-of-plane moments) often accompany bending moments, this book does not address torsional connections – torsion requires very specialized considerations such that it is difficult to generalize the approach in a book such as this. The basic concept of a MRC is shown in Figure 8.29. Continuity of moment is effected by the transfer of a couple of axial forces. Because precast connections are usually erected as pinned joints, it is only end moments from imposed loads which are carried by the MRC.

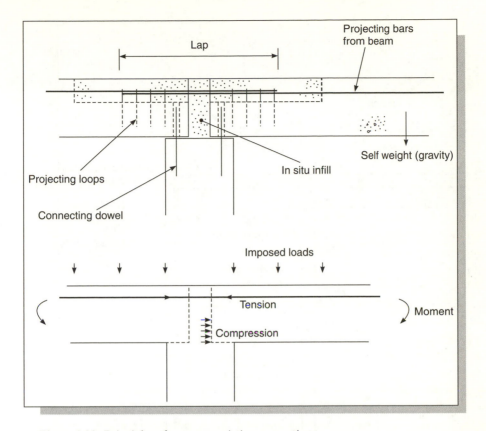

Figure 8.29: Principles of moment resisting connections.

Great care must be taken in detailing and constructing connections to be moment resisting, and the site operative should not be given the choice of whether to insert or not a vital element required to make the connection. For example, if additional bolts to those required for temporary stability are necessary to form a MRC, it is possible the site operative may omit them or insert under strength bolts. A similar situation arises where welding is required to form an MRC.

The locations where such connections may be made are summarized in Figure 8.30. These connections are used mainly to:

1 Stabilize and to increase the stiffness of portal and skeletal frames;

2 Reduce the depth of flexural frame members;

3 Distribute second order moments into beams and slabs, and hence reduce column moments; and

4 Improve resistance to progressive collapse.

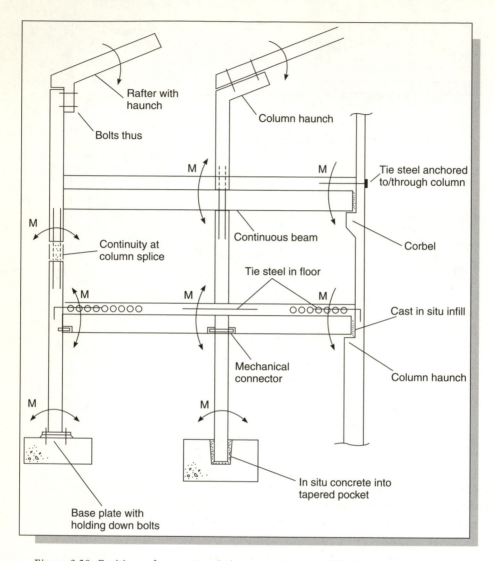

Figure 8.30: Positions of moment resisting connections in skeletal structures.

Not all precast frames lend themselves to having moment resisting connections. They may be considered only where there are a sufficient number of columns sharing the loads. It is also expected that the foundation will be encastre, which may not always be possible, especially on recycled land. The sway stiffness of the frame should not violate the drift criterion (usually sway deflection < height/500). The design of the floor diaphragm may also influence the decision to use moment connections, particularly if the diaphragm cannot be achieved over large distances. In such cases moment resisting frames would be required in all bays.

Figure 8.31: Moment continuity reinforcement placed in precast U-beams.

Moment resisting connections should be proportioned such that ductile failures will occur and that the limiting strength of the connection is not governed by shear friction, short lengths of weld, plates embedded in thin sections, or other similar details which may lead to brittleness. Deep spandrel beams with ample space for this purpose are specified as moment resisting frames, whilst the interior frames connections are all pinned-jointed shallow beams. Figure 8.31[11] shows how precast U-beams may be used to form moment connections by making the trough continuously reinforced across the column line. A similar approach is made using post-tensioning in the trough of a precast inverted-tee beam as indicated in Figure 8.32. In these cases, the connections are designed as pin jointed for self-weight loads, and moment resisting for imposed floor and horizontal loads. The sagging bending moments induced by horizontal sway loads are, by comparison, smaller than the hogging moments caused by gravity loads, but still, they must be considered in the design.

8.7.2 Structural elements in a moment resisting connection

MRC may be classified according to:

1 Generic type, e.g. mechanical or physical; and

2 Function within a structure, e.g. rigid foundation, frame action.

If such connections are to be used purposefully, either in reducing sagging moments in beams, or increasing the global strength and stiffness of the frame, a moment of resistance of at least 50–100 kNm is usually required. If the moment capacity is less than this, then it is probably better to design the connection as

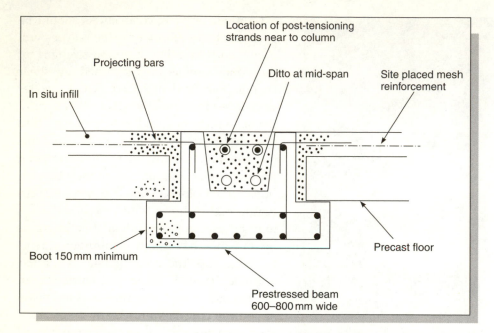

Figure 8.32: Moment continuity made by post-tensioning in the trough of a precast inverted-tee beam.

pinned-jointed. The methods used to achieve these capacities involve one or more of the following methods:

1 *Grouting* to projecting rebars, steel sections or similar. The grout may contain coarse aggregates and additives, such as expanding agents or epoxy resins, etc. An intermediate *high bond* sleeve may be used to reduce anchorage lengths.

2 *Bolting* between steel sections, plates, etc. The bolts may be friction bolts if shear forces are present. *Saw-tooth* plate washers may achieve similar means.

3 *Threaded bars* to couplers, cast-in sockets, or to nuts and plate washers. The bars may be threaded rebars, threaded *bright drawn* or *black* dowels, or long length bolts (e.g. holding down bolts).

4 *Welding* to steel sections, plates, rebars, etc.

Grouted joints for moment resistance – are formed by casting of a small quantity of in situ grout (or concrete) around projecting reinforcement (or other structural projections). The design for compression is dealt with in Section 8.3. Straight bars are lapped to resist tension. The bars should be tied together length-wise such that a full anchorage bond length is provided. Where high strength concrete is required ($f_{cu} > 50\,\text{N}/\text{mm}^2$), special mortars (e.g. epoxy-based grouts) or fibre reinforced

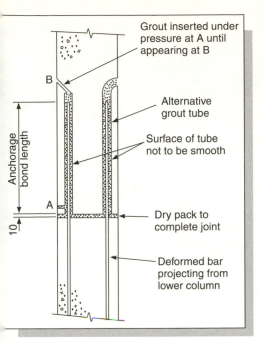

Grout inserted under pressure at A until appearing at B

Alternative grout tube

Surface of tube not to be smooth

Dry pack to complete joint

Deformed bar projecting from lower column

B

A

Anchorage bond length

10

Figure 8.33: Moment continuity by lapping of rebars in grouted sleeves.

concrete is used. Bond lengths to deformed HT bars of around 10 bar diameters are possible with high strength steel fibre reinforced concrete. Despite the full anchorage provided for the bars embedded in the precast and in situ concrete, bond stresses will ultimately break down close to the interface and the two halves of the joint may be considered separately. A tensile crack resulting from elastic deformation in the bar and slippage is formed in the interface and the joint's tension deformability may be calculated from values given in Table 8.3.

The main problem with vertical lapping is to ensure that the in situ infill grout forms a full and positive bond with the steel bars (Figure 8.33). Pressurized grout may be inserted through a hole beneath the level of the lap, and the appearance of the grout at a vent hole above the top of the lap is used as an indication of complete filling. The annulus should be at least 6 mm clear on all sides of the bars. The grout should be non-shrinkable and sufficiently flowable to allow pressure grouting. A 2:1 sand:cement mix containing a proprietary expanding agent is used to give a 24-hour strength of 20 N/mm^2 and a 28-day strength of around 60 N/mm^2. Note that the grading of the sand is crucial to the ultimate strength of this grout; medium zone sand with at least 50 per cent passing a 600 μm sieve is recommended. Grouting may be carried out by gravity pouring, but the annulus must be vented, or be sufficiently large in diameter, e.g. 50–60 mm for a 25 mm bar, to prevent air pockets forming. Superplasticizers are recommended to help grout flow as vibration cannot be used. The sand should be medium-coarse graded, but no coarse aggregate is used.

Bolted joints for moment resistance – are used to resist tension and shear forces, mainly at column foundations, column splices and at some types of beam connections. Fully anchored cast-in steel sections are necessary to generate large tensile capacity; barrel sockets are not adequate. Anchorage failures are sudden and brittle. To avoid this the anchorage length should be sufficient to enable the full strength of the concrete to form the conical plug as shown in Figure 8.34a. Load capacity is the product of the area of an assumed 45° failure surface times the tensile splitting stress of the concrete. Where the spacing between the bolts is less than the embedment length, the failure surfaces overlap and the gross capacity is reduced. The failure cone is replaced with a truncated pyramid with 45° vertices.

Where HT bolts are used, e.g. yield strength $>450\,\text{N}/\text{mm}^2$, the anchorage head requires enlargement. This is usually achieved using a so-called 'plate washer', of approximate size $100 \times 100 \times 8\,\text{mm}$. Groups of closely spaced bolts are connected through surface plates or rolled sections, e.g. Figure 8.34b. Here the bolts are designed not to carry the tensile (or shear) force, but to transfer forces to the steel insert

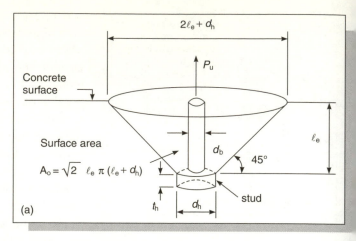

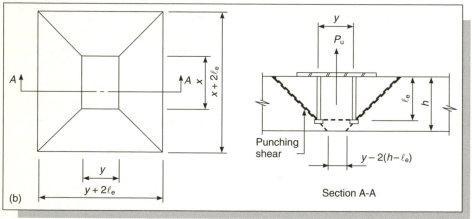

Figure 8.34: Failure zones in single and multiple bolted connections.

which itself is designed to interact with the concrete to much greater capacity possible than with bolts.

Threaded rebars for moment resistance – are used extensively in column splices and in certain types of beam connections, where the rebars are subjected to tension, or combined tension and shear. The shear capacity of projecting rebar (=a 'dowel' in this context) is a function of the shear strength of the rebar, the lever arm to the load (i.e. causing dowel bending), and the resistance of the concrete beneath the bar. In tension, the design is essentially a combination of bolt design, where the threaded portion is designed in the same manner as for bolts, and rebar anchorage, where a full anchorage bond length is provided. High tensile deformed bar is preferred in order to minimize the anchorage length.

Threaded rebars are inexpensive to make and may, with sturdy templates, be positioned accurately in the mould. Match casting may be used to locate the two

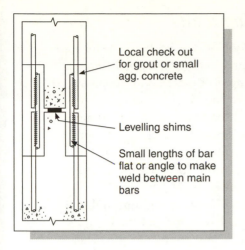

Figure 8.35: Continuity made by welding projecting plates.

Labels in figure:
Local check out for grout or small agg. concrete

Levelling shims

Small lengths of bar flat or angle to make weld between main bars

mating bars on site. If threaded couplers are used, the limiting strength should be that of the rebar. Projecting threads must be protected against corrosion and being damaged on site wrapping in waxed tape or similar. The bars should be enclosed within 180° U-bars or 90° links at not more than 40 mm cover from their end (25 mm cover is preferred). Corner bars should be enclosed by a link.

Welded joints for moment resistance – are found mainly in confined or heavily reinforced areas where the joint length is to be minimized, and immediate structural stability is required (or preferred). Welding is used to connect elements through projecting rebars, fully anchored steel plates or rolled steel sections, etc. The joint is often made using an intermediate bar or plate, as shown in Figure 8.35. Welding may equally be specified for vertical as well as horizontal connections. Projecting bars or steel sections should be confined by links or U-bars as specified above. Referring to Figures 8.23b and 8.35, the design of the weld is based on a weld size of 0.2ϕ (ϕ = diameter of bar) giving a circumferential contact breadth of 0.4ϕ. If the intermediate bar is fully stressed, and the two welds are equal in size, then:

$$p_{weld}0.4\phi l_w = \frac{f_y}{\gamma_{ms}}\frac{\pi\phi^2}{4} \qquad 8.24$$

Thus, the total length of weld, including two run outs, is:

$$l_w = \frac{f_y\pi\phi}{1.6p_{weld}\gamma_{ms}} + 0.4\phi \qquad 8.25$$

Moment resisting connections made by welding have the obvious advantage of being fully rigid and structurally continuous. The size of weld and adjoining elements may be designed to ensure that no failure occurs in the weld. The effects of heat dissipation, expansion and subsequent contraction of the recipient elements, and movement of the weld piece during welding should be considered when specifying the weld. Tolerances should be allowed in the positioning of the steel in the separate elements to give the site operative the opportunity to place an intermediate bar between the bars and place a well fashioned bead. Underhand welding should be avoided if possible, and the weld should be inspected afterwards. Grade E43 or E51 electrodes are used. The yield stress of mild steel ($f_y = 250\,\text{N/mm}^2$) is used for the connecting rebars, even though HT deformed bars may be specified.

8.7.3   *Floor connections at load bearing walls*

Horizontal joints in load bearing walls occur at floor and foundation levels. Primary forces in the joint are due to vertical compression from upper storey panels and horizontal shears from floor plate diaphragm effects. Connections at wall supports require careful detailing particularly if the floor units are supported within the breadth of the walls, and large wall loads are imposed. Some hollow core units (e.g. >250 mm deep) may require strengthening to prevent web buckling by filling the voids to a depth coincident with the edges of the walls. Double-tee units may require rib end closure pieces to form a vertical end diaphragm.

Referring to Figure 8.36 the bearing length l_s should be 75 mm minimum so that the clamping force N (acting in the vicinity of the precast slab and not the in situ infill) may generate a frictional force $F = \mu N$ over a sufficient contact length. The wall thickness should therefore be at least 200 mm, allowing a 50 mm wide gap for in situ concrete infill. It is assumed that the lever arm from the bearing ledge to the centroid of the tie steel bars is $0.8d$, and to the centre of bearing pressure is $0.67\,l_s$.

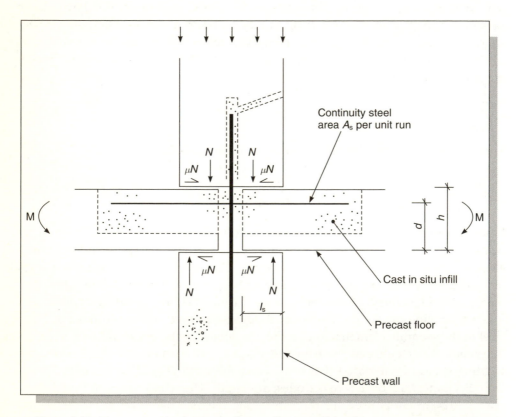

Figure 8.36: Moment continuity across floor–wall connections.

The tensile capacity of the concrete is ignored. The moment capacity of such connections is given by:

$$M = \mu Nh + 0.67 l_s N + 0.95 f_y A_s 0.8d \qquad 8.26$$

where μ = coefficient of friction, taken as 0.7,

f_y = yield stress in tie bars of area A_s,

d = effective depth to tie bars from bearing ledge, and $d > 0.5h$

Example 8.11
Calculate the ultimate moment of resistance in the floor slab to wall connection shown in Figure 8.37. The ultimate axial force from the upper wall is 500 kN/m run. Check the compression limit of the infill concrete in the 50 mm gap between the ends of the floor units. Check the limiting moment of resistance of the floor slab itself.

Use $f_{cu} = 40\,\text{N/mm}^2$, $f'_{cu} = 25\,\text{N/mm}^2$, $f_y = 460\,\text{N/mm}^2$ and $\mu = 0.7$.

Solution
Consider a 1 m width of slab.

Compressive stress beneath the upper wall $= \dfrac{500 \times 10^3}{1000 \times 200} = 2.5\,\text{N/mm}^2$

Clamping force $N = 2.5 \times 75 \times 1000 \times 10^{-3} = 187.5\,\text{kN}$

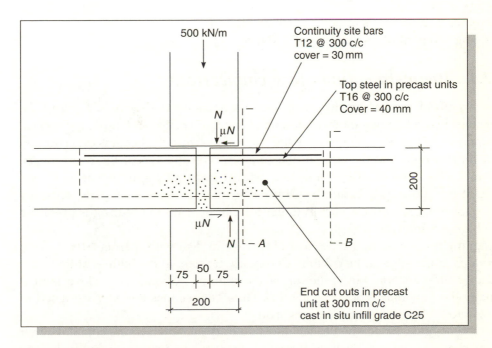

Figure 8.37: Detail to Example 8.11.

Plane A

At the edge of the bearing $A_s = 377\,mm^2/m$

$d = 200$ – cover 30 – bar radius $6 = 164\,mm$

$$M_R = [(0.7 \times 187.5 \times 10^3 \times 200) + (187.5 \times 10^3 \times 0.67 \times 75)$$
$$+ (0.95 \times 460 \times 377 \times 0.8 \times 164)] \times 10^{-6} = 57.3\,kNm \qquad (using\ Eq.\ 8.26)$$

Flexural horizontal compressive force in infill $= \dfrac{57.3 \times 10^3}{0.8 \times 164} = 436.6\,kN$

Corresponding stress assuming rectangular stress block depth $0.4d =$ $\dfrac{436.6 \times 10^3}{1000 \times 0.4 \times 164} = 6.6\,N/mm^2 < 0.45 f_{cu} = 11.25\,N/mm^2.$

Plane B

In the precast floor unit $A_s = 670\,mm^2/m$

$$F_s = 0.95 \times 460 \times 670 \times 10^{-3} = 292.7\,kN$$

$$X = \dfrac{292.7 \times 10^3}{0.45 \times 40 \times 1000 \times 0.9} = 18\,mm$$

$$z = 152 - 0.45 \times 18 = 144\,mm$$

$$M_R = 292.7 \times 10^3 \times 144 \times 10^{-6} = 42.1\,kNm < 57.3\,kNm$$

at edge of bearing.

Therefore, connection is critical in slab as desired.

Hogging moment of resistance $= 42.1\,kN/m$ width.

8.7.4 Beam-to-column face connections

The structural mechanism for beam end connections is shown in Figure 8.38. Consider first a hogging moment described in Figure 8.38a. At failure, the vertical component of the bending moment and shear force from the beam is concentrated in a contact region at the edge of the column, typically 1/5 of the depth of the column. This zone must be reinforced against horizontal splitting using closed links at not more than 25 mm beneath the seating. A small steel plate, typically $150 \times 150 \times 12\,mm$ in size, cast in to the column beneath the bearing is preferred to a highly reinforced region.

Sagging moments, Figure 8.38b are less easy to deal with, particularly if uplift develops at the edge of the column. Continuity tie steel in the bottom of the beam must be fully anchored over the top of the column. A bolted or welded joint is usually the only means of achieving this. Thus, connections are often designed as pin-jointed where sagging moments arise.

In the hogging mode, the tie steel must be fully anchored to the column, either by mechanical devices, such as threading into cast-in inserts, or by anchorage

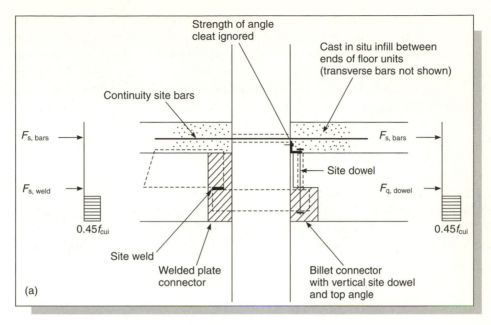

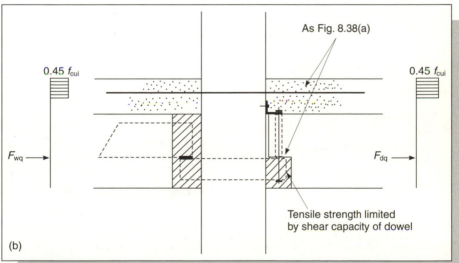

Figure 8.38: Moment resisting beam-to-column connections in (a) hogging mode; and (b) sagging mode.

bonding through grouted sleeves, etc. It is not sufficient to continue the tie steel around the column. The type of flooring is not important in this mode providing that the tie bars are positioned in confined concrete and interface shear links are provided to resist the force in the tie bars – T12 at 300 mm spacing is found to be adequate generally. Full scale testing[8] has shown that if the tie steel is not fully anchored to the column it achieves only about 25 per cent of yield value, and that the capacity of the connection is equal to that of the beam end connector. Two of the most popular types of beam end connections and the results of experimental tests to determine the moment of resistances are shown in Figure 8.39.[8,9] The most favourable situation is to design the connection to resist hogging moments only, and to class the sagging mode as pinned. In all but high sway load cases, the hogging moment resulting from gravity beam loads will dominate, and the connection may never experience sagging moments. The hogging moment of resistance for these connections is calculated as follows:

8.7.4.1   Welded plate connector

A thin plate is anchored to the beam using large diameter rebars, typically 25 mm HT. The plate is site welded to a projecting steel billet. Expansive infill concrete is used to fill the gap – left side of column in Figure 8.38a. Providing that the bars are

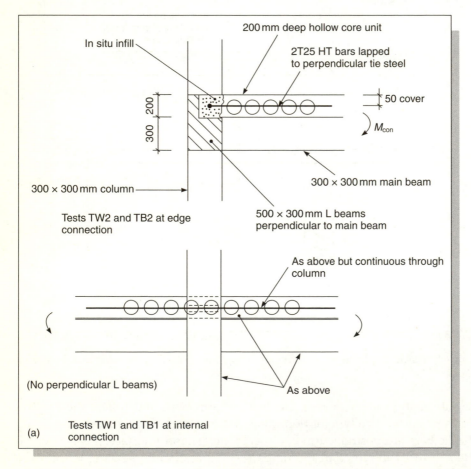

Figure 8.39: Results and details of moment resisting beam-to-column connection tests by Gorgun.

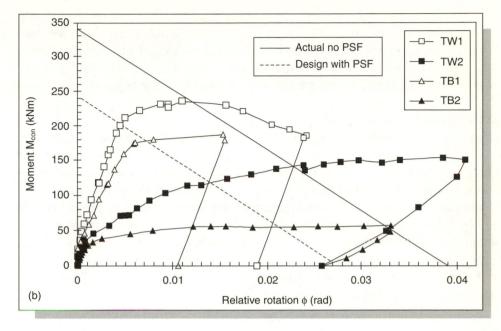

Figure 8.39 (continued): Results and details of moment resisting beam-to-column connection tests.[8]

fully anchored to the column, or are continuous through the column, the tie steel bars are fully stressed at the ultimate limit state. The beam plate is fully anchored such that the weld at the billet is also fully effective. Because of the end preparations made to the narrow plate, the weld size is a full 20 mm triangular fillet of 80 mm effective length (i.e. the normal rules for run-outs do not apply here). The compressive strength of the concrete at the bottom of the beam is limited by the strength f_{cui} of the infill concrete. The contribution of the solid steel billet is ignored. Then:

$$F_c = 0.45 f_{cui} b \, 0.9X \qquad\qquad 8.27$$

$$F_s = f_{yw} l_w t + 0.95 f_y A_s \qquad\qquad 8.28$$

and

$$F_c = F_s$$

hence the NA depth X may be determined, giving the lever arms z_1 and z_2 to the tie steel and weld, respectively. Then:

$$M_R = 0.95 f_y A_s z_1 + f_{yw} l_w t z_2 \qquad\qquad 8.29$$

If the bars are not suitably anchored to the column, the tie stress in Eq. 8.29 should be limited to $0.25f_y$. Horizontal interface shear links should be provided between the beam and floor slab. They should be capable of resisting the force $0.95f_yA_s$. It is suggested that the links should be distributed over a distance beyond the end of the connector equal to $1.5d$, where d = effective depth of the beam.

8.7.4.2 Steel billet connector

A threaded rod or dowel is site fixed through a hole in the beam and supporting steel billet and secured to a steel angle (or similar) at the top of the beam. Right side of column in Figure 8.38a. The annulus around the billet is site grouted. If the tie steel is fully anchored as described above the tie steel bars are fully stressed at the ultimate limit state. The shear strength of the vertical dowel is ignored due to the negligible strength of the bolted angle. The moment due to shear force in the (same) vertical dowel of area A_{sd} is fairly small owing to the nearness of the dowel to the compression zone. The compressive strength of the concrete at the bottom of the beam is limited by the strength f_{cui} of the narrow grouted joint. The contribution of the steel billet is ignored. Then:

$$F_c = 0.45f_{cui}b\,0.9X \qquad\qquad 8.30$$

$$F_s = 0.95f_yA_s + p_qA_{sd} \qquad\qquad 8.31$$

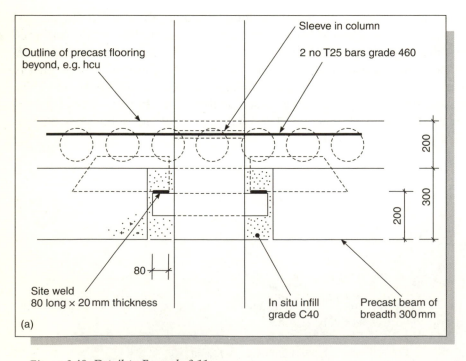

Figure 8.40: Detail to Example 8.11.

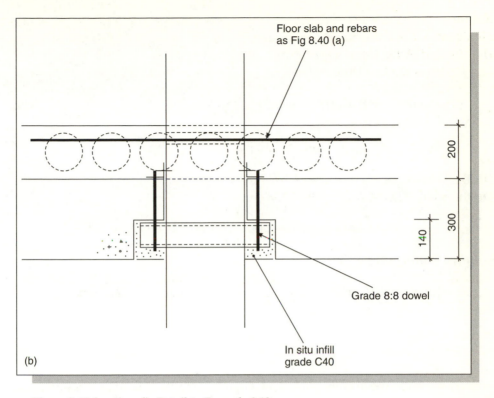

Figure 8.40 (continued): Detail to Example 8.12.

and

$$F_c = F_s$$

hence X and z are determined as before. Finally:

$$M_R = 0.95 f_y A_s z_1 + p_q A_{sd} z_2 \qquad\qquad 8.32$$

Example 8.12

Calculate the hogging moment of resistance of the beam-column connection shown in Figure 8.40a and b for the welded plate and billet connectors. In both cases, continuity reinforcement is positioned in the gap between the ends of floor slabs and above the centre line of the beam. Transverse tie steel would be present but is not shown here.

Use $f_{cui} = 40\,\text{N/mm}^2$, $f_y = 460\,\text{N/mm}^2$, p_q for grade 8:8 bolts $= 375\,\text{N/mm}^2$, $p_w = 215\,\text{N/mm}^2$ and cover to top steel $= 50\,\text{mm}$.

Solution

In both cases $d = 500 - 50 - 13 = 437\,\text{mm}$

(a) Welded Plate Connector

Weld length actually 80 mm.

$$F_s = \left[(0.95 \times 460 \times 982) + (215 \times 80 \times 20/\sqrt{2}) \times 10^{-3}\right] = 429.1 + 243.2 = 672.3\,\text{kN}$$

$$X = \frac{672.3 \times 10^3}{0.45 \times 40 \times 300 \times 0.9} = 138\,\text{mm} < 0.5d$$

z_1 to the bars $= 438 - 0.45 \times 138 = 376\,\text{mm}$
z_2 to the weld $= 200 - 0.45 \times 138 = 138\,\text{mm}$

$$M_R = 429.1 \times 0.376 + 243.2 \times 0.138 = 194.9\,\text{kNm}$$

(b) Billet Connector

$$F_s = [(0.95 \times 460 \times 982) + (375 \times 201) \times 10^{-3}] = 429.1 + 75.4 = 504.4\,\text{kN}$$

$$X = 104\,\text{mm}$$

$$M_R = 429.1 \times 0.391 + 75.4 \times 0.093 = 174.8\,\text{kNm}$$

Comparison with the results shown in Figure 8.39b of full scale experiments[8] having the same geometry are instructive. The failure moment in the tests using welded plate connector was 237 kNm compared with 194.9 kNm in Example 8.11, and for the billet connectors were respectively 190 kNm and 188 kNm. The tests reached greater values because the yield stress in the rebars used in the experiments was around 540 N/mm². (The sloping lines in Figure 8.39b are the so-called *beam-lines*, above which the capacity of the connection should exceed. PSF means 'partial safety factors', such as 1.5 for concrete and 1.05 for rebars, and therefore the requirements with PSFs will be less than without. The *beam-line* concept is introduced in Section 9.1.)

References

1 Prestressed Concrete Institute, Design Handbook, 4th edn, PCI, Chicago, USA, 1992.
2 Clarke, J. L. and Simmonds, R. M., Tests on Embedded Steel Billets for Precast Concrete Beam – Column Connections, Technical Report No. 42.523, Cement & Concrete Association, Wexham Springs, UK, August 1978, 12p.
3 Leonhardt, F., Vorlesungen über Massivbru-Zweiter Teil, Sonderfälle der Bemessung im Stahlbetonbau, Lectures about Fireproof Construction, Second Part, Special Cases of Design in Reinforced Concrete, 1975.
4 Baker, A. L. L. and Yu, C. W., Research to Investigate the Strength of Floor-to-Outside Wall Joints in Precast Concrete, The Stability of Precast Concrete Structures, Department of Environment and CIRIA Seminar, Ref. B387/73, March 1973.
5 Tharmartnam, K., Structural Behaviour of External Horizontal Joints in Large Panel Buildings, PhD thesis, University of London, 1972.

6 Bljuger, F., *Design of Precast Concrete Structures*, Ellis Horwood, Chichester, UK, 1988, 296p.
7 Mahdi, A. A., Moment Rotation Effects on the Stability of Columns in Precast Concrete Structures, PhD thesis, University of Nottingham, 1992.
8 Görgün, H., Semi-rigid Behaviour of Connections in Precast Concrete Structures, PhD Thesis, University of Nottingham, United Kingdom, 1997.
9 Elliott, K. S., Davies, G., Görgün, H. and Adlparvar, M. R., The Stability of Precast Concrete Skeletal Structures, *PCI Journal*, 43(2), 1998, pp. 42–57.
10 Same as 7.1.
11 Guidelines for the Use of Precast Concrete in Buildings, Study Group of the New Zealand Concrete Society and National Society of Earthquake Engineering, Christchurch, New Zealand, 1991, 174p.

9 Beam and column connections

9.1 Types of beam and column connections

This chapter is concerned with the major structural connections between beams, columns and foundations in skeletal frames. The importance of these connections to the behaviour of precast structures, both in the temporary construction phase and in service, cannot be over stated. It is not a coincidence that the feature which distinguished the patent of precast frames developed in the 1960s and 1970s was the beam-to-column connector – not the total connection, as defined in Section 8.1, but the physical connector itself.

The main types of connections are:

- beam-to-column face (Figure 9.1);
- beam-to-column head (Figure 9.2);
- column base plate to foundation (Figure 9.3); and
- column to pocket foundation (Figure 9.4).

Unlike cast in situ concrete work, the design philosophy for precast connections concerns both the structural requirements and the chosen method of construction. In many instances, the working practices in the factory may dictate connection design! Design philosophy depends on several factors, some of which may seem unlikely to the inexperienced, as follows:

Figure 9.1: Beam-to-column face connections. Asymmetrically loaded beams require couple connections to prevent beam twisting.

Figure 9.2: Beam-to-column head connections. A bearing pad and grouted dowels are also present.

Figure 9.3: Column-to-foundation connection using extended steel base plate.

Figure 9.4: Column-to-foundation connection using grouted pocket.

- the stability of the frame. Unbraced frames require moment resisting foundations, whereas braced frames do not. Braced frames may contain pin-jointed column splices;

- the structural layout of the frame: The number and available positions of columns and bracing elements may dictate connection design;

- moment continuity at ends of beams: Cantilevered beams always require moment resisting end connections (or otherwise beam continuity) whereas beams simply supported at both ends do not. Unbraced frames up to a certain height may be designed using rigid (or semi-rigid) end connections;

- fire protection to important bearings and rebars;

- appearance of the connection and minimizing structural zones, e.g. 'hidden' connections must be designed within the dimensions of the elements, whereas 'visible' connections are outside the elements;

- ease and economy of manufacture;

- the requirements for temporary stability to enable frame erection to proceed (=the need for immediate fixity/stability), e.g. torsional restraint at the ends of beams during floor erection;

- site access, or lack of it, may influence frame stability, and hence connection design;

- the chosen method(s) of making joints, e.g. grouting, bolting, welding, and the type of bearing(s) used; and

- the capabilities of hoisting and lifting plant.

In this context, 'design' means not only the selection of appropriate dimensions and materials for the connection devices, but understanding the nature of the force paths through the connected members and effects of volumetric changes they undergo. Whereas the effects of thermal, creep and shrinkage movement etc. in cast in situ concrete are intrinsically resisted by the minimum area of reinforcement, these have to be specifically allowed for in precast connection, either by 'enabling' (=allow to happen without damage) or 'restraining' (=prevent by adequate stiffness and strength). Anything in between will result in the damage situation shown in

Figure 9.5. Force transfer from tying forces and concentrated support bearings requires special detailing of the contact zones to avoid the development of stress concentrations and inevitable cracking.

All connections must have adequate strength, stiffness and ductility. The requirements for the mechanical behaviour of different types of connections depends on their intended purpose, and as explained in Figure 9.6, may differ widely whilst being perfectly suited to their need. In connection A, a large elastic stiffness may be required for cyclical loading, whilst the ductility is not important because there is no danger of overload in the connector. In B, non-linear deformation may be satisfactory if the connection is concerned only with strength. In C, low stiffness with post-yield ductility may be required if excessive deformations are acceptable. In all cases, the structural behaviour of the connection should exceed that of the connected member as shown in Figure 9.7. Connection X is a suitable connection because its deformation capacity is greater than that required by the connected member (dashed line, known as the *beam-line*). The residual

Figure 9.5: Damage to beam end due to lack of fit and/or spurious hogging moments.

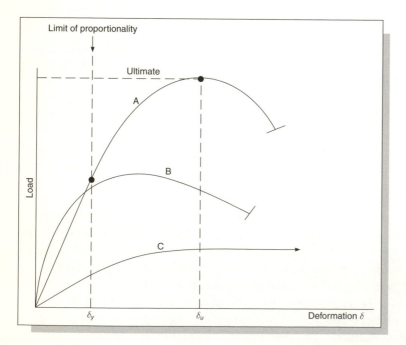

Figure 9.6: Schematic representation of behaviour of different types of connections.

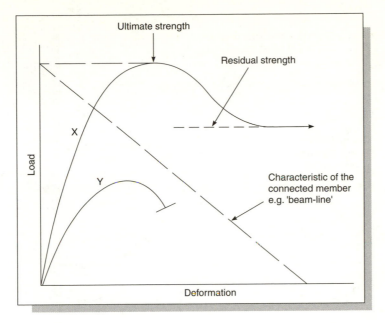

Figure 9.7: Relationship between structural behaviour of connections with respect to the member to which it is attached.

strength, rather than the actual ultimate strength, is often used in design. But Y is not a satisfactory connection because failure takes place in a brittle manner prior to matching the requirements of the member.

All connections must have a mechanical tensile force capacity, even compression and especially shear connections. Contact bond and friction is not allowed. All members must therefore have embedded anchors (Chapter 8). Even if tension is not present in the structural model, tensile capacity is provided for the purposes of robustness under abnormal loading conditions (Chapter 10).

Ductile capacity, δ_u/δ_y in Figure 9.6, is achieved by increasing the strength of brittle parts of connections. Brittle parts of the connections are well known to engineers, such as short dowels in shear, short bolts in tension, welds, congested reinforcement zones and confined rebar anchor lengths. Having established the rationale for the design of connections in the previous chapter, this chapter deals with the detailed design of beam-to-column and column-to-foundation connections.

9.2   Beam-to-column connections

Beam-to-column connections are the most important connections in precast skeletal frames. They are thought of by the profession at large as being difficult to specify, design and construct, especially those which are hidden within the

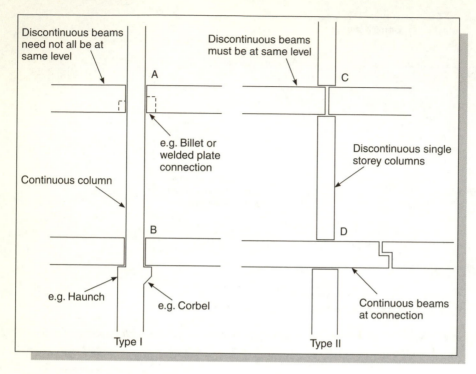

Figure 9.8: Types of beam-to-column connections.

beam. They dictate the manner of the beam in flexure – by controlling deflections and structural floor zones, and of the column in terms of frame stability and column buckling capacity.

There is a broad division shown in Figure 9.8 where:

I   the vertical member is *continuous* (both in design and construction terms) and horizontal elements are connected to it;

II   the vertical member is *discontinuous* (only in construction terms) and the horizontal elements are either structurally continuous or separate across the junction.

Type I connections fall into two further categories:

A   hidden connections, for which there is an enormous range, some of which are shown in Figures 9.9 to 9.12;

B   visible connections, such as shallow and deep corbels or nibs, shown in Figures 9.13 to 9.15.

These connections are designed mainly as simply supported for the reasons stated in Section 8.6. They are required to carry large shear forces V at the end of the beam and to transmit V into the column via an eccentric bearing. A beam–column

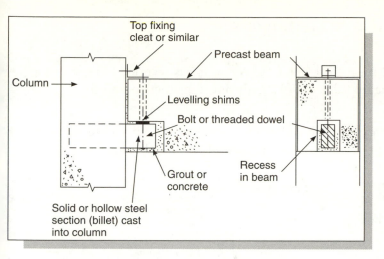

Figure 9.9: Billet beam-to-column hidden connection.

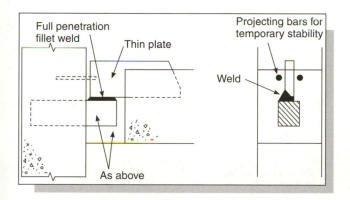

Figure 9.10: Welded plate beam-to-column hidden connection.

connection may comprise one, or more beams, typically two or three. They may be connected to the column at a common level, or may be displaced vertically depending on the versatility of the connector inside the column. Hidden connectors are more suited to multi-beam connections than corbels because of the difficulties in casting corbels on three or four faces. Hidden connectors usually comprise two main parts – a beam unit and a column unit. Some connectors introduce a third part linking the beam and column units to avoid having mould penetrations – in Figure 9.11, see how the column in the cleat connector may be cast in an un-punctured mould.

The end of the beam may be fully recessed, Figure 9.16, or if the beam is wide enough have a narrow pocket (Figure 9.17). Structurally, there is little difference between the two. However the pocketed end is easier to reinforce because the bars reach the end of the beam and therefore pass beyond the reaction point, whereas in the recessed end they terminate some distance from the reaction point.

Type II connections also fall into two categories:

C the ends of beams are simply supported and dowelled at the column head (Figure 9.18).

D continuous beams are supported and dowelled at the column head (Figure 9.19). A beam–beam half joint is made some distance from the face of the column, or else the beam forms a balcony cantilever.

Although there is continuity across the beam, especially in Type D (which is the whole purpose of this connection), Type II connections are pinned with regards to beam–column action. They offer no resistance to side sway.

9.2.1 Hidden connections to continuous columns – Type IA

To explain the structural mechanism, a billet type connector is selected. The structural model is shown in Figure 9.20a and 9.20b during construction and in service as follows:

At X: to transfer the shear force at the end of the reinforced (or uncracked prestressed) beams by a combination of vertical shear links and/or bent up bars, as shown in Figure 9.21, or into a prefabricated steel section, called a 'shear box', shown in Figure 9.22.

At Y: to ensure adequate shear capacity in the plane of the physical discontinuity between beam and column by either a projecting (solid or hollow) steel section, or a gusseted angle or T cleat bolted into anchored sockets.

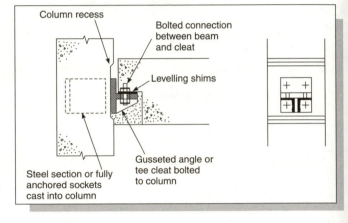

Figure 9.11: Cleat beam-to-column hidden connection.

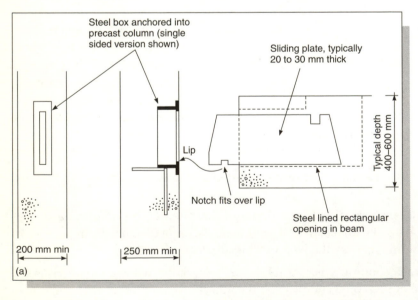

Figure 9.12a: Sliding plate beam-to-column hidden connection in detail.

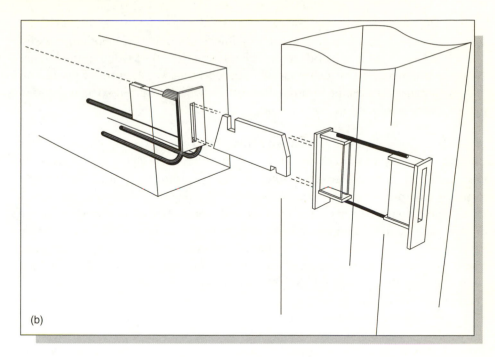

Figure 9.12b: Sliding plate beam-to-column hidden connection in isometric view.

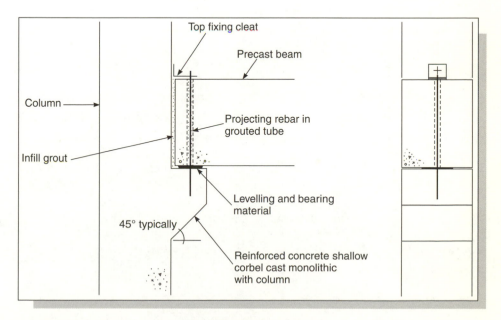

Figure 9.13: Shallow corbel beam-to-column visible connection.

At Z: to transfer the compressive loads into the reinforced concrete column. The effects of horizontal bursting forces (Section 8.3.2), both above and below the connection in the case of eccentrically loaded columns, are taken care of by using closely spaced links, see Figure 9.23. Column anchorages are generally either fully anchored cast-in-sockets, or steel box or H-section cast-inserts.

The least favourable position between contact surfaces is considered, taking into account the accumulation of frame and element tolerances. The gap between the precast elements is filled in situ using flowable mortar or grout containing a proprietary expanding agent. Figure 9.24 shows the result of shrinkage and other building movements if this operation is not correctly carried out. In some instances, particularly where the cover distance to the surface of the nearest steel insert exceeds about 50 mm, small diameter links are spot welded or otherwise attached to the inserts to form a small cage in grouted up recess. The grout, which has a minimum design strength of 30 N/mm^2 provides up to a two-hour fire protection and durability protection to the connection.

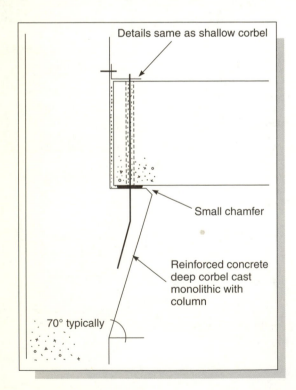

Figure 9.14: Deep corbel beam-to-column visible connection.

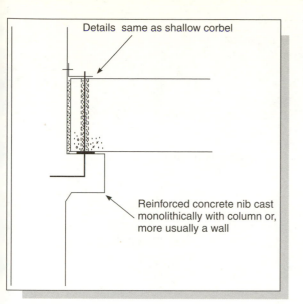

Details same as shallow corbel

Reinforced concrete nib cast monolithically with column or, more usually a wall

Figure 9.15: Nib beam-to-column (or wall) visible connection.

Figure 9.16: Recessed beam end connection to rectangular beam.

Figure 9.17: Pocketed beam end connection to inverted-tee beam.

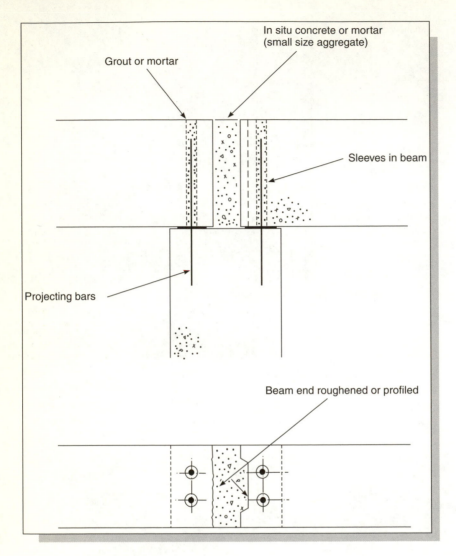

Figure 9.18: Beam-to-column head connection.

In some types of connectors, a vertical dowel (or two dowels side by side) passes through a tube in the beam and holes (circular or slotted) in the column insert (e.g. Figure 9.9). The diameter of the tube depends on the diameter of the dowel, but is typically 40–50 mm for 16–25 mm dowels. The edge cover distance to the tube is 25 mm minimum. Two or three small diameter loops, typically R8 or R10, pass around the tube to prevent localized splitting. The tube is filled in situ using expansive grout.

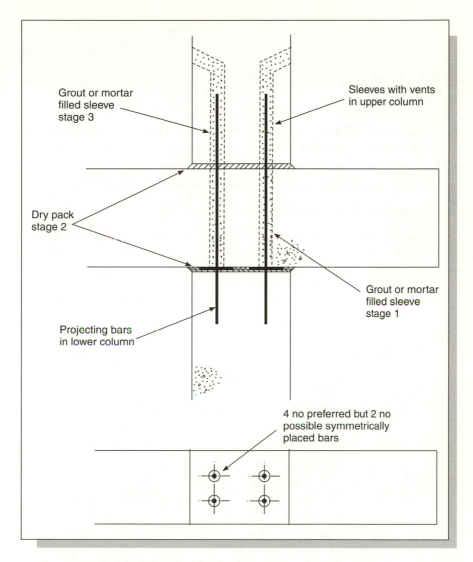

Figure 9.19: Continuous beam to discontinuous column connection.

9.2.2 Column insert design at Z

A 'column insert' is the name used to describe a steel section that is embedded in to precast columns in order to transfer shear and axial forces, and sometimes bending and torsion moments to the column. There are many types of inserts including:

- Universal column or beam (UC, UB)

- rolled channel, angle, or bent plate

- rolled rectangular or square hollow section (SHS, RHS)
- narrow plate
- threaded dowels or bolts in steel or plastic tubes
- bolts in cast-in steel sockets.

The insert may be either solid or tubular (Figure 9.9, 9.10) or cast-in sections (Figure 9.11, 9.12). The minimum breadth b_p of an insert is 50 mm. The minimum thickness of steel is taken as 6 mm for rolled sections and 4 mm for box sections providing that the insert is sealed to the passage of air and moisture. Inserts are classified as:

- 'wide sections', i.e. when b_p is in the range $75 \, \text{mm} < b_p < 0.4b$;

- 'thin plates', which include thin-walled rolled sections with wall thickness less than $0.1b$, or 50 mm. In general, additional bearing surfaces are required in thin section connectors; and

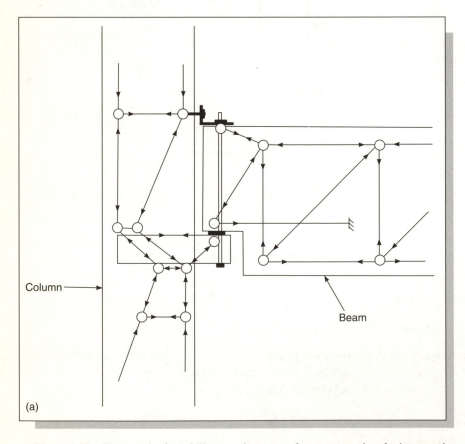

(a)

Figure 9.20a: Force paths for a billet type beam-to-column connection during erection.

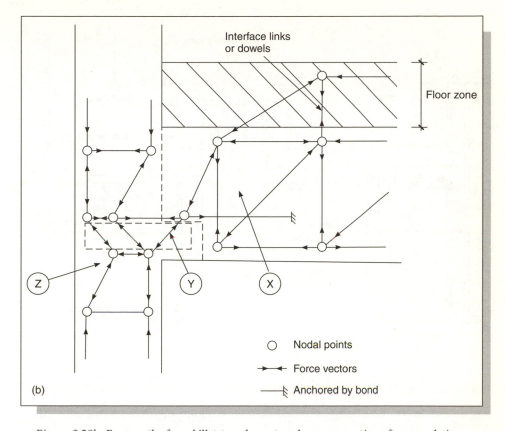

Figure 9.20b: Force paths for a billet type beam-to-column connection after completion.

- 'broad sections', $0.4b < b_p < (b - 2 \times \text{cover})$, where the cover distance to the sides of the insert is small enough to cause concern over shear cracking, and as a consequence the permissible stresses in the joint are reduced.

In the formulae above, b is the breadth of a rectangular beam of the upstand breadth of an inverted-tee beam. In the case of L beams, if the breadth of the upstand is such that a compressive strut may develop fully in the upstand, b may be taken as the breadth of the upstand. Otherwise, b is the breadth of the boot. The depth of the insert, plus cover distance to the soffit, should not be more than about half of the boot depth – detailed design examples will demonstrate the limitations here.

The *Institution of Structural Engineers Manual* on connections[1] proposes a method for determining the load and moment capacity of prismatic sections, and this has

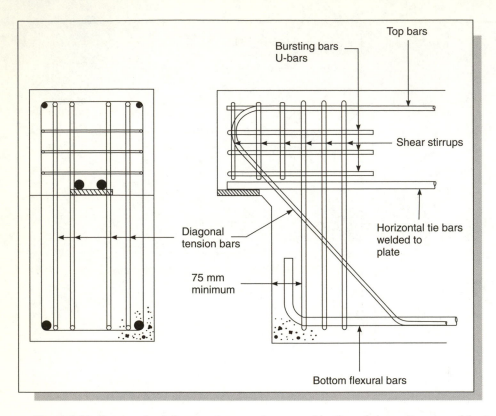

Figure 9.21: Beam end reinforcement cage using a combination of shear stirrups and bent-up bars.

found favour with many design engineers. An ultimate bearing stress of $0.8f_{cu}$ is used providing that the concrete directly above and below the column insert is confined using closely spaced links and the main longitudinal reinforcement is not interrupted by the insert. The centre-line of the insert should not be less than $1.5b_p$ from the edge of the column. The insert must lie within the column reinforcement cage.

In Figure 9.25, bearing is assumed to be uniform over limited areas of the insert and, at ultimate capacity of the insert, to be limited

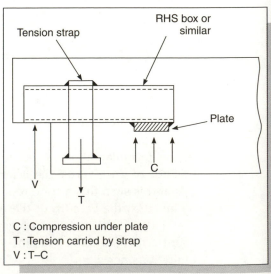

C : Compression under plate
T : Tension carried by strap
V : T–C

Figure 9.22: Prefabricated shear box.

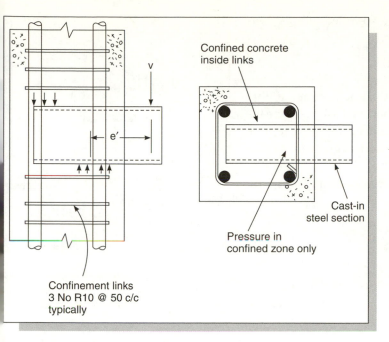

Figure 9.23: Principle of column insert design.

Figure 9.24: Shrinkage and other movement cracks around the edges of grouted up recesses.

to $0.8f_{cu}$ or f_b according to Eq. 8.2. The line pressure consists of two components – one part to react the vertical load V, and a second part to produce a couple to react the bending moment. Pressure in the cover concrete is ignored at the ultimate load.

The shear span L_1 must be taken assuming that the load V acts at the centre line of the bearing area plus an allowance for tolerances as given in Section 8.2. Then the pressure zone L_2 is given as:

$$L_2 = \frac{V}{0.8f_{cu}b_p} \qquad 9.1$$

The moment in the steel insert in the vertical plane zz is:

$$M_{zz} = V(L_1 + 0.5L_2) \qquad 9.2$$

Since the moment inside the column at zz is:

$$M_{zz} = 0.8f_{cu}b_pL_3(L_4 - L_2 - 0.5L_3 - 0.5L_3) \qquad 9.3$$

then

$$L_3(L_4 - L_2 - L_3) = L_2(L_1 + 0.5L_2) \qquad 9.4$$

and L_3 can be found by solving a quadratic equation. Note that L_4 is exclusive of the cover concrete to the links; $L_4 = h - 2 \times$ cover. To avoid overlapping stresses, the so-called '90 per cent rule' is used. Check that:

$$L_2 + 2L_3 \leq 0.9L_4 \qquad 9.5$$

so that the bearing surfaces do not overlap. If this relationship is not satisfied either increase f_{cu}, increase L_4, increase b_p or provide additional reinforcement or steel plates etc. welded to the insert as given in Section 9.2.3.

The maximum compressive force occurs below the insert and is given as:

$$F = 0.8f_{cu}b_p(L_3 + L_2) \qquad 9.6$$

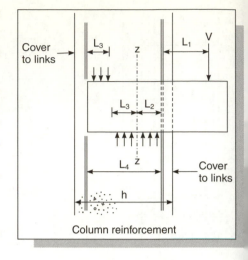

Figure 9.25: Definitions of dimensions and stress zones in column insert design.

The horizontal bursting force is calculated from end block theory to give:

$$F_{bst} = \zeta F \qquad 9.7$$

where ζ is the bursting force coefficient given in Table 8.1. The area of confinement steel A_{bst} is:

$$A_{bst} = \frac{f_{bst}}{0.95f_{yv}} \qquad 9.8$$

If the zone of pressure (i.e. $L_2 + L_3$) underneath the insert is small and located near to the front (i.e. nearest to the load) end of the insert, the area A_{bst} should be provided only by one leg of each of the confinement links. The links should be of the **closed** variety with proper anchorage. This is because the bursting forces will be present only in the front face of the column. Bursting forces will not affect the rear of the column until the pressure zone extends sufficiently far along the insert. Although there is a gradual increase in bursting forces in the column faces the following is used:

- If $L_2 + L_3 + $ cover $< h/3$, all A_{bst} to be provided by one leg of the links;
- If $h/3 < L_2 + L_3 + $ cover $< h/2$, 2/3 of A_{bst} to be provided by one leg of the links;
- If $L_2 + L_3 + $ cover $> h/2$, 1/2 A_{bst} to be provided by one leg of the links.

Confinement steel is placed above the insert to cater for the force:

$$F_{bst} = \zeta 0.8 f_{cu} b_p L_3 \qquad\qquad 9.9$$

A_{bst} is determined as above. It is likely that $L_3 < h/3$ in which case A_{bst} is provided by one leg of a link.

To design the steel insert, the section is assumed to bend about plane zz such that the plastic modulus is:

$$S > \frac{M_{zz}}{p_y} \qquad\qquad 9.10$$

and the web area is:

$$2dt > \frac{V}{0.6p_y} \qquad\qquad 9.11$$

In the case of rectangular hollow sections web buckling and bearing is avoided by filling the hollow with concrete or grout, either in the factory or during the grouting operation on site. (Note: Eq. 9.10 is quite conservative as the real behaviour is likely to be strut action, rather than bending.)

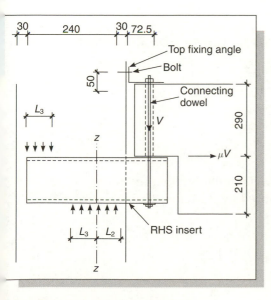

Figure 9.26: Detail to Example 9.1.

Example 9.1. *Column steel billet insert design*
Figure 9.26 shows a detail of rolled hollow section steel billet used to support a 500 mm deep× 300 mm wide precast concrete beam. The maximum depth of the RHS is 150 mm. Given that the ultimate beam reaction is 210 kN, calculate the size of the required billet, the confinement reinforcement, and the vertical threaded dowel bar.

Use $f_{cu} = 50\,\text{N/mm}^2$, $f_y = 460\,\text{N/mm}^2$, $p_y = 275\,\text{N/mm}^2$, $f_{yb}(\text{dowel}) = 450\text{N/mm}^2$, $f_{yb}(\text{bolts}) = 195\,\text{N/mm}^2$. Cover to all steel = 30 mm.

Solution
Try RHS $b_p = 100\,\text{mm} \times 150\,\text{mm}$ deep

Line pressure is the lesser of
$$f_b = 1.5 \times 50/1 + 2(100/300) = 45\,\text{N/mm}^2$$
$$\text{or} \quad 0.8 \times 50 = 40\,\text{N/mm}^2$$

$$L_2 = \frac{V}{p} = \frac{210 \times 10^3}{4000} = 52.5\,\text{mm}$$

From Figure 9.26a, $a_{max} = 72.5\,\text{mm}$

$$M_{zz} = 210 \times 10^3 \times (72.5 + 30 + 52.5/2) = 27 \times 10^6\,\text{Nmm}$$
$$L_4 = 300 - 2 \times \text{cover } 30 = 240\,\text{mm}$$

Also $M_{zz} = 4000L_3(240 - 52.5 - L_3) = 27 \times 10^6$
Solving $L_3 = 48.5\,\text{mm}$, check $L_2 + 2L_3 = 149.5 < 0.9 \times 240 = 216\,\text{mm}$
Total vertical force beneath insert $= 4000 \times (52.5 + 48.5) \times 10^{-3} = 404\,\text{kN}$
Bursting coefficient $\xi = 0.22$ for $b_p/b = 100/300 = 0.3$

$$A_{bst} = \frac{0.22 \times 404 \times 10^3}{0.95 \times 460} = 204\,\text{mm}^2$$

$L_2 + L_3 = 101\,\text{mm} <$ column $h/2$, then A_{bst} refers to one leg only
Use 2 no. T12 (226) at 50 mm spacing beneath insert.

Vertical force above billet $= 4000 \times 48.5 \times 10^{-3} = 194\,\text{kN}$

$$A_{bst} = 93\,\text{mm}^2$$

Use 1 no T12 (113) above insert.

RHS design
$M_{max} = 27 \times 10^6\,\text{Nmm}$ (ignore μV as this is resisted by surface friction over full contact area)

$$S_{xx} > 27 \times 10^6/275 = 98.2 \times 10^3\,\text{mm}^3$$

$$2dt > 210 \times 10^3/165 = 1273\,\text{mm}^2$$

Use $150 \times 100 \times 6.3$ RHS ($111\,\text{cm}^3$, $1890\,\text{mm}^2$).

Connecting dowel (Bending is ignored as the dowel is fully grouted in)
Horizontal force $H = \mu V = 0.4 \times 210 = 84\,\text{kN}$ is carried by dowel in double shear.
Maximum shear force $= 0.5 \times 84 = 42.0\,\text{kN}$

$$\text{Area} > \frac{42 \times 10^3}{0.6 \times 450} = 156\,\text{mm}^2$$

Bearing into 6.3 mm thick RHS

$$\text{Bolt diameter} > \frac{(42 \times 10^3)}{(460 \times 6.3)} = 14.5\,\text{mm}$$

Use M16 grade 8:8 dowel (210) in 50 mm diameter tube $(=16 + 2 \times 13.5$ tolerance + extra).

Top fixing angle

Horizontal force $= 42.0\,\text{kN}$

Assume upstand leg $= 100\,\text{mm}$ and thickness 12 mm

Maximum moment in angle $= 42.0 \times 10^3 \times (50 - 12) = 1596 \times 10^3\,\text{Nmm}$

$$S > \frac{1596 \times 10^3}{275} = 5804\,\text{mm}^3$$

$$\text{Shear requirement: } bt > \frac{42.0 \times 10^3}{165} = 254\,\text{mm}^2$$

Use 120 × 120 × 12 grade 43 angle × 165 wide (5940 mm³, 1980 mm²).

Bolts to column

$$T = 42.0\,\text{kN}$$

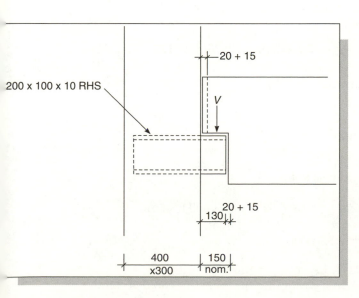

Figure 9.27: Detail to Example 9.2.

200 x 100 x 10 RHS

20 + 15

V

20 + 15

130

400

x300

150

nom.

$$\text{Area} = \frac{42.0 \times 10^3}{195} = 215\,\text{mm}^2$$

Use 2 no. M16 bolts (314).

Example 9.2. *Capacity of column billet insert*

Calculate the limiting capacity of the billet connector shown in Figure 9.27. Clearance to the end of the beam may be taken as 20 mm. The construction tolerance for the connection may be taken as 15 mm.

Use cover to all steel or rebars as 40 mm. $f_{cu} = 50\,\text{N/mm}^2$ and $p_y = 275\,\text{N/mm}^2$

Solution

Distance a from face of column to centre of bearing (see Figure 9.27)

$$a = 35 + \frac{130 - 35}{2} = 83 \, \text{mm}$$

$$L_4 = 400 - 2 \times \text{cover } 40 = 320 \, \text{mm}$$

Limiting stress zones $L_2 + 2L_3 = 0.9L_4$

Therefore,

$$L_3 = 144 - 0.5L_2 \tag{1}$$

Let line pressure $p = 0.8 f_{cu} \, b_p$ (it will be shown to cancel out in the subsequent analysis). Moments z–z at end of zone L_2 are:

$$M_{zz} = pL_2(83 + 40 + 0.5L_2) = 123pL_2 + 0.5pL_2^2 \tag{2}$$

and

$$M_{zz} = pL_3(L_4 - L_2 - L_3) = pL_3L_4 - pL_3L_2 - pL_3^2 \tag{3}$$

Substitute Eq. 1 into Eq. 3 to give

$$M_{zz} = 0.25pL_2^2 - 0.5pL_4L_2 + p(144L_4 - 20\,736) \tag{4}$$

Eq. 4 = Eq. 2 & divide through by p gives the quadratic $0.25L_s^2 + (123 + 0.5L_4)L_2 - (144L_4 - 20\,736) = 0$

Hence, $L_2 = 83.4 \, \text{mm}$ and $L_3 = 102.3 \, \text{mm}$.

Resolve vertically $V = pL_2 = 4000 \times 83.4 \times 10^{-3} = 333.6 \, \text{kN}$

Check capacity of RHS

From standard tables $S_{xx} = 252\,000 \, \text{mm}^3$

$$M_R = 275 \times 252\,000 \times 10^{-6} = 69.3 \, \text{kNm}$$

$$M_{zz} = pL_2(123 + 0.5L_2) \text{ where } p = 0.8 \times 50 \times 100 = 4000 \, \text{N/mm}$$

$$M_R = M_{zz} \text{ gives } 69.3 \times 10^6 = 492\,000L_2 + 2000L_2^2$$

Hence, $L_2 = 100.1$ which is greater than L_2 from the concrete stress analysis and is therefore not critical.

Check shear capacity of RHS

$$2dt = 4000 \, \text{mm}^2$$

$$V_R = 165 \times 4000 \times 10^{-3} = 660 \, \text{kN} > 333.6 \, \text{kN}$$

Therefore, not critical

Maximum capacity = 333.6 kN

Alternatively, the ultimate load capacity V_u may be calculated directly from a knowledge of p ($=0.8f_{cu}b_p$) and L_4 ($=h - 2 \times$ cover) for given values of L_1. Setting the gap between stress zones L_3 equal to $0.1L_4$

$$0.9L_4 = 2L_3 + L_2$$
$$pL_3(L_3 + 0.1L_4) = pL_2(L_1 + 0.5L_2)$$

Hence, $V_u = pL_2 = p(0.9L_4 - 2L_3)$ leading, by substitution and expansion, to:

$$V_u = p\left[\sqrt{4L_1(L_1 + L_4) + 1.99L_4^2} - 2L_1 - L_4\right] \qquad 9.12$$

e.g. for $L_4 = 240 \, \text{mm}$ and $L_1 = 85 \, \text{mm}$, and letting $p = 4000 \, \text{N/mm}$
$V_u = 4000 \times 64.47 = 257.9 \times 10^3 \, \text{N} = 257.9 \, \text{kN}$
Carrying out the reverse operation according to Eqs 9.1–9.4 gives:

$$L_2 = \frac{257.9 \times 10^3}{4000} = 64.5 \, \text{mm}$$

$$M_{zz} = 257.9 \times 10^3 \left(85 + \frac{64.5}{2}\right) = 30.24 \times 10^6 \, \text{Nmm}$$

$$M_{zz} = 4000L_3(240 - 85 - L_3)$$

Then, $L_3 = 76.2 \, \text{mm}$

Example 9.3

Determine the area of bursting reinforcement beneath the billet in Example 9.2. Use $f_y = 250 \, \text{N/mm}^2$.

Solution

$$b_p/b = 100/300 = 0.33$$

$$\zeta = 0.22 \, \text{(from Table 8.1)}$$

$$F = 4000(83.4 + 102.3) \times 10^{-3} = 742.8 \, \text{kN}$$

$$F_{bst} = 0.22 \times 742.8 = 163.4 \, \text{kN}$$

$$A_{bst} = \frac{163.4 \times 10^3}{0.95 \times 250} = 688 \, \text{mm}^2$$

Length of stress zone $L_2 + L_3 +$ cover $= 225.7\,\text{mm} = 0.56h > 0.5h$
Therefore, leg of link $= 1/2 \times 688 = 344\,\text{mm}^2$
Use 4 no. R12 at 50 mm spacing (452).

9.2.3 Additional reinforcement welded to inserts

Where the insert lies close to the top of the column, such that the restraining force $0.8\,f_{\text{cu}}b_{\text{p}}L_3$ cannot develop at its end due to shear failure to the side of the remote end of the insert, reinforcement should be welded to the sides of the insert. The bars are anchored by full bond development as shown in Figure 9.28. This occurs if:

$$0.8f_{\text{cu}}b_{\text{p}}L_3 > b_{\text{p}}d_{\text{t}}v_{\text{ult}} \qquad\qquad 9.13$$

where d_{t} is the distance from the top of the insert to the uppermost link, and v_{ult} is the maximum ultimate shear strength – the lesser of $0.8\,\sqrt{f_{\text{cu}}}$ or $5.0\,\text{N}/\text{mm}^2$. Generally if d_{t} is less than about 150 mm then it is certain that a designer will NOT allow the insert to rely on the holding down pressure $0.8\,f_{\text{cu}}b_{\text{p}}$, for the simple reason that it is likely that in such thin sections the limiting bearing stress would be about $0.3f_{\text{cu}}$. Referring to Figure 9.28:

$$M_{zz} = 0.8f_{\text{cu}}b_{\text{p}}L_2(L_1 + 0.5L_2) \qquad\qquad 9.14$$

$$M_{zz} = 0.8f_{\text{cu}}b_{\text{p}}L_3(d - \text{cover} - L_2 - 0.5L_3) \qquad\qquad 9.15$$

hence L_3 may be determined. Check $L_2 + L_3 < 0.7L_4$ so that a couple may be generated. The steel required to replace the up-thrust is:

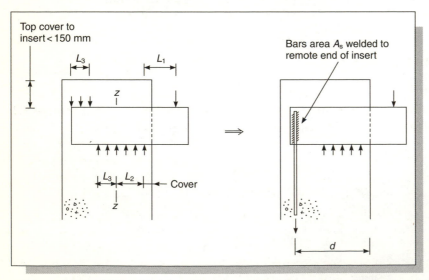

Figure 9.28: Additional holding down tie welded to insert at top of column.

$$A_s = \frac{0.8f_{cu}b_pL_3}{0.95f_y} \qquad 9.16$$

Where the top cover to the insert is greater than 200 mm, at least 3 no. closed links may be provided and the concrete well compacted, thus allowing the bearing pressures to develop.

If the strength capacity of an insert is less than that required due to overlapping pressure around the insert, one remedy is to add extra bars to both the front and rear of the insert. The bars can act in tension or compression both above and below the insert, and can develop their ultimate yield strength by the action of bond as shown in Figure 9.29. Although deformed bar is used in order to keep the anchorage bond lengths to a minimum, the ultimate stress in the bar is taken as the value used for mild steel reinforcement, i.e. 250 N/mm². In order to develop a full bond strength the bars should be positioned with a centroidal cover distance of at least 50 mm, and be enclosed within confining links of area not less than $0.5A_s$. The weld can be deposited on both sides of the bar, i.e. effective thickness $= 2 \times$ throat thickness $= 1.4 \times$ weld leg. The weld design should be according to Section 8.4.5.

The moment of resistance using this reinforcement increases from M in the unreinforced case, to M_r. The corresponding vertical shear increases from V to V_r. Referring to Figure 9.29 and resolving vertically before the introduction of the bars $V = 0.8f_{cu}b_pL_2$, and taking moments:

$$V(L_1 + 0.5L_2) = 0.95f_yA'_s(d - d') \qquad 9.17$$

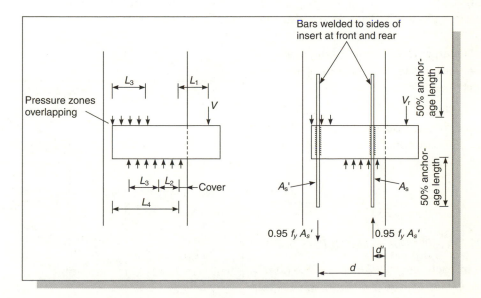

Figure 9.29: Additional reinforcement welded to insert.

and $A_s = A'_s$ because the vertical force V is resisted by the concrete alone. When V is increased to V_r then:

$$V_r = 0.8f_{cu}b_pL_2 + 0.95f_y(A_s - A'_s) \qquad 9.18$$

and

$$(V_r - V)(d + L_1 - \text{cover}) = 0.95f_yA_s(d - d') \qquad 9.19$$

where $A_s > A'_s$ and L_2 is assumed to remain unchanged from the previous case.

Example 9.4. *Column billet with additional welded bars*
The billet connector shown in Figure 9.30 is subjected to an ultimate load of 200 kN acting at a maximum distance of 75 mm from the face of the column. Check if additional bars welded to the insert are necessary, and if so, design those bars. Assume the RHS is adequate.

Use $f_{cu} = 50\,\text{N/mm}^2$, $f_y = 250\,\text{N/mm}^2$, $p_{weld} = 215\,\text{N/mm}^2$, cover $= 40\,\text{mm}$

Solution
Line pressure $p = 0.8 \times 50 \times 100 = 4000\,\text{N/mm}$
$L_2 = V/p = 200 \times 10^3/4000 = 50\,\text{mm}$
$M_{zz} = 200\,000\,(75 + 40 + 0.5 \times 50) = 28.0 \times 10^6\,\text{Nmm}$. Without additional bars,
$M_{zz} = pL_3(L_4 - L_2 - L_3) = 4000L_3\,(220 - 50 - L_3) = 28.0 \times 10^6$
Solving $L_3 = 70\,\text{mm}$, if $v_{ult} = 0.8\sqrt{f_{cu}} = 5.05$ or $5.0\,\text{N/mm}^2$, Eq. 9.13 gives $4000 \times 70 > 100 \times 150 \times 5.0$. Therefore, additional bars are needed at remove end of insert.
Edge distance to additional bars $= 40 + 10$
links $+$ radius say $16 = 66\,\text{mm}$. Therefore, $d = 234\,\text{mm}$.

$M_{zz} = 4000L_3(234 - 40 - 50 - 0.5L_3)$
$= 28.0 \times 10^6\,\text{Nmm}$. Hence, $L_3 = 61.9\,\text{mm}$
Force in additional bars $= 4000 \times 61.9 \times 10^{-3} = 247.6\,\text{kN}$.

$$A_s = \frac{247.6 \times 10^3}{0.95 \times 250} = 1042\,\text{mm}^2$$

Use 2T32 (1608) bars welded to RHS insert.

Force in each bar $= 247.6/2 = 123.8\,\text{kN}$
Use 6 mm weld size on each side of bar (i.e <0.2 diameter).

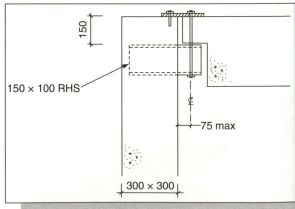

150 × 100 RHS
150
75 max
300 × 300

Figure 9.30: Detail to Example 9.4.

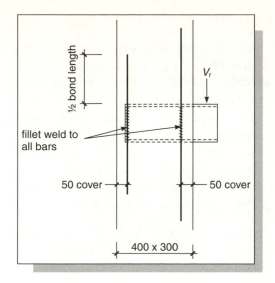

Figure 9.31: Detail to Example 9.5.

Therefore,

Leg length $= (123.8 \times 10^3)/(2 \times 0.7 \times 6 \times 215) = 68\,\text{mm} + 2 \times 6\,\text{mm}$ run outs $= 80\,\text{mm}$

Use 6 mm CFW × 80 mm long to each side of bar.

Example 9.5. *Capacity of column billet insert with additional bars*

Calculate the capacity of the billet connector in Example 9.2 if 2 no. R20 additional bars are welded to the sides of the billet as shown in Figure 9.31. Check the diameter of the bars at the remote end of the insert.

Use $f_y = 250\,\text{N/mm}^2$. Cover to additional bars $= 50\,\text{mm}$

Solution

From Example 9.2, $L_1 = 83 + 40 = 123\,\text{mm}$

$$d = 400 - \text{cover } 50 - \text{bar radius (say) } 10 = 340\,\text{mm}$$
$$d' = 50 + 10 = 60\,\text{mm}$$

$$(V_r - V)(340 + 83) = 0.95 \times 250 \times 628 \times 280 \;(\textit{using Eq. 9.19})$$
$$V_r - V = \underline{98.7\,\text{kN}}$$

Total capacity $= 333.6$ (from Example 9.2) $+ 98.7 = \underline{432.3\,\text{kN}}$

Force in bars at remote end $= (0.95 \times 250 \times 628 \times 10^{-3}) - 98.7 = 50.45\,\text{kN}$

Therefore, $A_s = (50.45 \times 10^3)/(0.95 \times 250) = 212\,\text{mm}^2$

Use 2 no. R12 bars (226).

9.2.4 *Narrow plate column inserts*

Narrow-plate inserts are steel plates not thicker than $0.1b$, where b is the width of the column, or 50 mm, which ever is least, set vertically in the column to support a beam as shown in Figure 9.32. When loaded, such plates tend to produce a splitting effect in the columns under the bearing surface, unless they are supported by transverse bearing plates or reinforcement welded to the sides of the plate. Small diameter links (T8 at 50 mm centres) are placed around the bars to confine the concrete immediately beneath the thin plate. The compressive bond resistance of the bars is used if there is an insufficient tension anchorage bond length available above the connection.

To calculate the maximum shear and moments on the plate assume that load V acts before. Referring to Figure 9.32, calculate the force F_{s2} in the steel as:

$$F_{s2} = VL_1/L_2 \qquad\qquad 9.20$$

and F_{s1} as:

$$F_{s1} = F_{s2} + V \qquad\qquad 9.21$$

Provide reinforcement A_{s1} and A_{s2} in the usual manner. Place additional column confinement links above and below the plate using $\zeta = 0.5$ for point loads in steel bars.

9.2.5 *Connections to columns using corbels*

A corbel is a short cantilever projection from the face of a column which supports a load bearing element on its upper horizontal ledge. It is used extensively in situ-

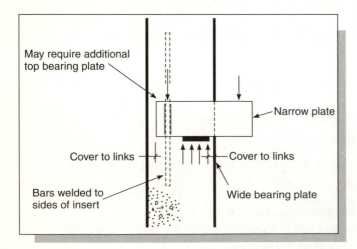

Figure 9.32: Narrow plate insert with additional reinforcement.

Figure 9.33: Corbel connections at Hartwell Arena car park, Finland (courtesy Betoni).

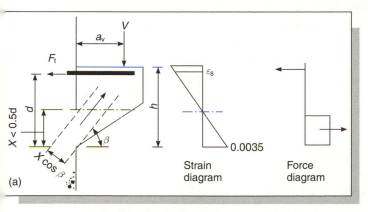

(a)

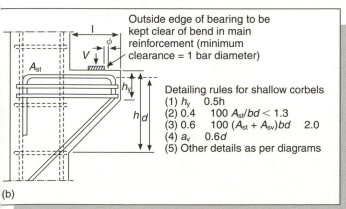

Outside edge of bearing to be
kept clear of bend in main
reinforcement (minimum
clearance = 1 bar diameter)

Detailing rules for shallow corbels
(1) h_y 0.5h
(2) 0.4   100 $A_{st}/bd < 1.3$
(3) 0.6   100 $(A_{st} + A_{sv})bd$ 2.0
(4) a_v 0.6d
(5) Other details as per diagrams

(b)

Figure 9.34: Principle of shallow corbel design.

ations where the visual aesthetics of connections is not important, for example in the parking structure shown in Figure 9.33, and where particularly heavy loads are to be transferred from beams to columns – also the situation shown in Figure 9.33!

In other situations it is often desirable to contain the connection within the overall depth of the beam. For reasons of appearance this calls for a 'shallow' corbel, defined in Figure 9.34. Shallow corbels are designed as short cantilevers, according to BS8110, Part 1, Clause 5.2.7, where the distance to the effective position of load action $a_v < 0.6d$. The depth of the face of the corbel should not be less than half the total depth. The width of a corbel must be less than about 500 mm, otherwise it qualifies as a continuous nib. Compared with nibs, which are used for lighter loads, say less than 150 kN, forces are distributed more evenly in a corbel thanks to the inclined soffit. The other difference between a corbel and a nib is that in a corbel closed links are provided horizontally to resist the shear force, whereas in a nib the links are only provided vertically to resist short bending (or strut-and-tie) actions.

9.2.5.1 Shallow corbels

In a shallow corbel design shear failure is first precluded. The ultimate shear stress $v = V/bd$ at the face of the column should not exceed the lesser of $v_{ult} = 0.8\sqrt{f_{cu}}$ or $5.0\,\text{N/mm}^2$. Shear reinforcement A_{sv} is required if $v > v_c(2d/a_v)$ such that:

$$A_{sv} = \frac{a_v b \left[v - v_c \left(\frac{2d}{a_v} \right) \right]}{0.95 f_{yv}} > \frac{0.4 a_v b}{0.95 f_{yv}} > 0.4\, bd/100 \qquad 9.22$$

$A_{sv} > 0.5A_s$ (where A_s is obtained from Eq. 9.27) ensures confinement of the concrete in the compressive strut. Thus, the minimum total area of steel $A_s + A_{sv} > 0.6\% \times$ concrete area. Horizontal links are placed in the upper two-thirds of the corbel. The bar diameter should not be more than 12 mm. The distance to the first link should not be more than 75 mm from the bearing surface.

The effective bearing area is confined to within the reinforced concrete zones, i.e. any net dimensions are exclusive of the concrete cover c. If the breadth of the corbel is given by b, the bearing length is given by:

$$a_{eff} = \frac{V}{0.4 f_{cu}(b - 2c)} \qquad 9.23$$

where V is the ultimate applied force.

The length of the seating ledge is determined from the sum of the distance to the centre of the load a_v plus half the bearing length a_{eff} plus an edge distance allowance. For HT bar with an inner bend radius of 3ϕ, the allowance is $5\phi + $ cover to bar; for mild steel bars it is $4\phi + $ cover. If the overall depth h of the corbel is at least $1.5a$, and the depth of the front face of the corbel is $0.5h$, the effective depth d is given as:

$$d = 1.5a_v + 0.75a_{eff} + 7\phi + 0.5c \qquad 9.24$$

Then

$$\beta = \tan^{-1}\frac{d - 0.5X}{a_v} \qquad 9.25$$

and

$$F_t = V/\tan\beta \quad \text{and} \quad F_c = V/\sin\beta \qquad 9.26$$

The tensile tie capacity F_{tR} is given as:

$$F_{tR} = A_s 0.95 f_y \qquad 9.27$$

$A_s > 0.4\%$ of the section at the face of the column, and the steel should be fully anchored at the far face of the column. The compressive capacity of the diagonal strut F_{cR} is given by:

$$F_{cR} = 0.4 f_{cu} b X \cos \beta \qquad 9.28$$

To prevent the concrete strain from exceeding 0.002 at ultimate then $X < 0.5d$. Additional horizontal U shape bars are added beneath the bearing to cater for the horizontal friction force μV.

Example 9.6. *Shallow corbel*
A 300×300 mm column supports a 300 mm wide precast beam on a single sided reinforced concrete corbel. The beam has pretensioning strands extending to the end of the beam. The gap between the end of the beam and column is, inclusive of all tolerances, 20 mm. Because the beam experiences end rotations due to flexure of 0.005 radians a neoprene bearing pad is to be used. Given that the ultimate end reaction from the beam is 200 kN, determine the size of the bearing pad, size of corbel and reinforcement required.
   Use $f_{cu} = 50$ N/mm², $f_y = 460$ N/mm², $f_y = 250$ N/mm² for stirrups, $f_{neoprene} = 10$ N/mm², $E_{neoprene} = 50$ N/mm². Cover to all steel = 30 mm.

Solution
Bearing pad
(Iteration is necessary due to beam-end rotation influencing final pad stresses)
   Without rotation effects, Area $> 200 \times 10^3/10 = 20\,000$ mm². Maximum width $b_p = 300 - 2$ cover $20 = 240$ mm less tolerances $= 220$ mm. Therefore, $b_l \geq 20\,000/220 = 91$ mm. Add about, say 30 per cent to this length to cater for end rotations. Try $200 \times 120 \times 10$ bearing pad (10 mm is considered to be a minimum thickness practically). Then $A = 24 \times 10^3$ mm² and $Z = 200 \times 120^2/6 = (480 \times 10^3)$ mm³. Additional eccentricity e due to end rotation $= \theta$.

$$e = \frac{\theta Z E b_\ell}{2Vt} = \frac{0.005 \times 480 \times 10^3 \times 50 \times 120}{2 \times 200 \times 10^3 \times 10} = 3.6 \text{ mm}$$

Maximum stress in pad $= (200 \times 10^3)/(24 \times 10^3) + (200 \times 10^3 \times 3.6)/(480 \times 10^3) = 9.8$ N/mm² < 10 N/mm².
Use $200 \times 120 \times 10$ bearing pad.

Corbel ledge and reinforcement

Ref. Figure 9.35.

Assume T12 mm bars in top of corbel

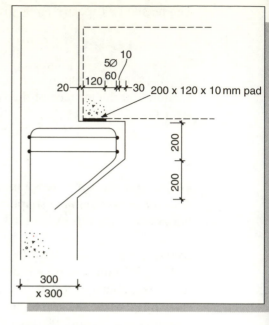

a = gap $20 + 120 + 5 \times$ diameter $12 +$ link $10 +$
     cover $30 = 240$ mm, round off to 250 mm

$h \geq 1.5a = 375$ mm, round off to 400 mm

$h_f \geq 0.5h = 200$ mm

$d = 400 - 30 - 6 = 364$ mm

Shear check

$v = (200 \times 10^3)/(300 \times 364) = 1.83 \, \text{N/mm}^2$
$< 0.8\sqrt{50} = 5.66$ or $5.0 \, \text{N/mm}^2$

Use 250 mm long ledge $\times$ 400 deep corbel with 200 deep face.

$$a_v = 20 + 120/2 + 3.6 = 84 \, \text{mm}$$

$$a_v/d = 0.22 < 0.6$$

$$X = 0.5d = 182 \, \text{mm}$$

Figure 9.35: Detail to Example 9.6.

$\beta = \tan^{-1}(364 - 0.5 \times 182)/84 = 73°$ (*using Eq. 9.25*)

$F_c = \dfrac{200}{\sin 73°} = 209.3 \, \text{kN}$ (*using Eq. 9.26*)

$F_{cR} = 0.4 \times 5.0 \times 300 \times 182 \times \cos 73° \times 10^{-3} = 319 \, \text{kN} > 209.3 \, \text{kN}$ (*using Eq. 9.28*)

$F_t = 200/\tan 73° = 61.1 \, \text{kN}$ (*using Eq. 9.26*)

Horizontal friction force $= 0.7 \times 200 = 140 \, \text{kN}$

Total $F_t = 61.1 + 140 = 201.1 \, \text{kN}$

$A_{st} = (201.1 \times 10^3)/(0.95 \times 460) = 460 \, \text{mm}^2$

Minimum area $= 0.4\% \times 300 \times 364 = 437 \, \text{mm}^2$

Use 4 no. T12 bars (452) at 70 mm spacing.

Force in each bar $= 201.1/4 = 50.3 \, \text{kN}$

Bend radius $r = 50.3 \times 10^3(1 + 2(12/58))/2 \times 50 \times 12 = 59 \, \text{mm} > 4 \times$ diameters assumed $= 48 \, \text{mm}$. Therefore, corbel length to increase by 11 mm to 251 mm as already provided.

Shear reinforcement

$$v = 1.83 \, \text{N/mm}^2$$

$$100A_s/bd = (100 \times 452)/(300 \times 364) = 0.41$$

$$v_c = 0.56 \, \text{N/mm}^2 \times 2d/a_v = 4.8 \, \text{N/mm}^2 > 1.83 \, \text{N/mm}^2$$

Provide minimum links $A_{sv} = (0.4 \times 110 \times 300)/(0.95 \times 250) = 56 \, \text{mm}^2$ or $0.5A_{st} = 252 \, \text{mm}^2$

Use 2 no. R10 (314) at 100 mm spacing from top of corbel.

Minimum total steel $= 0.6\% \times 300 \times 364 = 655 \, \text{mm}^2 < 452 + 314 = 766 \, \text{mm}^2$ provided.

9.2.5.2 Deep corbels

Deep corbels simulate inclined columns where $a_v < 0.2d$. A deep corbel is usually required because the capacity of the shallow corbel is insufficient. However, the problem of local bursting stresses at the bearing surface is just as important as before, and so mild steel spreader plates are often cast into the bearing surface. The tension steel A_s is welded to the steel plate to form a positive anchorage.

The same design equations are used for deep and shallow corbels except that the shear stress in the compressive region should not exceed $1.3 \, \text{N/mm}^2$ (BS8110

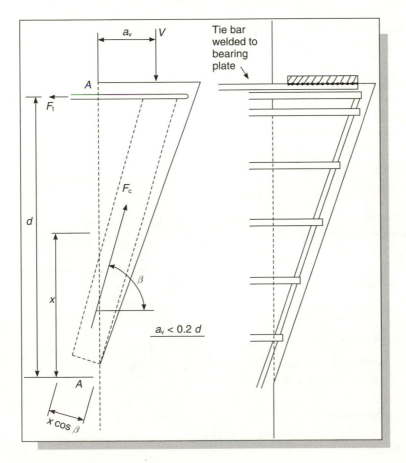

Figure 9.36: Principle of deep corbel design.

Part 1, Clause 5.3.7) and the compressive strut should be restrained using horizontal bars as shown in Figure 9.36. The *I Struct E Manual*[1] suggests that apart from A_s, main strut reinforcement A_{sc} should be provided such that:

$$A_{sc} = 0.5bX \cos \beta / 100 \qquad\qquad 9.29$$

and horizontal links should be uniformly distributed over the full depth of the corbel of 0.4 per cent of the concrete area. The size of the bars should be at least 1/4 of the diameter, and the spacing not exceeding 12 times the diameter of the main strut reinforcement.

9.3 Beam end shear design

Special attention is given to shear reinforcement near to the end connections. Stirrups and bent-up reinforcing bars are provided to ensure the transfer of shear in the critical region. Bent up bars may only be used to resist not more than one-half of the shear force – the remainder must be provided by vertical links. In some instances, a prefabricated shear box (Figure 9.22) partially or wholly replaces the stirrup cage. Shear boxes are used to:

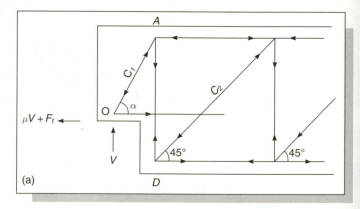

1 transfer the shear forces to a point in the beam where the stirrups are considered to be fully effective;

2 prevent bursting of the concrete at ends of the beam; and

3 provide a steel bearing plate with which to make a steel-to-steel connection to the supporting member.

There are two different design approaches depending on the depth of the end recess – or more correctly whether an inclined compressive strut, O–A in Figure 9.37a, can or cannot develop fully above the bearing in the nib.

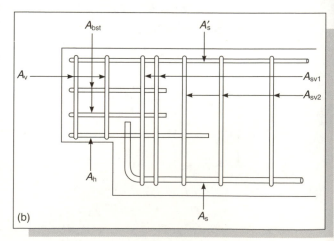

Figure 9.37: Principle of beam end design with shallow recess.

In both cases, end bursting reinforcement (Section 8.3.2) and horizontal friction reinforcement (Section 8.3.3) must be provided. A check to determine whether a steel bearing plate is necessary is carried out (Section 8.3.2). The plate, or short length of rolled steel angle, is fully bonded to the beam by virtue of tie bars welded to the inside surface of the plate (or angle). Further, from the recess truss action develops in the beam.

At the end of the recess, the ultimate shear stress is given as:

$$v = \frac{V}{bd_h} \le 0.8\sqrt{f_{cu}} \quad \text{or} \quad 5.0 \, \text{N/mm}^2 \qquad\qquad 9.30$$

where b = breadth of the beam, and d_h the effective depth of the nib.

Horizontal friction reinforcement A_h should be provided as shown in Figure 9.37b such that:

$$A_h = \frac{\mu V + F_t}{0.95 f_y} \qquad\qquad 9.31$$

where μ is defined in Table 8.2, and F_t is a stability tie force (Section 10.4) being carried through the connector. If the tie is carried elsewhere, for example by site bars, then $F_t = 0$.

At least two mild steel bars should be used for A_h and they should preferably be welded to the bearing plate as it is unlikely that full anchorage would be achieved by other means. If no plate is required a single bar should be formed into a U shape and placed directly over the bearing with 15–20-mm cover. The size of this bar should not be greater than 20-mm diameter in order to achieve the correct position. It is important that the internal radius of the bend should be checked for bursting forces. If more than one bar is required the vertical separation distance should not exceed 50 mm.

According to BS8110, Part 1 Eq. 50, the internal radius r of a bar of diameter ϕ that extends for more than 4 diameters beyond the point at which it is required to resist the full force F is given by:

$$r = \frac{F(1 + (2\phi/a_b))}{2f_{cu}\phi} \qquad\qquad 9.32$$

where in this case $F = 0.5\,(\mu V + F_t)$ because the U-bar has two legs extending into the beam. Also, a_b is the distance from the inside face of the bar to the nearest free surface, or the clear distance between bars.

Case 1 *Shallow Recess*
Where $a_v < 0.6 d_h$, truss action develops by providing adequate compression and tensile members inside the beam. The ultimate vertical force V is resolved into the

first compressive strut, assumed to be inclined at α to the horizontal as shown in Figure 9.37(a). The compression in the strut is:

$$C_1 = \frac{V}{\sin \alpha} \qquad 9.33$$

This compressive force is limited such that the depth of compressive zone, when projected onto a vertical plane, does not exceed $0.5\,d_h$. Then:

$$C_{1,\max} = 0.4f_{cu}b\,0.5d_h \cos \alpha \qquad 9.34$$

This strut must be anchored into a horizontal tie at the loaded area. The area of the horizontal tie, which is in addition to that given in Eq. 9.31 is:

$$A_h = \frac{V \cot \alpha}{0.95f_y} \qquad 9.35$$

At the top of the strut, point A, the force C_1 is resolved vertically to give:

$$A_{sv} = \frac{V}{0.95f_y} \qquad 9.36$$

and resolved horizontally to give compression steel:

$$A'_s = \frac{V \cot \alpha}{0.95f_y} \qquad 9.37$$

The second diagonal strut forms at an assumed inclination of 45° such that:

$$C_2 = \frac{V}{\sin 45°} \qquad 9.38$$

with a limiting value of $C_2 = 0.4f_{cu}b0.5\,d \cos 45° = 0.14f_{cu}bd$ $\qquad 9.39$

The area of longitudinal steel in the bottom of the beam is:

$$A_s = \frac{C_2 \cos 45°}{0.95f_y} = \frac{V}{0.95f_y} \qquad 9.40$$

These bars should be anchored for a distance of $12 \times$ diameter beyond the nodal point D in Figure 9.37.

The bars shown as A_v in Figure 9.37b require no design, unless a diagonal crack occurs in the nib and the compressive strut O–A is rendered ineffective. This is unlikely to happen providing the horizontal bars A_{bst} and A_h are in place. However, it is common practice to provide 2 no. links to form A_v of the same diameter as A_{sv}.

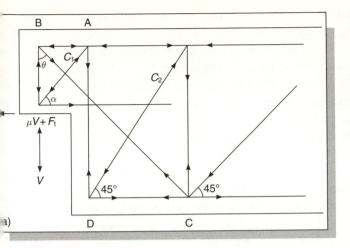

Case 2 *Deep Recess*

Where the depth of the recess is more than $2d/3$, or where $a_v > 0.6\,d_h$, or where d_h is less than about 200 mm, the structural behaviour changes from the pattern shown in Figure 9.37 to that shown in Figure 9.38. Here 50 per cent of the end reaction V is transferred to the body of the beam using a diagonal tie B–C, inclined at θ to the vertical, whilst the remainder is carried through the inclined strut. Remember that lateral bursting steel A_{bst} and horizontal tie steel A_h is still present.

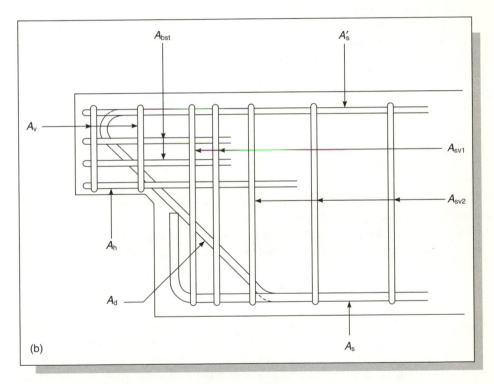

Figure 9.38: Principle of beam end design with deep recess.

The area of diagonal bars A_d is given by:

$$A_d = \frac{0.5V}{0.95f_y \cos\theta}$$

9.41

These bars must be fully anchored in both the top and bottom of the beam as shown in Figure 9.38b. The bend at the top is usually 135° and so the bending radius, Eq. 9.32, must be strictly obeyed because the force in the bar must continue around the top bend. The diagonal bar must also continue for a full anchorage length from the root of the recess. A small triangular chamfer, say $25 \times 25\,mm$, is often used to reduce the possibility of cracks developing at the internal corner.

The remaining $0.5V$ is carried by the first diagonal compressive strut, assumed to be inclined at α to the horizontal. This force is dealt with by using Eqs 9.33, 9.34 and 9.35, substituting $0.5V$ for V. The top bars A'_s must not be less than the minimum percentage steel for concrete in compression, and they must be surrounded by links up to the end of the beam in order to be fully effective. The bars should extend within a cover distance from the end of the beam and be lapped to the diagonal bars a distance equal to 1.4 times the basic lap length. Occasionally, the top bars are welded to the first vertical link to provide the necessary anchorage at the end of the cage.

To complete the shear cage the first full depth vertical links A_{sv1} designed using the truss action, shown in Figure 9.38, should be placed at one cover distance from the inside face of the recess. The area of the links is:

$$A_{sv} = \frac{0.5V}{0.95f_y}$$

9.42

Note that the concrete shear stress v_c is ignored. It is likely that pairs of links are required. The links should enclose the diagonal bars A_d, the hair-pin bars A_{bst} and the tie bars A_h. The vertical force in the links is equilibrated by a second compressive strut C_2, inclined at 45° to the horizontal. Equations 9.38 and 9.39 are used, again substituting $0.5V$ for V.

At the bottom of the inclined bar, the two separate actions are additive. Thus, the area of the longitudinal bar in the bottom of the beam is given by Eq. 9.40. At this same point, the vertical resolution of the diagonal forces calls for a second set of vertical links A_{sv2} positioned at one effective depth beyond C so that:

$$A_{sv2} = \frac{V}{0.95f_y}$$

9.43

Beyond this point the development of shear resistance is subsequently catered for by the links and concrete according to BS8110, Part 1, Clause 3.4.5.

Example 9.7. *Recessed beam end reinforcement*

A precast reinforced concrete beam is 500 mm deep × 300 mm wide. It is supported on steel billet inserts cast into 300×300 mm columns. The clear span of the beam is 5.7 m. The cross-section of the billet is 150 mm deep and 100 mm wide.

The total ultimate uniformly distributed loading on the beam (including self weight) is 70 kN/m (see Figure 9.39).

Given the following data, calculate the reinforcement required in the end of the beam and, if necessary, a bearing plate. Assume that stability tie forces F_t are carried elsewhere.

Use $f_{cu} = 40\,\text{N/mm}^2$, $f_y = 460\,\text{N/mm}^2$, $f_y = 250\,\text{N/mm}^2$ for welded bars, $p_y = 275\,\text{N/mm}^2$ for plate, $f_{weld} = 215\,\text{N/mm}^2$. Cover to all steel = 30 mm. Nominal gap between end of beam and column face = 20 mm. Bearing shims thickness = 10 mm.

Solution

Beam end reaction $V = 70 \times 6.0/2 = 210\,\text{kN}$

Beam length tolerance = $3 \times 5.7 = 17.1\,\text{mm}$

Construction tolerance = 10 mm between column faces

Total tolerance = 27.1 mm = 13.5 mm per end

Therefore, Maximum gap = 20 (nominal) + 13.5 = 35 mm rounded up.

Refer to Figure 9.39a. Let edge distance to grout tube = 25 mm and tube radius = 25 mm. Therefore, distance from end of beam to centre of tube = 50 mm.

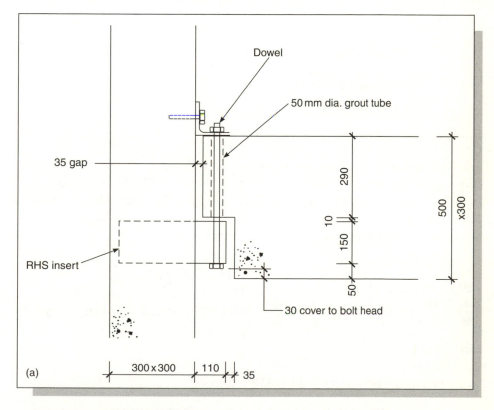

(a)

Figure 9.39: Detail to Example 9.7.

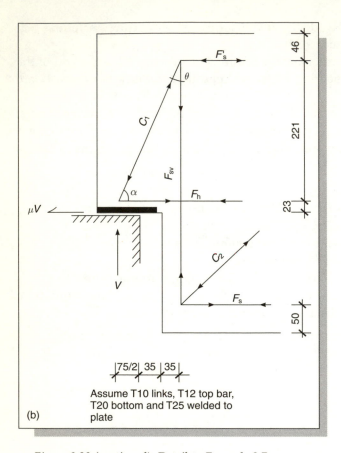

Figure 9.39 (continued): Detail to Example 9.7.

Target distance for centre of bearing to coincide with dowel position, $a = 35\,\text{gap} + 50$ to dowel $= 85\,\text{mm}$, $\pm 13.5\,\text{mm}$ tolerance $= 72$–$98\,\text{mm}$ range.

Try billet length $= 110\,\text{mm}$, then $a = (110 + 35)/2 = 72.5\,\text{mm}$ which is within the above range.

Maximum bearing length $= 110 - 35 = 75\,\text{mm}$

Shear stress at nib

$d_\text{h} = 290 - 10\,\text{plate} - 13\,\text{bar radius} = 267\,\text{mm}$

$$v = \frac{210 \times 10^3}{300 \times 267} = 2.62\,\text{N/mm}^2 < 0.8\sqrt{40} = 5.05 \text{ or } 5.0\,\text{N/mm}^2 \qquad (using\ Eq.\ 9.30)$$

Bearing plate

Try $b_\text{p} = 100\,\text{mm}$

Allowable bearing stress $= 0.8 \times 40 = 32\,\text{N/mm}^2$

$$b_1 = (210 \times 10^3)/(100 \times 32) = 65\,\text{mm} < 75\,\text{mm available.}$$

However, provide 120 mm wide × 100 mm long plate to allow for lateral and longitudinal tolerances.

Plate thickness.
Shear strength $= 0.6p_y$
Plate in double shear. Effective length $= 75$ mm

$$t > (210 \times 10^3)/(2 \times 0.6 \times 275 \times 75) = 8.5 \text{ mm}$$

Plate also subject to horizontal force $\mu V = 0.4 \times 210 = 84$ kN

$$t > (84 \times 10^3)/(120 \times 275) = 2.5 \text{ mm}$$

Use 120 × 100 × 10 plate grade 43.

Reinforcement Design
$$\alpha = \tan^{-1} 221/107.5 = 64° \quad \text{(Ref. Figure 9.39b)}.$$

$$C_1 = \frac{210}{\sin 64°} = 233.6 \text{ kN} \qquad (using\ Eq.\ 9.33)$$
$$C_{1,\max} = 0.4 \times 40 \times 300 \times 0.5 \times 267 \cos 64° \times 10^{-3} = 281 \text{ kN} \qquad (using\ Eq.\ 9.34)$$

Compressive strut capacity OK.

Horizontal bars welded to plate
Use HT deformed bars with mild steel stress

$$F_h = V \cot \alpha + \mu V + F_t = 210 \cot 64° + 0.4 \times 210 + 0 = 186.4 \text{ kN}$$
$$A_h = \frac{186.4 \times 10^3}{0.95 \times 250} = 784 \text{ mm}^2 \qquad (using\ Eqs\ 9.31\ and\ 9.35)$$

Use 2 no. T25 bars (981) welded to plate.

Bar length $= 34 \times$ diameter $= 34 \times 25 = 850$ mm from end of plate.
Force in each bar $= 186.4/2 = 93.2$ kN
Try 6 mm double sided fillet weld leg, $l_w = (93.2 \times 10^3)/(2 \times (6/\sqrt{2}) \times 215) = 51 + 12$ run outs $= 63$ mm < 75 mm available.
Use 6 mm CFW × 60 long.

Vertical stirrups

$$A_{sv} = \frac{210 \times 10^3}{0.95 \times 460} = 481 \text{ mm}^2 \qquad (using\ Eq.\ 9.36)$$

Use 3 no. T12 stirrups (678) at 50 mm spacing.

Top longitudinal reinforcement

$$A'_s = \frac{210 \times 10^3 \cot 64°}{0.95 \times 460} = 234\,\text{mm}^2 \qquad \text{(using Eq. 9.37)}$$

Use 2 no. T16 bars (402).
Length of bar from nodal point $= 35 \times$ diameter $= 35 \times 16 = 560\,\text{mm}$
Length of bar from end of beam $= 560 + 145 - 30 = 675\,\text{mm}$

Compression field

$$C_2 = 210/\sin 45° = 297\,\text{kN} \qquad \text{(using Eq. 9.38)}$$

$$C_{2,\text{max}} = 0.14 \times 40 \times 300 \times 450 \times 10^{-3} = 756\,\text{kN} \qquad \text{(using Eq. 9.39)}$$

Compressive strut 2 capacity OK.

Bottom longitudinal reinforcement

$$A_s = (210 \times 10^3)/(0.95 \times 460) = 481\,\text{mm}^2 \qquad \text{(using Eq. 9.40)}$$

Use 2 no. T20 (628) with full bend at end.

End lateral bursting

$$b_p/b = 100/300 = 0.33$$
$$\xi = 0.22 \text{ (from Table 8.1)}$$
$$F_{\text{bst}} = 0.22 \times 210 = 46.2\,\text{kN}$$
$$A_{\text{bst}} = 106\,\text{mm}^2$$

Use 3 no. T8 bars (150) at 75 mm spacing with first bar 10 mm above welded horizontal bar.
Leg length $= 34 \times 8 = 272\,\text{mm}$, provide also 106 mm² closed links in rib end.
Use 2 no. T8 closed links (100) at 50 mm spacing.

Example 9.8.   *Recessed beam end reinforcement*
Repeat Example 9.7 using a beam depth of 400 mm.

Solution
Bearing plate same as Example 9.7
nib depth $= 190\,\text{mm}$ (Ref. Figure 9.40a)

Shear stress at nib

$$v = (210 \times 10^3)/(300 \times 167) = 4.19\,\text{N/mm}^2 < 5.05\,\text{N/mm}^2$$

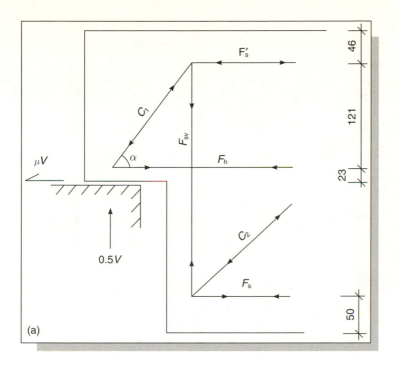

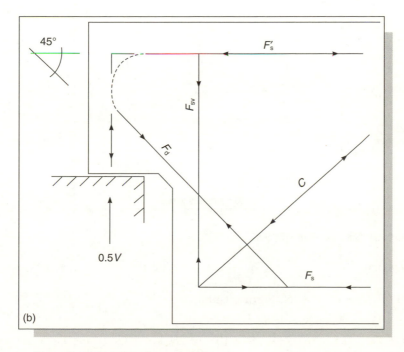

Figure 9.40: Detail to Example 9.8.

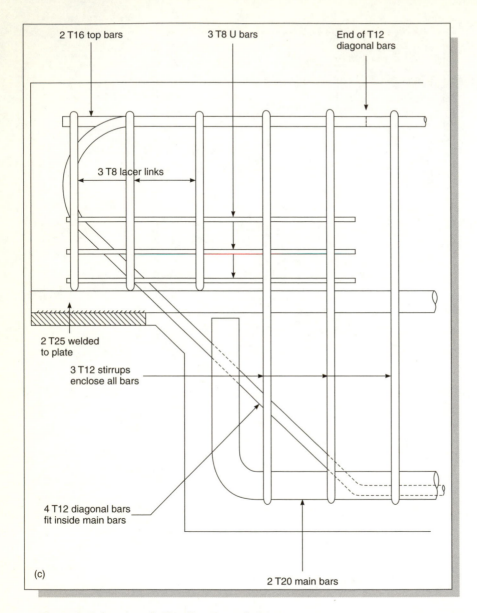

Figure 9.40 (continued): Detail to Example 9.8.

Reinforcement Design

$\alpha = \tan^{-1} 121/107.5 = 48.4°$ (Ref. Figure 9.40a)

$C_1 = 210/\sin 48.4° = 280.8 \, \text{kN}$ *(using Eq. 9.33)*

$C_{1,\text{max}} = 0.4 \times 40 \times 300 \times 0.5 \times 167 \cos 48.4° \times 10^{-3} = 266 \, \text{kN} > C_1$ *(using Eq. 9.34)*
Strut capacity at limit. Use diagonal bars to carry $0.5V$ (see Fig 9.40b).

$$C_1 = 0.5 \times 280.8 = 140.4\,\text{kN} < 266\,\text{kN}$$

Compressive strut 1 capacity OK.

Horizontal bars welded to plate

Use HT deformed bars (to reduce bond length) with mild steel stress $f_y = 250\,\text{N/mm}^2$

$$F_h = V\cot\alpha + \mu V + F_t = 0.5 \times 210\cot 48.4° + 0.4 \times 210 + 0$$
$$= 177.2\,\text{kN} \qquad (using\ Eqs\ 9.31\ and\ 9.35)$$
$$A_h = \frac{177.2 \times 10^3}{0.95 \times 250} = 746\,\text{mm}^2$$

Use 2 no. T25 bars (981) welded to plate as in Example 9.7.

Diagonal bars inclined at 45°

$$F_d = 0.5V/\sin 45° = 105/\sin 45° = 148.5\,\text{kN}$$
$$A_d = (148.5 \times 10^3)/(0.95 \times 460) = 340\,\text{mm}^2$$

Use even number of bars. Avoid using a bar on the centre line of the beam because of the space needed for grout tube to fixing dowel.
Use 4 no. T12 bars (452) at 55 mm apart.

Internal radius
$$F \text{ per bar} = 148.5/4 = 37.1\,\text{kN}$$
$$A_b = 55 - 12 = 43\,\text{mm}$$
$$r > \frac{37.1 \times 10^3(1 + 2(12/43))}{2 \times 40 \times 12} = 60\,\text{mm}$$

Anchorage length from root of nib $= 35 \times 12 = 420\,\text{mm}$

Remainder of the reinforcement is as Example 9.7. Figure 9.40c shows the completed cage.

Example 9.9. *Recessed beam end shear box*
Repeat Example 9.8 using a beam depth of 350 mm. If steel sections are required use $p_y = 275\,\text{N/mm}^2$ and $p_{\text{weld}} = 215\,\text{N/mm}^2$.

Solution
It is clear from Figure 9.40c it is impossible to position the diagonal reinforcing bar in the nib. Also, from Figure 9.41a, $\alpha = \tan^{-1} 71/107.5 = 33°$ for which inclined strut action is not possible (also by inspection with Example 9.8). The solution calls for the use of a prefabricated shear box as shown in Figure 9.41b.

To provide 30 mm cover to links (of say 10 mm diameter), maximum depth of steel box $= 140 - 30 - 10 = 100\,\text{mm}$. Therefore, try 100×100 SHS.

Try reinforcement hanger bars

Assuming reinforcement hanger bars of 25-mm diameter $L_1 = 37.5 + 35 + \text{cover } 30 + 12.5 = 115\,\text{mm}$, but allow extra say $5\,\text{mm}$ tolerances $= 120\,\text{mm}$ (Figure 9.41(b)).

Maximum pressure under box is lesser of:

$$f_b = \frac{1.5 \times 40}{1 + (2 \times 100/300)} = 36\,\text{N/mm}^2 \quad \text{or} \quad 0.8 \times 40$$
$$= 32\,\text{N/mm}^2$$

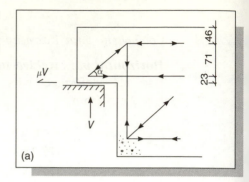

(a)

Therefore, line pressure under box $= 32 \times 100 = 3200\,\text{N/mm}^2$

Taking moments at hanger bars $210 \times 10^3 \times 120 = 3200\,L_3(L_4 - 0.5L_3)$, then $L_3L_4 - 0.5L_3^2 = 7875$. Minimum value of L_3 exists when $L_3 = 0.5L_4$ for which $L_3 = 125.5\,\text{mm}$ and $L_4 = 251\,\text{mm}$. In practice, this arrangement will give rise to very large strain gradients and a potentially small lever arm between the hanger bar and compression zone. Increase $L_4 = 297.5\,\text{mm}$ (making the overall length of shear box $= 297.5 + 152.5 = 450\,\text{mm}$). Then $L_3 = 27.8\,\text{mm}$

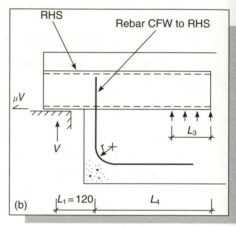

(b)

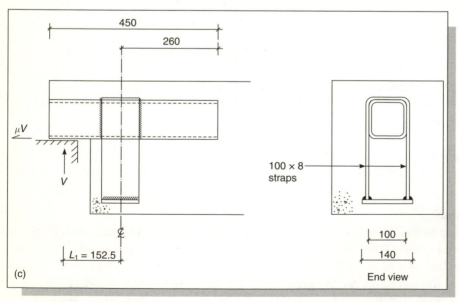

(c)

Figure 9.41: Detail to Example 9.9.

$$C = 3200 \times 27.8 \times 10^{-3} = 89.0\,\text{kN}$$

$$T = 210 + 89 = 299\,\text{kN}$$

$$A_s = (299 \times 10^3)/(0.95 \times 250) = 1259\,\text{mm}^2$$

Use 2 no. R32 (1608) bars welded to sides of box.

Check bend radius
Force in each bar $= 299/2 = 149.5\,\text{kN}$
$a_b = 100\,\text{mm}$ between bars

$$r = \frac{149.5 \times 10^3(1 + 2(32/100))}{2 \times 40 \times 32} = 96\,\text{mm} \qquad \textit{(using Eq. 9.32)}$$

Minimum depth beneath box to accommodate these bars $= 96 + 32 +$ link 10 cover 30 $= 168\,\text{mm} < 210\,\text{mm}$ available. This means that the straight height of the hanger bar beneath the box $= 42\,\text{mm} < 4 \times$ diameter (128). This renders the use of hanging bars not possible.

Try mild steel strap, 450 mm long
Assuming length of strap $= 100\,\text{mm}$ (Ref. Figure 9.41c).
$$L_1 = 37.5 + 35 + \text{cover } 30 + (100/2) = 152.5\,\text{mm}$$

RHS design
$$M_{\text{max}} = 210 \times 10^3 \times 152.5 = 32.0 \times 10^6\,\text{Nmm}$$

$$S_{xx} > (32.0 \times 10^6)/275 = 116.5 \times 10^3\,\text{mm}^3$$

$$2dt > (210 \times 10^3)/165 = 127.3\,\text{mm}^2$$

Use 100 $\times$ 100 $\times$ 10 RHS grade 43 steel (119 cm^3, 2000 mm^2).
Using $L_4 = 260\,\text{mm}$ (making overall length of shear box $260 + 152.5 + 37.5 = 450\,\text{mm}$), taking moments at centre line of strap:

$$210 \times 10^3 \times 152.5 = 3200L_3(260 - 0.5L_3)$$

$$L_3 = 41.9\,\text{mm}$$

$$C = 3200 \times 41.9 \times 10^{-3} = 134.1\,\text{kN}$$

$$T = 210 + 134.1 = 344.1\,\text{kN}$$

Area of 2 no. side straps $= 344.1 \times 10^3/275 = 1251\,\text{mm}^2$
Thickness $= 1251/2 \times 100 = 6.3\,\text{mm}$
Use 100 mm $\times$ 8 mm thick straps.

Bottom plate design
Length $= 100\,\text{mm}$

Width $= 100 + (2 \times 8) + 2 \times$ weld leg say $8 = 132$ mm, use 140 mm
Bearing capacity of concrete above plate $= 0.8 \times 40 \times 100 \times 40 \times 10^{-3} = 448$ kN > 344.1 kN required.
Thickness of plate (based on shear) $p_q = 0.6 \times 275 = 165$ N/mm^2
$$t = (344.1 \times 10^3)/(2 \times 100 \times 165) = 10.4 \, \text{mm}$$

Use 12-mm thick bottom plate.

Weld straps to bottom plate. Weld is deposited on both sides of both straps. Maximum weld length $= 100 - 2$ run out say $16 = 84$ mm

$$t_{\text{weld}} = (344.1 \times 10^3)/(0.7 \times 4 \times 215 \times 84) = 6.8 \, \text{mm}$$

Use 8 mm CFW $\times$ 100 mm long weld to RHS box.

Shear links between end of straps and end of box, and beyond end of box for 1 effective depth.
Main bar in bottom of beam 2 no. T20 (from Example 9.7)

$$d = 350 - \text{cover } 30 - \text{levels } 10 - \text{bar radius } 10 = 300 \, \text{mm}$$

$$v = (210 \times 10^3)/(300 \times 300) = 2.33 \, \text{N/mm}^2$$

$$100 A_s/bd = (100 \times 628)/(300 \times 300) = 0.7$$

$$v_c = 0.71 \, \text{N/mm}^2$$

$$A_{sv} = \frac{(2.33 - 0.71) \times 300}{0.95 \times 460} = 1.11 \, \text{mm}^2/\text{mm} = 555 \, \text{mm}^2/\text{m per leg}$$

Use T10 links at 140 mm spacing (560).

9.4 Column foundation connections

Connections to foundations, such as pad footings, pile caps, retaining walls, ground beams etc., are made in one of three ways:

1 base plate, Figure 9.42. The size of plate is either greater than the size of the column ('extended plate', see Figure 9.3), or equal to the column ('flush plate', see Figure 9.43);

2 grouted pocket, Figure 9.44 (see also Figure 9.4); and

3 grouted sleeves, Figure 9.45.

Although the base plate method is the most expensive of the three options it has the advantage that the column may be immediately stabilized and plumbed vertical by adjusting the level of the nuts to the holding down bolts. This is

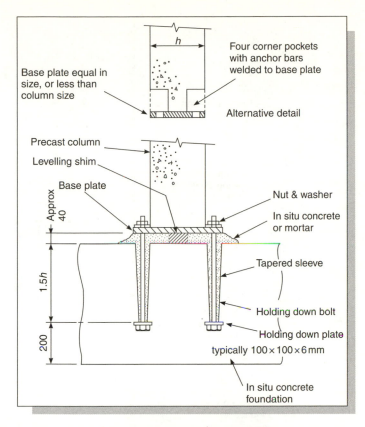

Figure 9.42: Column base plate connection details.

Figure 9.43: Column-to-foundation connection using flush steel base plate.

particularly important when working in soft ground conditions where temporary propping may not provide adequate stability alone.

All column–foundations connections may be designed either as pinned or moment resisting – the designer has the choice depending on the overall stability requirements of the frame. However, the normal grouted pocket method has inherent strength and stiffness providing a moment resisting connection de facto. The attitudes towards the choice in using base

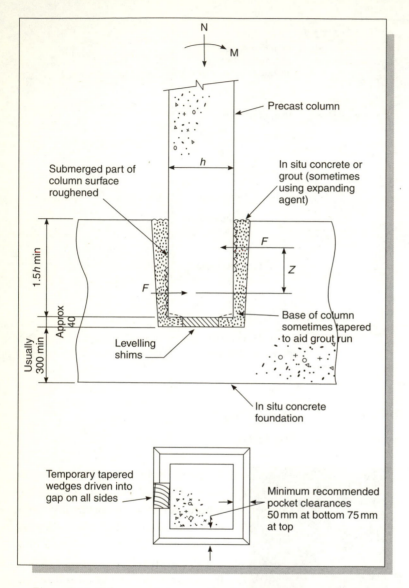

Figure 9.44: Column pocket connection details.

plates rather than pockets tend to be based more on production rather than structural decisions.

9.4.1   Columns on base plates

A moment resisting connection requires a sufficiently large lever arm z, shown in Figure 9.46, between the holding down bolts and the centroid of the compression

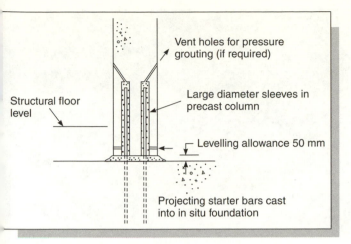

Vent holes for pressure
grouting (if required)

Large diameter sleeves in
precast column

Structural floor
level

Levelling allowance 50 mm

Projecting starter bars cast
into in situ foundation

Figure 9.45: Column grouted sleeve connection details.

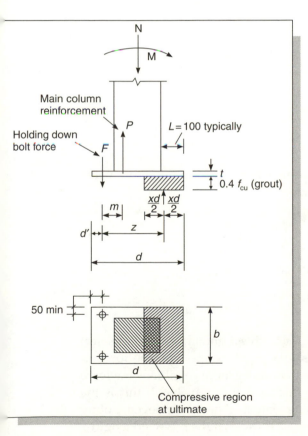

N

M

Main column
reinforcement

P

$L = 100$ typically

Holding down
bolt force

F

t
$0.4 f_{cu}$ (grout)

$\frac{xd}{2}$ $\frac{xd}{2}$

m

d'

z

d

50 min

b

d

Compressive region
at ultimate

Figure 9.46: Extended base plate design.

zone. To achieve this the base plate is usually (but not always) larger than the size of the column, projecting over two, three or four faces as necessary.

To fabricate the base plate reinforcing 'starter' bars are fitted through holes in the plate and fillet welded at both sides (Figure 9.47). Although HT bars are used in order to reduce the compression bond length the yield strength is $f_y = 250 \, \text{N/mm}^2$. Links, typically 2 or 3 T10 or T12 bars, are provided close to the plate at 50–75 mm spacing. The maximum projection L of the plate is, therefore, usually restricted to 100 mm, irrespective of size. 100 mm is also a minimum practical limit for detailing and site erection purposes.

Holding down bolts of grade 4:6 or 8:8 are used as appropriate. The length of the anchor bolt is typically 375–450 mm for 20–32 mm-diameter bolts. The bearing area of the bolt head is increased by using a plate, nominally $100 \times 100 \times 8 \, \text{mm}$. The bottom of the bolt is a minimum of 100 mm above the reinforcement in the bottom of the footing. Confinement reinforcement (in the form of links) around the bolts is usually required, particularly where narrow beams and/or walls are used and where the edge distance is less than about 200 mm. The steel is designed on the principle of shear friction but should not be less than 4 no. R 8 links at 75 mm centres placed near to the top of the bolts. Anchor loops are usually provided around the bolts in order to achieve the full strength of the bolt if the horizontal edge distance is less than about 200 mm. The gap between the plate and foundation is filled using in situ concrete or mortar of grade C30 to C40 depending on the design – although $f_{cu} = 40\,\text{N/mm}^2$ is normally specified.

The design method considers the equilibrium of vertical forces and overturning moments. Two methods are used depending on whether the bolts achieve tension or not. Referring to Figure 9.46 and resolving vertically. If $F > 0$:

$$F + N = 0.4f_{cu}bXd \qquad 9.44$$

where $Xd = $ compressive stress block depth.

Figure 9.47: Starter bars welded to extended base plate.

Taking moments about centre line of compressive stress block

$$M = F(d - d' - 0.5Xd) + N(0.5d - 0.5Xd) \qquad 9.45$$

also, $M = Ne$, such that

$$\frac{N(e + 0.5d - d')}{0.4f_{cu}bd^2} = X\left(1 - \frac{d'}{d}\right) - 0.5X^2 \qquad 9.46$$

from which X and F may be calculated.

The size of the base plate is optimized with respect to obtaining a minimum value for d when a solution exists for Eq. 9.46 as follows:

$$\frac{2N(e + 0.5d - d')}{0.4f_{cu}bd^2} = \left(1 - \frac{d'}{d}\right)^2 \qquad 9.47$$

Letting $x = N/0.4f_{cu}b$ the solution to Eq. 9.47 is:

$$d_{min} = 0.5x + d' + \sqrt{(0.5x + d')^2 - d'^2 + 2x(e - d')} \qquad 9.48$$

Substituting d_{min} from Eq. 9.48 into Eq. 9.46 gives (the inevitable answer) $Xd = (d - d')$, i.e. the pressure zone extends to a point in line with the force in the holding down bolts. However, it is likely that under these conditions the force F will be very large and the resulting plate thickness unacceptable. If this is the case increase b (=decrease x) until a reasonable plate thickness is achieved, noting that the minimum b is equal to the column breadth and that the projected length L should not be less than about 80–100 mm.

Returning to the solution of Eq. 9.46, if $X > N/0.4f_{cu}bd$, then F is positive. Assume N number bolts each of root area A_b and ultimate strength f_{yb} to be providing the force F, then:

$$A_b = \frac{F}{Nf_{yb}}$$ 9.49

For grade 4:6 bolts use $f_{yb} = 195\,N/mm^2$, and $f_{yb} = 450\,N/mm^2$ for grade 8:8 bolts.

The thickness of the base plate is the greater of:

$$t = \sqrt{0.8f_{cu}L^2/p_y}$$ (based on compression side)   9.50

or

$$t = \sqrt{4Fm/bp_y}$$ (based on tension side)   9.51

where L = overhang of plate beyond column face, m = distance from centre of bolts to centre of bars in column, and p_y = yield strength of the plate = $275\,N/mm^2$ for steel grade 43.

If $X < N/0.4f_{cu}bd$, then F is negative and the above Eqs 9.44 and 9.45 are not valid. The analysis simplifies to the following:

$$X = 1 - \frac{2e}{d}$$ 9.52

and

$$N = f_c bXd$$ 9.53

because the infill concrete not fully stresses to $0.4f_{cu}$. Equation 9.50 is modified to:

$$t = \sqrt{\frac{2f_cL^2}{p_y}}$$ 9.54

Pinned-jointed footings can be designed by decreasing the in-plane lever arm. Base plates using two bolts on one centre-line, or four bolts closely spaced also give the desired effect.

Base plates equal to or smaller than the column are used where a projection around the foot of the column is structurally or architecturally unacceptable. The holding-down bolt group is located in line with the main column reinforcement. The base plate is set flush with the bottom of the precast column and small pockets, typical 100-mm cube (for access purposes), leave the plate exposed at each corner, or on opposite faces as shown in Figure 9.48.

Non-symmetrical base plates are used in situations where the overhang of the plate is not possible on one or two sides, as shown in Figure 9.49. The plate overhang must allow at least three holding down bolts to be positioned. The force(s) in the bolt(s) create a couple with the compression under the plate. The design analysis proceeds in a similar manner to that of columns centralized on symmetrical plates, except that now an additional eccentricity exists for the axial load. Referring to Figure 9.49a for the three-sided plate with a clockwise moment $M = Ne$ the equilibrium of moments is, by taking moments about the centre line of the compression stress block, given as:

$$Ne = 2F(d - d' - 0.5Xd) + N(0.5h - 0.5Xd)$$

9.55a

Substituting $2F = 0.4f_{cu}bXd - N$, and simplifying, gives (a modified version of Eq. 9.46)

$$\frac{N(e + d - d' - 0.5h)}{0.4f_{cu}bd^2} = X\left(1 - \frac{d'}{d}\right) - 0.5X^2$$

9.56a

Figure 9.48: Flush base plates for 4 bolts (foreground) and 2 bolt (left) connections.

from which X and F may be calculated.

If the moment is anticlockwise, then referring to Figure 9.49b, the equilibrium is given as

$$Ne = F(d - d' - 0.5Xd) + N(d - 0.5h - 0.5Xd) \qquad 9.55b$$

and

$$\frac{N(e + 0.5h - d')}{0.4f_{cu}bd^2} = X\left(1 - \frac{d'}{d}\right) - 0.5X^2 \qquad 9.56b$$

Example 9.10. *Base plate design*

A 400 mm deep × 300 mm wide column is subjected to an ultimate axial force of 2000 kN and a moment about its major axis of 300 kNm. Design an extended base

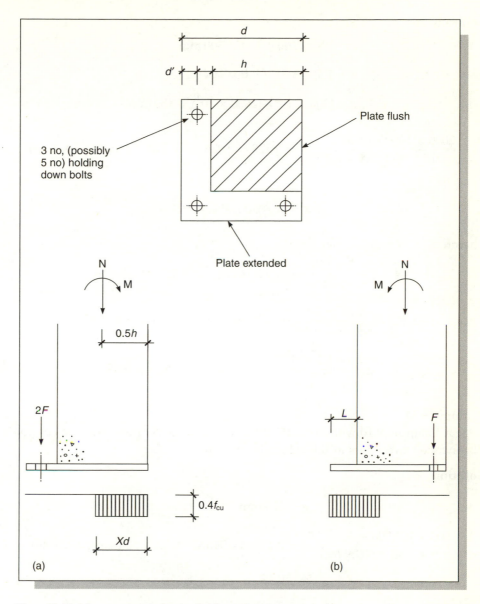

Figure 9.49: Non symmetrical extended base plates.

plate if the projection length is 100 mm and the distance from the edge of the plate to the holding down bolts is 50 mm.

$$\text{Use } f_{cu} = 40\,\text{N/mm}^2, \, p_y = 275\,\text{N/mm}^2.$$

Solution

$$b = 300 + 100 + 100 = 500\,\text{mm}$$

$$d = 400 + 100 + 100 = 600\,\text{mm}$$

$$e = (300 \times 10^6)/(2000 \times 10^3) = 150\,\text{mm}$$

$$\frac{2000 \times 10^3(150 + 300 - 50)}{0.4 \times 40 \times 500 \times 600^2} = 0.278 = 0.916X - 0.5X^2 \qquad (using\ Eq.\ 9.46)$$

$$X = 0.383$$
$$X < N/0.4f_{cu}bd = 0.417$$

Therefore, F is negative, set $F = 0$

$$X = 1 - \frac{2 \times 150}{600} = 0.5 \qquad (using\ Eq.\ 9.52)$$

$$f_c = \frac{2000 \times 10^3}{500 \times 0.5 \times 600} = 13.3\,\text{N/mm}^2 \qquad (using\ Eq.\ 9.53)$$

$$t = \sqrt{\frac{2 \times 13.3 \times 100^2}{275}} = 31\,\text{mm} \qquad (using\ Eq.\ 9.54)$$

Example 9.11.   *Base plate design*
Repeat Example 9.10 with $N = 600\,\text{kN}$ using $f_{yb} = 450\,\text{N/mm}^2$ and the centroidal distance to starter bars in the column $= 50\,\text{mm}$

Solution

$$e = 500\,\text{mm}$$

$$\frac{600 \times 10^3(500 + 300 - 50)}{0.4 \times 40 \times 500 \times 600^2} = 0.156 = 0.916X - 0.5X^2 \qquad (using\ Eq.\ 9.46)$$

$$X = 0.19$$

$$X > N/0.4f_{cu}bd = 0.125,\ \text{therefore, } F \text{ is positive}$$

$$F = 0.4 \times 40 \times 500 \times 0.19 \times 600 \times 10^3 - 600 = 312\,\text{kN} \qquad (using\ Eq.\ 9.44)$$

$$A_b > \frac{312 \times 10^3}{450} = 693\,\text{mm}^2 \qquad (using\ Eq.\ 9.49)$$

Use 2 no. M24 grade 8.8 holding down bolts (706).

Plate thickness

$$t > \sqrt{\frac{0.8 \times 40 \times 100^2}{275}} = 34.1\,\text{mm} \qquad \text{(using Eq. 9.50)}$$

$$t > \sqrt{\frac{4 \times 312 \times 10^3 (50 + 50)}{500 \times 275}} = 30.1\,\text{mm} \qquad \text{(using Eq. 9.51)}$$

Use 600 × 500 × 35 mm base plate grade 43.

Example 9.12.   *Optimized base plate depth*
A base plate is used to support a 400 mm deep × 300 mm wide column subjected to an ultimate axial force of 1500 kN and an ultimate moment about the major axis of 360 kNm. Optimize the size of the base plate with respect of depth for the minimum possible breadth and determine the magnitude of the force in the holding down bolts. Calculate the thickness of the base plate, stating whether this is an economical solution.

Use $f_{cu} = 40\,\text{N/mm}^2$, $p_y = 275\,\text{N/mm}^2$. Centroidal cover distance to bars in the column = 50 mm. Edge distance to holding down bolts = 50 mm.

Solution
Minimum breadth of plate = 300 mm

$$e = (360 \times 10^6)/(1500 \times 10^2) = 240\,\text{mm}$$
$$x = (1500 \times 10^3)/(0.4 \times 40 \times 300) = 313\,\text{mm}$$

$$d = 156.5 + 50 + \sqrt{206.5^2 - 50^2 + 118\,940} = 606\,\text{mm} \qquad \text{(using Eq. 9.48)}$$
$$Xd = 606 - 50 = 556\,\text{mm}$$
$$F = 0.4 \times 40 \times 300 \times 556 \times 10^{-3} - 1500 = 1168.8\,\text{kN} \qquad \text{(using Eq. 9.44)}$$

Plate overhang $L = 606 - 400/2 = 103\,\text{mm}$
$$m = 53 + 50 = 103\,\text{mm}$$

Plate thickness

$$t = \sqrt{\frac{0.8 \times 40 \times 103^2}{275}} = 35\,\text{mm} \qquad \text{(using Eq. 9.50)}$$

$$t = \sqrt{\frac{4 \times 1168.8 \times 10^3 \times 103}{300 \times 275}} = \underline{76\,\text{mm}!} \qquad \text{(using Eq. 9.51)}$$

Clearly, this is not an economical solution.
The reader should repeat the exercise using larger values for b until t, based on the tension side, is equal to 35 mm.

9.4.2   Columns in pockets

This is the most economical solution from a precasting point of view, but its use is restricted to situations where fairly large in situ concrete pad footings can easily be constructed. The precast column requires only additional links to resist bursting pressures generated by end bearing forces, and a chemical retarding agent to enable scabbling to expose the aggregate in the region of the pocket. In cases where the column reinforcement is in tension, the bars extending into the pocket must be fully anchored by bond (BS8110, Part 1, Clause 3.12.8). In order to reduce the depth of the pocket to a manageable size these bars may need to be hooked at their ends.

The concrete foundation is cast in situ using a tapered box shutter to form the pocket. The gap between the pocket and the column should be at least 75 mm at the top of the pocket. The pocket is usually tapered 5° to the vertical to ease the placement of concrete or grout in the annulus. This gives rise to a wedge force equal to $N \tan 5°$, where N is the ultimate axial load in the column. The precast column requires only additional links to resist bursting pressures generated by end bearing forces using $\zeta = 0.11$ (ζ is defined in Section 8.3.2 and given in Table 8.1).

Vertical loads are transmitted to the foundation by a combination of skin friction (between column and in situ infill) and end bearing. It is not instructive to know the proportion of the load transmitted by either of these mechanisms, only that the total load is transferred to the foundation. To increase skin friction shear keys may be formed in the sides of the pocket or on the sides of the column to transfer axial load by the action of shear wedging. Shear keys should conform to Figure 8.17.

If overturning moments are present half of the skin friction is conservatively ignored due to possible cracking at the precast/in situ boundary. Ultimate load design considers vertical load transfer by end bearing based on the strength of the gross cross-sectional area of the reinforced column and equal area of non-shrinkable concrete or grout. The design strength of the expansive infill is usually $f'_{cu} = 40\,\text{N/mm}^2$. The failure mode may be by diagonal-tension shear across the corner of the pocket in which case links are provided around the top half of the pocket. Another mode of failure is crushing of the in situ concrete in the annulus. This is guarded against by using an ultimate stress of $0.4f'_{cu}$ working over a width equal to the precast column only, i.e. ignoring the presence of the third dimension.

Bruggeling[1] propose that the depth of the pocket D is related to the ratio of the moment M and the axial force N as follows:

$$\text{If } e = M/N < 0.15h \quad \text{then} \quad D > 1.2h$$
$$\text{If } e = M/N > 2.00h \quad \text{then} \quad D > 2.0h$$

9.57

In Figure 9.50, the force F acts such that a couple F_Z is generated over a distance the greater of:

$$z = (D - 0.1D)/2 = 0.45D$$

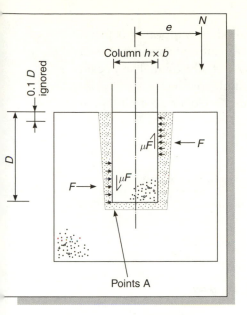

Figure 9.50: Pocket foundation design.

or

$$z = (D - \text{cover})/2 \qquad 9.58$$

This is because the top $0.1D$ of the pocket is ignored within the cover zone. Referring to Figure 9.50 and taking moments about A:

$$N\,e - \mu Fh - 0.45FD = 0 \qquad 9.59$$

Then

$$F = \frac{Ne}{\mu h + 0.45D} < 0.4f'_{cu}b(0.45D) \qquad 9.60$$

The analysis is for uniaxial bending only. There is no method for dealing with biaxial bending, although the method for dealing with biaxial bending in columns may be adopted here, i.e. an increased moment in the critical direction is considered as uniaxial moment.

The total depth of the foundation is equal to the pocket depth plus the plinth depth. There is no analysis to determine the plinth depth because when the annulus is filled the design of the foundation is based on the total depth, not the plinth depth. There is no punching shear because compression is transferred directly to the foundation. However, punching shear is present prior to the hardening of the infill, i.e. in the temporary construction phase. To this end, the plinth depth is made nominally equal to the smaller dimension of the column up to a maximum depth of 400 mm. As an approximate guide the total depth should be such that a 45° load distribution line can be drawn from the edge of the column to the bottom corner of the foundation. Thus, if the breadth of the foundation is B and the breadth of the column is b, the depth of the foundation should be approximately $(B - b)/2$. (see Figure 9.51).

The design of the foundation itself is according to standard reinforced concrete practise.

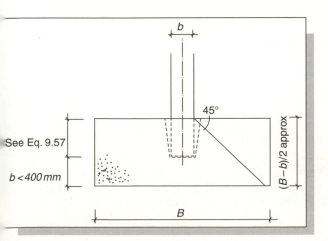

Figure 9.51: Determination of depth of foundation.

Example 9.13.   *Column pocket foundation*

Design a column pocket foundation connection required to support a 300 mm wide by 400 mm deep precast column subjected to an ultimate axial force of 2000 kN and an ultimate moment (about the major axis) of 100 kNm. The pocket taper is 5°. Determine the required strength f'_{cu} of the infill concrete.

Use $f_{cu} = 50\,\text{N/mm}^2$ for the column, $f_y = 460\,\text{N/mm}^2$. Cover to foundation bars = 50 mm.

Solution

$$e = (100 \times 10^6)/(2000 \times 10^3) = 50\,\text{mm} < h/6 = 400/6 = 67\,\text{mm}$$

Therefore, no tension in column bars

Pocket depth

$$e = 50\,\text{mm} < 0.15 \times 400 = 60\,\text{mm}$$

$$D = 1.2h = 480\,\text{mm} > 1.5b \qquad (\textit{using Eq. 9.57})$$

$$F = \frac{2000 \times 10^3 \times 50}{0.7 \times 400 + 0.45 \times 480} \times 10^{-3} = 201.6\,\text{kN} \qquad (\textit{using Eq. 9.60})$$

Pocket breadth $b = 300\,\text{mm}$

$$f'_{cu} = \frac{201.6 \times 10^3}{0.4 \times 300 \times 0.45 \times 480} = 7.8\,\text{N/mm}^2 \quad (\textit{using Eq. 9.60})$$

use $f'_{cu} = 40\,\text{N/mm}^2$ (for rapid early strength)

Confinement rebar in column

$$F_{bst} = 0.11 \times 2000 = 220\,\text{kN}$$

$$A_{bst} = (220 \times 10^3)/(0.95 \times 460) = 503\,\text{mm}^2/2\ \text{legs} = 252\,\text{mm}^2$$

Use 4 no. T10 (314) links at 50 mm centres.

Pocket reinforcement

$$H = N \tan 5° = 2000 \tan 5° = 175\,\text{kN}$$

Force across pocket due to moment = 201.6 kN

$$A_{sv} = (175 + 201.6) \times 10^3/0.95 \times 460 = 862\,\text{mm}^2/2\ \text{legs} = 431\,\text{mm}^2$$

Use 4 no. T12 (452) links at 80 mm centres around top of pocket.
Provide 3 no. T12 hanger bars to support links.

Example 9.14. *Concrete pad foundation*
Determine the size of the in situ concrete pad foundation for the column-to-foundation connection in Exercise 9.13. Assume that the ultimate load and moment are based on $\gamma_f = 1.2$. Calculate the reinforcement in the foundation. Use $f_{cu} = 30 \, \text{N/mm}^2$. Cover to bars $= 50 \, \text{mm}$. Ground bearing pressure $= 500 \, \text{kN/m}^2$.

Solution
Try depth of foundation of 1.0 m
Net bearing pressure $= 500 - 24 \times 1.0 = 476 \, \text{kN/m}^2$

$$N/BH + 6M/BH^2 < 476 \, \text{kN/m}^2$$

Let $B = H$ because M is small

$$2000/1.2B^2 + 6 \times 100/1.2B^3 = 476 \, \text{kN/m}^2$$

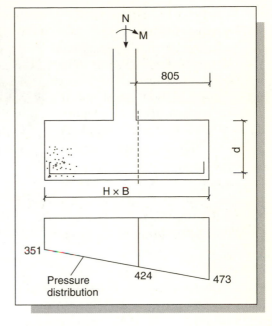

Figure 9.52

from which $B \geq 2.01 \, \text{m}$
Bearing pressures $= 412 \pm 61 = +473$ and $+351 \, \text{kN/m}^2$
Depth of foundation for 45° load spread from edge of column $= \dfrac{2010}{2} - \dfrac{400}{2} = 805 \, \text{mm}$
Plinth depth $= 805 - 480 - 75$ level allowance $= 250 \, \text{mm} < 300 \, \text{mm}$ but say OK.

Moment about face of column $= 424 \times 2.01 \times 0.805^2/2 + 49/2 \times 2.01 \times 2/3 \times 0.805^2 = 276.1 + 21.3 = 297.4 \, \text{kNm}$

$$M_{ult} = 1.2 \times 297.4 = 357 \, \text{kNm}$$

$$B = 2010 \, \text{mm}$$

$$d = 850 - 50 - \text{say } 12 = 788 \, \text{mm}$$

$$K = \frac{357 \times 10^6}{30 \times 2010 \times 788^2} = 0.01$$

Therefore, $z/d = 0.95$

$$A_s = (357 \times 10^6)/(0.95 \times 460 \times 0.95 \times 788) = 1092 \, \text{mm}^2$$

Use 6 no. T16 (1206) at 400 centres.

9.4.3 Columns on grouted sleeves

One of the most popular (and easily the most economical) column foundation detail is the grouted sleeve (see Figure 9.45). Starter (or waiting) bars projecting from the foundation pass into openings, usually circular sleeves, in the column. The annulus around the bars is afterwards filled with (gravity or pressurized) expansive flowable grout of strength equal to that of the column, but not usually less than $f_{cu} = 40\,\text{N/mm}^2$.

Figure 9.53: Corrugated steel sleeves used for grouted sleeve connection.

The annulus must be 6 mm nominal. If the annulus around the bar is quite large, say more than 10–15 mm, the sleeve can be gravity fed, otherwise pressure grouting must be used. The corrugated pressed sheet sleeves shown in Figure 9.53 are large enough to enable gravity filling. The thickness of the material is around 1 mm. The corrugations increase the bond strength (by wedging action) and may be left inside the column. If the sleeve is smooth it should be withdrawn. Figure 9.54 shows the manufacture of columns using sleeves. The upper ends of the

Figure 9.54: Preparation of reinforcement cage and sleeves for column shown in Figure 9.51

sleeves are open and flush to the face of the column (the white plugs prevent concrete ingress during pouring).

The column is positioned onto packing shims which provide a fixing tolerance of around 40 mm. The gap at the bottom of the column is site filled using mortar (or concrete containing small size aggregate ≤6 mm) of compressive strength equal to that of the column. The joint possesses most of the advantages (confinement of concrete, thin dry packed joint, continuity of HT reinforcement, easy to manufacture and fix) and few of the disadvantages (fully compacted grout in the sleeves) associated with precast construction methods. The column must remain propped until the grout has hardened. However, props usually remain in position until the first floor beams and slabs have been placed.

The design procedure is the same as for prismatic reinforced concrete columns. The assumption is that a full bond is provided to the starter bars enabling their full strength to be developed. The starter bars are placed in the corners of the column to maximize the effective depth. However, this means that the main reinforcement in the column must be placed inside the starter bars, and this becomes the critical design situation. Attempts are made to position the main reinforcement at the edge of the column with respect to the major axis of the column, and further from the edge with respect to the minor axis. The effective depth to the reinforcement in the minor axis is therefore:

$$d = h - \text{cover} - \text{column link} - \text{column bar} - \text{space say 10} - \text{radius of sleeve}$$

This is typically $h - 110$ mm.

Example 9.15. *Grouted sleeve foundation connection*
A precast concrete column 400 mm deep × 300 mm wide carries an ultimate force of 1500 kN and a major axis moment of 300 kNm. The column is to be founded on pressure grouted sleeves using withdrawn sleeves. Determine the starter bars required for this connection.

Use $f_{cu} = 50 \, \text{N/mm}^2$, $f_y = 460 \, \text{N/mm}^2$, cover to 10 mm links in the column = 30 mm.

Solution
Guess the diameter of the starter bar = 32 mm.

Diameter of sleeve = 32 + 6 + 6 = 44 mm, use 50-mm internal diameter.

Effective depth to starter bars to major axis = 400 − 30 − 10 − 50/2 = 365 mm.

$$N/bh = 12.5$$
$$M/bh^2 = 6.25$$
$$d/h = 365/400 = 0.91$$

BS8110, Part 3, Chart 49

$$A_{sc} = 2.4\% \ bh = 2880 \, \text{mm}^2$$

Use 4 no. T32 bars (3216).
Original guess for size of bars OK.

Reference

1   Institution of Structural Engineers, Structural Joints in Precast Concrete, London, August 1978, 56p.

10 Ties in precast concrete structures

10.1 Ties in precast concrete structures

It is often said that the two greatest problems with precast concrete skeletal structures are: (a) the avoidance of joints in critical locations; and (b) how to design the structure to prevent progressive collapse.

Item (a) has been thoroughly dealt with in Chapters 8 and 9. This chapter deals with item (b). It introduces a method of placing continuous steel reinforcing ties to prevent progressive collapse, and examines why ties are needed and the resulting structural mechanisms in the event of accidental loading. There is probably no need to explain what 'progressive collapse' is, this is clearly shown in Figure 10.1.

The need for ties is therefore obvious. No structural element, whether it be precast concrete, stone, steel, aluminium, or timber, should be designed and constructed such that a loss of bearing, stability or load capacity would cause total failure of that element, and more importantly extensive damage and failure of the entire structure. If the latter causes damage which is disproportionate to the cause (Figure 10.1) in such a manner that the failure of one element leads to the progressive failure of others, e.g. a vehicle damages a column and this in turn causes the beams and floor slabs to fall, this is termed 'progressive collapse'. Structures which are not able to avoid this are not 'robust'. Following some rather dramatic failures of concrete structures in the 1960s and 1970s, e.g. Ronan Point (Figure 10.1), high alumina cement concrete beams, etc. BS8110 gives prominence to this by specifying the need for structural ties in all buildings in Part 1, Clause 2.2.2.2. It states...

Robustness. Structures should be planned and designed so that they are not unreasonably susceptible to the effects of accidents. In particular, situations should be avoided where damage to small areas of a structure or failure of a single element may lead to collapse of major parts of a structure.

The message to designers of elemental precast concrete structures is clear – ensure that the failure of an element does not cause failure of its neighbour, and its neighbour in turn. In pin-jointed structures where there is no moment continuity at the connections, structural continuity must in some way be designed into the structure. Note that this is not for ultimate limit strength, and therefore does not apply to Chapters 3 to 9. Structural continuity may be achieved in the elements and connections themselves, for example by the dowel connecting the beam to the column in Figure 9.9, or by additional site placed means. The problem with the former is that if the structure containing the dowel is subjected to accidental loading the dowel may fail due to a large horizontal shear force, and the structural continuity would be lost. If the latter method is used, and steel ties are placed continuously across the connection, as shown in Figure 8.38, then as long as the skeletal elements are anchored to the ties isolated failure cannot occur.

Figure 10.1: Progressive collapse of precast wall frame at Ronan Point.

BS8110, Clause 2.2.2.2 continues this theme and offers two solutions:

> Unreasonable susceptibilty to the effects of accidents may generally be prevented if the following precautions are taken.
> ...(b) All buildings are provided with effective horizontal ties (1) around the periphery; (2) internally; (3) to columns and walls.
> (c) The layout of the building is checked to identify any key elements the failure of which would cause the collapse of more than a limited portion close to the element in question.
> (d) Buildings are detailed so that any vertical load bearing element other than a key element can be removed without causing the collapse of more than a limited portion close to the element in question.

There are three proposals here – but items (b) and (d) are generically similar, i.e. continuous ties should be provided both vertically and horizontally. In fact Clause 2.2.2.2(d) offers a further alternative called the 'alternative load' path or 'bridging' method in which:

such (vertical) elements should be considered to be removed in turn and elements normally supported by the element in question designed to 'bridge' the gap...

Note how often the term 'element' appears in this clause. Code writers clearly had precast concrete structures in mind. Of course, properly detailed and constructed cast in situ concrete structures are inherently robust and require only minimal additional provisions to satisfy the robustness issue.

Quantitative data on tie forces and key elements' loading are given elsewhere in the code, and later in this chapter. However, the principles of how robustness is satisfied must first be addressed.

10.2   Design for robustness and avoidance of progressive collapse

The term 'progressive collapse' was first used following the partial collapse of a precast concrete wall frame at Ronan Point, London in 1968 (see Figure 10.1). The collapse is well documented.[1] A gas explosion in a corner room at the 18th floor level caused the connection between the floor and wall to fail locally. The corner wall peeled away from the floor slab leaving the slabs supported on two inner adjacent edges. Falling debris and a loss of bearing resulted in a portion of every floor being damaged. Site investigations found deficiencies in the manner in which the precast elements were tied to one another. Poor detailing and unsatisfactory workmanship was blamed for the disproportionate amount of damage. However, another most crucial factor was that there was no suitable design information to guide the designer towards a robust solution, and in the absence of such guidance engineers did not question the effect of a gas explosion, even though at the time some 400 gas explosions per year had caused structural damage.

Tests demonstrated that the connection was capable of resisting wind suction for which the wall panels had been designed. However, the connections lacked the strength and ductility to resist the blast force for sufficient time to enable venting to occur through doors and windows. Tests were also carried out to estimate the pressure at which failure at Ronan Point may have taken place, and as a consequence BS8110, Part 2, Clause 2.6 states...

Loads on key elements....an element and connection should be capable of withstanding a design ultimate load of $34 \, kN/mm^2$, to which no partial safety factors should be applied, from any direction. A horizontal member, or part of a horizontal member that provides lateral support vital to the stability of a key element, should also be considered a key element.

Not only should a wall be designed for this pressure, but the connections from the wall to the floor slab, and the diaphragm forces in the floor too, particularly as the blast pressure is localized. This clearly had severe implications for using hollow core floor units as floor diaphragms without a structural topping where the provision of coupling bars became very onerous.

The outcome of this was that key elements are infrequently designed. Some specialized structures and isolated facade elements requiring resistance against bomb damage have been designed in this manner, but generally whole structures are not.

The 'bridging' method, or so called 'alternative path' method specifies walls, beams and columns, or parts thereof, which are deemed to have failed. The remaining structure is analysed for the removal of each element. The elements in the remaining structure are called 'bridging elements'. At each floor level in turn (including basement floors), every vertical load bearing member, except for key elements, is sacrificed and the design should be such that collapse of a significant part of the structure does not result. BS8110 does not quantify the term 'significant part of the structure'. The method is not widely used because of the implicit necessity to provide additional reinforcement, most of which is designed to act in catenary, and which is permitted in the fully tied method.

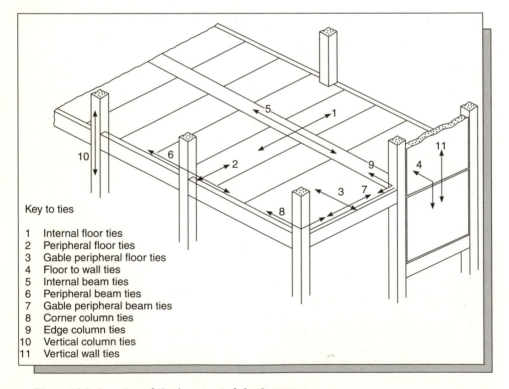

Key to ties

1   Internal floor ties
2   Peripheral floor ties
3   Gable peripheral floor ties
4   Floor to wall ties
5   Internal beam ties
6   Peripheral beam ties
7   Gable peripheral beam ties
8   Corner column ties
9   Edge column ties
10 Vertical column ties
11 Vertical wall ties

Figure 10.2: Location of ties in precast skeletal structures.

In the tied method, BS8110, Part 1, Clauses 2.2.2.2 and 3.12.3 set out the requirements as to when and how the ties should be used, and this de facto avoids using key and bridging elements. Structural continuity between elements is obtained by the use of horizontal floor and vertical column and wall ties positioned as shown in Figure 10.2. These are as follows:

- horizontal internal and peripheral ties, which must also be anchored to vertical load bearing elements;

- vertical ties.

Horizontal ties are further divided into floor and beam ties (Figure 10.3):

- floor ties, to provide continuity between floor slabs, or between floor slabs and beams;

- internal and peripheral beam ties, to provide continuity between main support beams; and

- gable peripheral beam ties, to provide continuity between lines of main support beams.

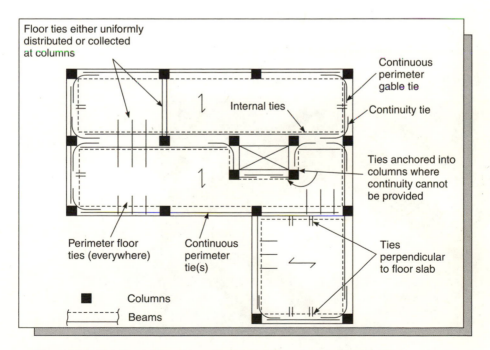

Figure 10.3: Details and locations of horizontal floor ties in a precast floor.

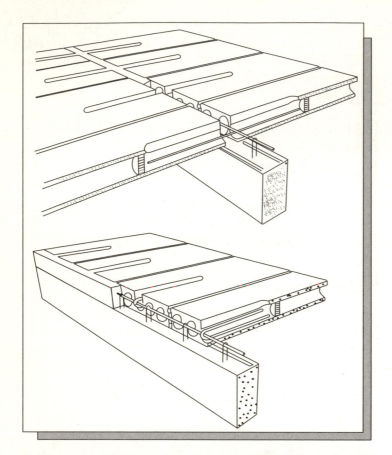

Figure 10.4: Continuity floor ties in hollow core slabs (courtesy FIP).

Figure 10.5: Continuity of beam ties at corner columns.

Figure 10.6: Corner recess in hollow core slabs to permit the correct placement of ties at corners.

The best way of providing floor ties is to provide a continuous ring of reinforcement around each bay of floor slabs bounded by beams. Floor ties must span over supporting beams, either directly as a single bar as shown in Figure 10.4, or, if the slots in slabs are not coincident, lapped to beam ties as L shape bars. Beam ties must span past columns, either through sleeves or pass on either side of the column. Beam ties at corner columns must also be continuous – Figure 10.5 shows a possible solution where ties rest on ledges made into the corners of floor slabs, Figure 10.6, in order to achieve the correct bend radius of about 500 mm.

10.3   The fully tied solution

In addition to the aforementioned clauses, specific requirements relating to ties in precast concrete structures are given in BS8110, Part 1, Clause 5.1.8. These are satisfied either by using individual continuous ties provided explicitly for this purpose in in situ concrete strips, or using ties partly in the in situ concrete and partly in the precast elements.

The structural model is as follows. In the event of the complete loss of a supporting column or beam at a particular floor level, the floor plate at this level and the next level up must resist total collapse by acting in catenary (=chain link action) as shown in Figure 10.7. At the moment, the accident occurs an alternative load path for the floor beams which were previously supported by the damaged member may not be immediately available. If a column is removed the tie forces over the beams must be mobilized. The column that is directly above the damaged unit carries the beam end reactions of the beams at this level as a tie in vertical suspension. If a beam support is lost the floor ties act in catenary. With increasing deformation a new equilibrium state will develop as shown in Figure 10.7 where the deflection reaches a critical value Δ_{crit}. If the deflection exceeds Δ_{crit} the tie steel will either fracture or debond in the adjacent spans. This behaviour is applicable to uniformly distributed loads in the beam at the floor level under consideration, and to point loads from the column above this level.

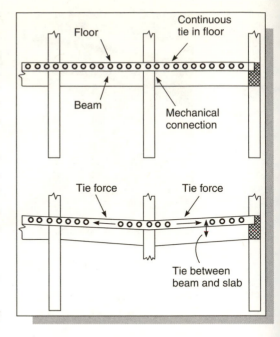

If the sagging deflected shape of the beam is deduced from uniformly distributed loading (udl), and the tie steel is elastic-perfect-plastic, it can be shown that for a characteristic udl w acting on a beam of length $2L$ (=two spans L affected by the loss of an internal column) the catenary force T is given as:

$$T = 2wL\left(\frac{0.208L}{\Delta_{\text{crit}}} + \frac{0.25\Delta_{\text{crit}}}{L}\right) \qquad 10.1$$

and for a point load P acting at mid span of a beam of length $2L$:

$$T = 0.5P\sqrt{\left(\frac{L}{\Delta_{\text{crit}}}\right)^2 + 1} \qquad 10.2$$

Figure 10.7: Catenary action between precast elements.

In this equation, P is the characteristic axial force in the column from only the next storey above, i.e. $w(L/2 + L/2)$ because catenary action is taking place at all subsequent floors above the level under consideration.

In a 3D precast concrete orthogonal slab field the tie forces above will act in two mutually perpendicular directions, namely along the beam as a 'beam tie' and along the slabs as 'floor ties'. The floor ties will only act in the vicinity of the beam where the bearing is lost, which could be estimated as $L/2$ from the ends. Δ_{crit} is related to the strain in the tie steel, which is a function of the type of reinforcement and detailing, and can only be obtained by testing. Test results (Ref. 8.4) show that just prior to failure $\Delta_{\text{crit}} \approx 0.2L$. An alternative, dynamic approach to the same problem is given by Engström.[2]

The above equations may be used to determine catenary tie forces and assist in understanding the derivation of the stability tie forces that are given in BS8110.

Tie bars are either high tensile deformed bar using a design strength of $460\,\text{N/mm}^2$ ($\gamma_m = 1.0$), or helical prestressing strand using a design strength of 1580 to $1760\,\text{N/mm}^2$. The strand is laid unstressed, but stretched tightly. It is considered continuous if it is correctly lapped. The lap length for deformed bar is usually taken as 44 diameters for $f_{\text{cu}} = 25\,\text{N/mm}^2$. The lapping length for strand is based on the transmission lengths given in BS8110, Part 1,

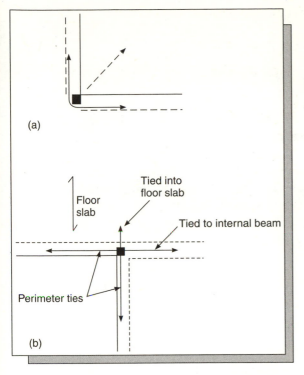

Figure 10.8: Ties at re-entrant corners: (a) Resultant force problem; and (b) Solutions for continuity.

Clause 4.10.3. Using $f_{cu} = 25\,\text{N/mm}^2$ (in place of f_{ci}) and $K_t = 360$ for 7-wire drawn strand, a length of $72 \times$ diameter is found. This does not agree with some test results and so to err on the side of caution a lap length of 1200 mm is used for 12.5-mm diameter strand. The position of the lap is staggered. The tie, of diameter ϕ must be embedded in in situ concrete at least $2(\phi + h_{agg} + 5\,\text{mm})$ wide. In most instances, this means that the in situ concrete must be at least 60 mm wide to accommodate the space occupied by aggregate and bars.

A problem occurs at re-entrant corners, shown in Figure 10.8a. The resultant force from the perimeter tie steel is pulling outwards into unconfined space. One cannot rely on the shear capacity of the column to restrain this force. Two alternatives are possible, as follows:

1 The tie steel continues through or past the side of the column as though it were an edge column as shown in Figure 10.8b. Cast-in couplers (or similar) may be used to anchor the tie steel to the walls.

2 A structural topping containing a steel fabric may be used in this localized area. To avoid increasing the overall depth of the floor a shallower precast floor unit can often be used.

Example 10.1. *Estimation of tie forces (not BS8110 values)*
An internal column carries a symmetrical arrangement of two beams, which in turn carry a symmetrical arrangement of floor slabs. The column grid is 6×6 m. The characteristic floor loading is $10.0\,\text{kN/m}^2$. If the central column was to be removed in an accident calculate the magnitude of the beam and slab tie force required to establish a new equilibrium sagging position.

Solution

$$w = 10 \times (6.0/2 + 6.0/2) = 60.0\,\text{kN/m}$$
$$L = 6.0\,\text{m}$$
$$\Delta_{\text{crit}} = 0.2 \times 6.0 = 1.2\,\text{m}$$

$$T = 2 \times 60 \times 6 \times (1.04 + 0.05) = 785\,\text{kN} \qquad (using\ Eq.\ 10.1)$$

$$P = 60 \times \left(\frac{6.0}{2} + \frac{6.0}{2} \right) = 360\,\text{kN}$$

$$T = 0.5 \times 360 \times 5.099 = 918\,\text{kN} \qquad (using\ Eq.\ 10.2)$$

Total tie force $T = \underline{1703\,\text{kN}}$.

Consider that the tie force may be distributed equally between the beam tie and the floor ties owing to the rectangular grid dimensions.

Beam tie $= 1703/2 = \underline{851\,\text{kN}}$

Floor ties $= 1703/2 = 851\,\text{kN}$ distributed over 6 m wide floor $= \underline{142\,\text{kN/m run}}$.

(Note these values are without partial safety factors etc. and may be tentatively compared by the reader with the tie forces given in Section 10.4.)

10.4   Tie forces

The basic horizontal tie force F_t is given in BS8110, Part 1, Clause 3.12.3 as the lesser of:

$$F_t = (20 + 4n)\,\text{kN/m} \quad \text{or} \quad 60\,\text{kN/m width} \qquad\qquad 10.3$$

where n is the number of storeys including basements. F_t is considered as an ultimate value and is **not** subjected to the further partial safety factor of $\gamma_f = 1.05$ given in BS8110, Clause 2.4.3.2.

If the total characteristic dead (g_k) + live(q_k) floor loading is greater than $7.5\,\text{kN/m}^2$ and/or the distances (l_r) between the columns or walls in the direction of the tie is greater than 5 m, the force is modified as the greater of:

$$F'_t = F_t \frac{g_k + q_k}{7.5} \cdot \frac{l_r}{5.0} \qquad\qquad 10.4$$

or

$$F'_t = 1.0\,F_t \qquad\qquad 10.5$$

10.4.1   Horizontal floor and beam ties

Internal floor ties parallel with the span of the flooring are either distributed evenly using short lengths of tie steel anchored by bond into the opened cores of the hollow core floor units, or grouped in full depth in situ strips at positions coincident with columns. This relies on an adequate pull out force generated by tie steel cast into in situ concreted hollow cores (see Section 8.5). A typical configuration is to use 12-mm diameter HT bar or 12.5-mm diameter helical strand at 600 mm centres, i.e. 2 bars per 1200 mm wide slab. For greater quantities, it is better to decrease the bar spacing rather than to increase the size or number of bars. At no time should two bars be placed in one core.

Ties along gable edge beams are placed into the broken out cores of slabs at intervals varying between 1.2 m and 2 m. The success of the ties relies on adequate anchorage of small loops cast into in situ concrete which itself is locked into the bottom of the hollow core. Generous openings (say 300 mm long) should be made in the floor slabs to ensure that any projecting tie steel in the beam may be lapped without damage to the slab or tie bar.

The magnitude of the tie forces F'_t between precast hollow core slabs is as given by Eq. 10.4 or 10.5. The area of steel is:

$$A_s = \frac{F'_t \gamma_m}{f_y} \qquad 10.6$$

where $\gamma_m = 1.0$ for reinforcement (BS8110, Part 1, Clause 2.4.4.2).

Continuity between beams is provided across the line of columns by calculating the magnitude of the internal and peripheral tie force F'_t according to Eq. 10.4 or 10.5. The tie force in the beam $F'_{t,\text{beam}}$ is the summation of all the internal tie forces across the span L of the floor, i.e.:

$$F'_{t,\text{beam}} = F'_t L \qquad 10.7$$

If the spans of the floor slabs on either side of the beam are different, L_1 and L_2 the summation of half of each span, i.e. $0.5\,(L_1 + L_2)$, is used to replace L in equation 10.7. In an edge beam $L_2 = 0$. Where the floor slabs to one side of the beam are spanning parallel with the beam, and on the other side are spanning on to the beam with a span L_1, the beam tie force is taken as:

$$F'_{t,\text{beam}} = 1.0 F'_t + 0.5 L_1 F'_t \qquad 10.8$$

Where the floor slabs are spanning parallel with the span of an edge beam, a nominal tie force $1.0\,F'_t$ need only be provided. The area of steel is calculated according to Eq. 10.6.

The internal tie bars should be distributed equally either side of the centre line of the beam, be separated by a distance of at least 15 mm to ensure adequate bond (assuming 10 mm size aggregate), and be positioned underneath projecting loops from the beam. Peripheral tie bars should be similarly positioned, and, according to BS8110, Part 1, Clause 3.12.3.5, be located within 1.2 m of the edge of the building. This is not usually a problem except in the case of cantilevers where there is no beam nearer than 1.2 m from the end of the cantilever. In this case, the tie must be located in a special edge beam, which is cast for the sole purpose of providing the peripheral tie.

The success of the tie beams depends largely on detailing. Figure 10.9 illustrates the concept; two tie bars are fixed on site and pass underneath the projecting

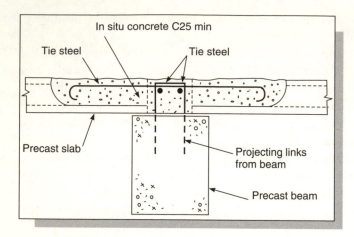

Figure 10.9: Continuity details at internal beams and floors.

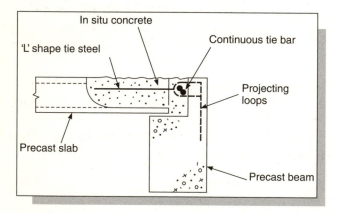

Figure 10.10: Continuity details at external beams and floors.

reinforcement loops. The hooked bars are fully anchored into the hollow cores of the slab. Finally, the tie bars pass through small sleeves preformed in the columns, or pass by the side of the column. Figure 10.10 shows how continuity may be satisfied at external positions.

10.4.2   Horizontal ties to columns

BS8110, Part 1, Clause 3.12.3.6 specifies that external columns (internal column ties are not required) should be tied horizontally in a direction into the structure at each floor and roof level with ties capable of developing a force $F_{t,col}$ equal to the greater of:

1   the lesser of 2.0 F_t or 0.4h F_t where h is the storey height in metres; or

2   0.03N, i.e. 3 per cent of the vertical ultimate axial force carried by the column at that level using $\gamma_f = 1.05$.

Corner columns should be tied in to the structure in two mutually perpendicular directions with the above force. Ties in a direction parallel with beams may be provided at columns in one of two ways. If there is no positive tie force between the beam and column, loose tie steel should be placed on site and be fully anchored. The net cross-section of any threaded bar (to the root of the thread) is used in calculating the area of the tie bar. Resin anchored reinforcement may also be used, but this involves the use of proprietary materials and techniques. The manufacturer's specification with regard to the size of hole and insertion of the resin should be followed.

The second method is to design the beam-to-connector so that the horizontal shear capacity may be used to provide the necessary tie force from the column to

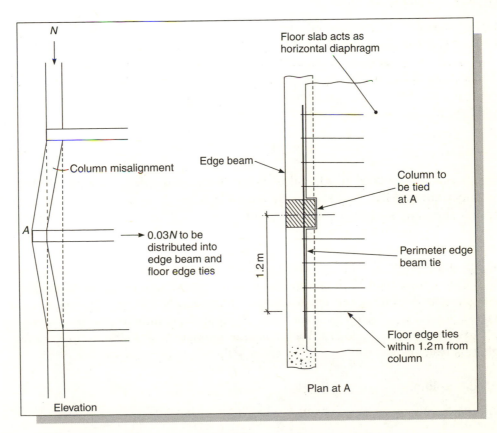

Figure 10.11: Provision of column-to-floor ties via beam and floor ties.

the beam. This imposes additional axial forces in the beam which must be dealt with as explained in Section 9.3. There must be a positive no-slip horizontal connection between the beam and column. This is achieved by surrounding any mechanical connectors, such as bolts or dowels, with in situ concrete. The site workmanship should be especially supervised in these situations because evidence of poor compaction can easily be covered over.

Ties in a direction perpendicular to edge beams do not have to be connected directly to columns. The tie force may be distributed to the edge floor ties that lie within 1.2 m of the column. This is explained in Figure 10.11. The assumption is that if the floor slab is acting as a horizontal diaphragm, then the edge column cannot be displaced outwards without mobilizing the floor plate. If the floor is tied to the edge beams, and this in turn is tied to the column as given above, the column tie must be secured. The floor tie force in this locality must therefore be $F_t = F_{t,\text{col}}/2.4\,\text{kN/m}$. This replaces the floor tie force F_t' obtained from Eq. 10.4.

10.4.3   Vertical ties

BS8110, Part 1 calls for continuous vertical ties in all buildings capable of resisting the tensile force given in Clause 3.12.3.7. The vertical tie force capacity is calculated from the summation of the ultimate beam reactions N only at the floor level immediately above where the tie is designed, not the total load from every floor above the tie. This means the reinforcement provided in the column, and in any splice within that column, should be at least:

$$A_s = \frac{N}{f_y} \qquad\qquad 10.9$$

The loading on the floor may be taken as the characteristic dead load plus 1/3 of the superimposed live load (unless the building is being used for storage where the full superimposed load is taken). γ_f for gravity loads may be taken as 1.05.

Example 10.2.   *Stability ties in a precast structure*
Design the stability ties in the form of helical strand for the six-storey structure for which a floor plan is shown in Figure 10.12. The storey height is 3.5 m. The characteristic floor dead and live loads are 5.5 and 5.0 kN/m², respectively. There is no structural topping on the floor slabs. The beam-to-column connectors are not able to carry stability tie forces. The maximum ultimate axial forces in the various columns (calculated using $\gamma_f = 1.05$) are as follows:

- edge column   5000 kN
- corner column   3000 kN
- internal column   8000 kN

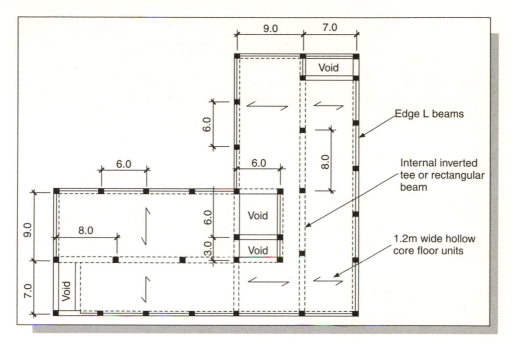

Figure 10.12: Detail to Example 10.2, (Dimensions in m).

Use $f_y = 1750\,\mathrm{N/mm^2}$ for longitudinal ties and $f_y = 460\,\mathrm{N/mm^2}$ for floor ties, $f_{cu} = 25\,\mathrm{N/mm^2}$ for infill concrete, $\gamma_m = 1.0$.

Solution

$$F_t = 20 + 4 \times 6 = 44\,\mathrm{kN/m\ run} \qquad (\textit{using Eq. 10.3})$$

Floor ties for 9 m span floors

$$F'_t = 44 \times 9/5 \times 10.5/7.5 = 111\,\mathrm{kN/m} \qquad (\textit{using Eq. 10.4})$$
$$A_s = 111 \times 10^3/460 = 241\,\mathrm{mm^2/m\ run} \times 1.2$$
$$= 290\,\mathrm{mm^2 per\ 1.2\,m\ wide\ floor\ unit} \qquad (\textit{using Eq. 10.6})$$

Use T12 bars at 400 mm centres (339).
Anchorage length $= 44 \times 12 = 528\,\mathrm{mm}$

Provide 600 mm long milled slot into hollow cores.

Floor ties for 7 m span floor
Pro-rata $7/9 \times$ previous case $= 225\,\mathrm{mm^2}$ per 1.2 m wide unit

Use same as for 9 m floor so that positions of bars coincide.

Perimeter tie to edge beam supporting 9 m long floors

$$F'_t = 44 \times 6/5 \times 10.5/7.5 = 74\,\text{kN/m}$$

Supported floor $= 9.0/2 = 4.5\,\text{m}$ span

$$F_{t,\text{beam}} = 74 \times 4.5 = 333\,\text{kN} \qquad (using\ Eq.\ 10.7)$$

$$A_s = (333 \times 10^3)/1750 = 190\,\text{mm}^2$$

Use 2 no. 12.5 mm diameter helical strand (188).
Ditto to 7 m long floors

Gable end tie to 9 m long beam

$$F'_t = 111\,\text{kN/m} \times 1.0\,\text{m nominal} = 111\,\text{kN}$$
$$A_s = 111 \times 10^3/1750 = 63\,\text{mm}^2$$

Use 1 no. 12.5 mm strand (94).
Ditto to 7 m long gable beam

Internal tie to main Spine Beam

$$F'_t = 44 \times 8/5 \times 10.5/7.5 = 98.6\,\text{kN/m}$$

Supported floor $= 9.0/2 + 7.0/2 = 8.0\,\text{m}$

$$F_{t,\text{beam}} = 98.6 \times 8.0 = 789\,\text{kN} \qquad (using\ Eq.\ 10.7)$$
$$A_s = 789 \times 10^3/1750 = 450\,\text{mm}^2$$

Use 4 no. 15.2 mm strands (552).
Lap length to all strand $= 100 \times$ diameter
Therefore, Lap to 12.5/15.2 mm diameter $= 1250/1520\,\text{mm}$, respectively.

Bar bending radius
For 12.5 mm diameter strand $F = 1750 \times 94 = 164.5 \times 10^3\,\text{N}$

$$a_b = 50\,\text{mm}$$
$$f_{cu} = 25\,\text{N/mm}^2$$

$$r = \frac{164.5 \times 10^3 \left(1 + \dfrac{2 \times 12.5}{50}\right)}{2 \times 25 \times 12.5} = 395\,\text{mm} \qquad (using\ Eq.\ 9.32)$$

Use 400 mm radius.
For 15.2 mm diameter strand $F = 1750 \times 138 = 241.5 \times 10^3\,\text{N}$

$$a_b = 50\,\text{mm}$$
$$r = 510\,\text{mm}$$

Use 510 mm radius.

Prepare corners of hcu with 500×500 mm triangular recess as per Figure 10.6.

Internal tie along front of central core area

End of tie cannot be continuous at columns at corner of voids. Therefore, use HT bar passing through sleeve in column and secured by plate washer.

$$F'_t = 44 \times 6/5 \times 10.5/7.5 = 74 \, \text{kN/m}$$

Supported floor $= 3.0/2 = 1.5 \, \text{m}$

$$F_{t,\text{beam}} = 74 \times 1.5 = 111 \, \text{kN} \qquad (using \ Eq. \ 10.7)$$
$$A_s = 111 \times 10^3/460 = 241 \, \text{mm}^2$$

Use 1 no. T20 bar threaded to M20 (245).

Column ties

Edge column

$$N = 5000 \, \text{kN}$$

$$\begin{aligned} F_{t,\text{col}} &= 2.0 \times 44 = 88 \, \text{kN} \\ &0.4 \times 3.5 \times 44 + 61.6 \, \text{kN} \\ &0.03 \times 5000 = \underline{150 \, \text{kN}} \end{aligned}$$

In the direction of the beam $A_s = 86 \, \text{mm}^2 < 188 \, \text{mm}^2$ provided above.

In the direction perpendicular to the beam $A_s = 150 \times 10^3/460 \times (1.20 + 1.20) = 136 \, \text{mm}^2/\text{m}$ run $< 282 \, \text{mm}^2$ provided above by T12 at 400 c/c.

Corner column

$$N = 3000 \, \text{kN}$$

$$\begin{aligned} F_{t,\text{col}} &= 2.0 \times 44 = 88 \, \text{kN} \\ &0.4 \times 3.5 \times 44 + 61.6 \, \text{kN} \\ &0.03 \times 3000 = \underline{90 \, \text{kN}} \end{aligned}$$

In the direction of the beams $A_s = 52 \, \text{mm}^2 < 94 \, \text{mm}^2$ provided to the gable beam above.

Internal column – has no requirement.

References

1   HMSO, Report of the inquiry into the collapse of flats at Ronan Point, Canning Town, London, 1968.
2   Engström, B., Connections Between Precast Components, *Nordisk Betong*, Journal of the Nordic Concrete Federation, No. 2–3, 1990, pp. 53–56.

index